Lecture Notes in Physics

Volume 1000

The series Lecture Notes in Physics (LNP), founded in 1969, reports new developments in physics research and teaching - quickly and informally, but with a high quality and the explicit aim to summarize and communicate current knowledge in an accessible way. Books published in this series are conceived as bridging material between advanced graduate textbooks and the forefront of research and to serve three purposes:

- to be a compact and modern up-to-date source of reference on a well-defined topic;
- to serve as an accessible introduction to the field to postgraduate students and non-specialist researchers from related areas;
- to be a source of advanced teaching material for specialized seminars, courses and schools.

Both monographs and multi-author volumes will be considered for publication. Edited volumes should however consist of a very limited number of contributions only. Proceedings will not be considered for LNP.

Volumes published in LNP are disseminated both in print and in electronic formats, the electronic archive being available at springerlink.com. The series content is indexed, abstracted and referenced by many abstracting and information services, bibliographic networks, subscription agencies, library networks, and consortia.

Proposals should be sent to a member of the Editorial Board, or directly to the responsible editor at Springer:

Dr Lisa Scalone
lisa.scalone@springernature.com

Roberta Citro • Maciej Lewenstein •
Angel Rubio • Wolfgang P. Schleich •
James D. Wells • Gary P. Zank
Editors

Sketches of Physics

The Celebration Collection

Springer

Editors

Roberta Citro
Department of Physics "E.R.Caianiello"
University of Salerno
Fisciano, Italy

Angel Rubio
Theory Department
Max Planck Institute for the Structure and
Dynamics of Matter
Hamburg, Germany

James D. Wells
Physics Department
University of Michigan–Ann Arbor
Ann Arbor, MI, USA

Maciej Lewenstein
Quantum Optics Theory
Institute of Photonic Sciences
Castelldefels (Barcelona), Spain

Wolfgang P. Schleich
Institute of Quantum Physics
University of Ulm
Ulm, Germany

Gary P. Zank
CSPAR & SPA
University of Alabama in Huntsville
Huntsville, AL, USA

ISSN 0075-8450   ISSN 1616-6361   (electronic)
Lecture Notes in Physics
ISBN 978-3-031-32468-0   ISBN 978-3-031-32469-7   (eBook)
https://doi.org/10.1007/978-3-031-32469-7

This Springer imprint is published by the registered company Springer Nature Switzerland AG
The registered company address is: Gewerbestrasse 11, 6330 Cham, Switzerland

Paper in this product is recyclable.

Lecture Notes in Physics: The Formative Years

In the mid-1960s, Springer-Verlag—until then predominantly a German language publishing house—wanted to become one of the leading publishers worldwide for scientific literature in mathematics and physics. A big step was the formation of an associated firm, Springer-Verlag New York, on July 29, 1964. Shortly thereafter, on September 1st, a new office was opened in the Flatiron Building in the heart of Manhattan. Immediately, Klaus Peters, head of the mathematics editorial in Heidelberg, who over the next 15 years nearly single-handedly brought Springer to the forefront of mathematical publishing, began to intensify contacts with American mathematicians. In the same year, he founded the *Lecture Notes in Mathematics* as a way to distribute preprints that usually were sent only to a handful of closer colleagues, rarely to a wider audience. Klaus was convinced that this would be regarded positively in the mathematics community worldwide, in particular because it was timeliness rather than typesetting that would be the hallmark of the series. At the time, no one expected that the series would become so successful and respected. As a matter of fact, it contributed enormously to the rapid recognition of Springer-Verlag as an international high-quality science publisher. In December 1969, *Lecture Notes in Physics* commenced publication with its first three volumes. In 1973, the *Lecture Notes in Computer Science* were founded and many more lecture notes series followed.

In the mid-60s, computer composition was still some 15 years away. Photo composition was not yet used for scientific books. Manuscripts were submitted to the publisher, typed on mechanical typewriters with formulae inserted by hand. These manuscripts needed careful copyediting, often including grammar and language editing, before they could be sent to the printer. Printing was done in hot type and the mathematical typesetting via hand composition. This process, all together, was time-consuming—but the outcome was far superior to most modern book production. It also required well-trained personnel. For the rapidly growing company, both this long processing time and a shortage of qualified copy editors and production editors were limiting factors, and so the directors of the company immediately supported the idea to create the less formal lecture notes series as a compliment to its expanding traditional book publishing.

The 1960s saw a rapid growth of the mathematical and physical sciences as never before experienced. Its protagonists had to give priority to their research work and barely one of them found the time to write a textbook or a carefully written monograph. To teach students and to present their scientific findings in some detail, authors produced mimeographed notes of their lectures in manuscript form using typesetter composition with handwritten formulae. (Donald Knuth's *TeX*, Leslie Lampert's *LaTeX*, and all their derivatives were not readily available to authors before the 1980s.) From those notes, often sketchy in style, students learned of recent developments, and it was this huge set of preprints that paved the way for newcomers to modern research. And so the idea was born: Why not, instead of waiting only for full-fledged and well-composed book manuscripts, should the publisher not act as a distributor for some of the most valuable preprints as well. All that was needed was at hand at Springer-Verlag: an advertising department, contacts to photo typesetters, contacts to libraries and bookstores. So, "let's start" was the motto at the time.

Ten years ago the series *Lecture Notes in Physics* was founded with two goals in mind: Firstly, very good lectures, especially those devoted to themes of actual research, should become accessible to a wider audience, and secondly, proceedings of conferences should appear quickly and at a lower price. In both cases, however, preference was given to genuine "Lecture Notes", i.e., contrary to concise articles in scientific journals written for the expert, the contributions to this series should present many different aspects of the problem in question and should be intelligible for a wider audience quite often including students or workers from other fields of research. At the same time "Lecture Notes" also means that the form of a text might be tentative and less accurate than usually required for a textbook or a monograph. However, although tentative in form, a high scientific standard in correctness and in selection of the material presented is asked for.

The above text had been written in 1979. In the ten years since the founding of the series, 111 volumes had already appeared and it continued to grow; only five years later, Volume 222 was published. At that time the series was already well-established and fully accepted in the physics community. What is not mentioned in the above-cited published aims and scope for the series is the following remarkable agreement between the publisher and the scientists: Firstly, the list price of the books should be low and, in return, the authors waive their honorarium; secondly, the authors should not be bound by contractual restrictions concerning the publication of their material in other form. Indeed, quite a few of the expository lecture notes were later published as carefully worked-out monographs. This agreement created a special publisher-author relationship. In retrospect, it seems clear that the success of the series is to no small extent due to the early involvement of most esteemed and leading physicists, both as authors and editors.

Indeed, special mention has to be made and special praise has to be given to the founding editors Jürgen Ehlers, Klaus Hepp, and Hans-Arwed Weidenmüller. Early discussions between these three scientists and the publisher shaped the concept of *Lecture Notes in Physics*; also extremely helpful were the encouraging discussions with Klaus Peters who brought to those meetings his experience as a science publisher, in particular of the already widely known *Lecture Notes in Mathematics*. Ehlers, Hepp, and Weidenmüller were highly respected in their scientific communities, and they could solicit excellent manuscripts right from the beginning. In 1975 Johannes Zittartz, and in 1977 Rudolf Kippenhahn, joined the editorial board of the series. The rapid development of various branches of physics in the 1960s and 1970s—general relativity awoke from its dormancy, in sub-nuclear physics, the big accelerator laboratories produced a host of important results culminating in the standard model, the rapid growth of telescope technology for ever-wider spectral regions made astronomy an ever-busier field of research in physics, and the emergence of modern mathematical physics pushed statistical physics to its center stage—made this expansion of the board necessary.

In retrospect, this rapid publication of new physics, albeit expository but otherwise not yet available in traditional book form, made Springer-Verlag known within a very short time as an international publisher. It gave Springer-Verlag's acquisition editors access to a fairly large pool of potential authors, referees, and science advisors, but there was a price to pay. In the 15 formative years under scrutiny in this article, certainly more than 2.3 million copies of the Lecture Notes in Mathematics and Lecture Notes in Physics have been printed. As was the custom in those early days, the complete print-run for sales over many years to come was delivered upon appearance of the individual lecture notes volume and that caused considerable storage problems in the warehouse. The rapid growth of the company implied that the sales department had to supply booksellers all over the world—a significant handling problem. But there also emerged serious problems in the editorial approach, concerning the very aims and scopes of the lecture notes series. At the heart of the program had been the teaching and introduction of newcomers to recent research. Long and well-done surveys from seminar notes and topical symposia were therefore also welcomed as lecture notes to serve this purpose; hence, carefully edited proceedings were admitted. But by 1985, proceedings outnumbered monographic volumes in the textitLecture Notes in Physics by about 2 to 1, and it is not clear that the original pedagogical value of the series is consistently maintained in all of these proceeding publications. For years to come a satisfactory solution of this dilemma had not been found.

Now, more than fifty years after its creation one may conclude: the concept of lectures notes, including the continued publication of the *Lecture Notes in Physics*, is an important contribution to the scientific community.

Acquisition Editor for *Lecture Notes in Physics* in days of yore   Wolf Beiglböck

Lecture Notes in Physics: The Renaissance Years

I joined Springer-Verlag in 1999 as assistant to Prof. Wolf Beiglböck, and certainly as an admirer of Springer books in general. I even owned some volumes of *Lecture Notes in Physics* (LNP). These I had purchased as a PhD student through one or another of the then typical special sales (the most famous being the "yellow sale" of Springer mathematics titles), which from time to time would also be devoted to Springer's backlist in physics. I recall too that I wasn't always entirely happy with what I got from LNP—I wanted them as self-study introductions on topics I wasn't very familiar with (i.e., almost everything as a PhD student. . .), and they sometimes disappointed by not delivering what they seemed to promise. Little did I suspect then that I would soon be given the chance to take action on this.

As it turned out, this was a time at which the LNP had almost fallen victim to its own success: While the annual output was impressive, typically 30 or more new releases, the proceedings volumes had come to far outnumber the introductory monographs that were originally the focus of the series. What's more, due to an inflation of conferences and workshops worldwide, the scientific value of proceedings dropped correspondingly, mostly due to the increasingly superficial and repetitive nature of such contributions, exceptions notwithstanding. In 1991, the LNP had already been split into a proceedings and monograph series, but the monograph series, with far fewer volumes and necessarily appearing as a new series on the market, had not gained as much visibility as the former combined series or, for that matter, the now separate LNP proceedings series. For the latter, the number of copies sold per volume dropped rapidly as the market was being swamped with proceedings from all publishers.

After some deliberation, Wolf Beiglböck and I came to the conclusion that the best way forward was to merge the two LNP subseries again and to slowly but steadily phase out the proceedings in favor of either monographic contributions or edited volumes that were invitation-only and under the tight editorial control of the volume editors. Indeed, just stopping the proceedings subseries overnight would have left the LNP below any reasonable critical mass that was required by its then still very many subscribers. The last volume of the monograph subseries, volume m73, was thus published in 2003 and the proceedings subseries dropped its name continuing, as of this moment, once more simply as THE *Lecture Notes in Physics*. It would be too easy to say that the rest is history, as it actually took almost a decade to re-establish a balance between monographic and edited volumes, though traditional proceedings had been completely banned by then. Rare exceptions were made for school proceedings or carefully selected material from topical workshops, which are typically characterized by a small number of contributions with greater depth and the inclusion of sufficient introductory or background material, respectively.

It was also around the time of the merger of the subseries that the essentials of the current aims and scope of the series were shaped:

> The series *Lecture Notes in Physics* (LNP), founded in 1969, reports new developments in physics research and teaching—quickly and informally, but with a high quality and the explicit aim to summarize and communicate current knowledge in an accessible way. Books published in this series are conceived as bridging material between advanced graduate textbooks and the forefront of research and to serve three purposes:
>
> - to be a compact and modern up-to-date source of reference on a well-defined topic;
> - to serve as an accessible introduction to the field for postgraduate students and non-specialist researchers from related areas;
> - to be a source of advanced teaching material for specialized seminars, courses and schools.
>
> Both monographs and multi-author volumes will be considered for publication. Edited volumes should however consist of a very limited number of contributions only. Proceedings will not be considered for LNP.

Traditional proceedings were diverted from then on into the revamped book series *Springer Proceedings in Physics* , and the revised aims and scope tried also to set the LNP apart from other physics books series, such as the equally well-known *Springer Tracts in Modern Physics* which at that time were typically short monographs on niche subjects, often based on habilitation theses. When I eventually took over the re-combined series, the idea was again to reach out to experienced researchers as authors, yet often at a stage of their careers before they would find time to write full-scale textbooks or monographs. Also, together with the editorial board, we tried, albeit with varying success, to insist that volumes should not exceed 200–250 pages, so as to really remain geared toward first introductions for young researchers and non-specialists from related fields. When asked by colleagues how I would characterize the LNP, my favorite answer was always "as textbooks in search of a course."

It would not do justice to the changes occurring in scientific publishing at that time to attribute the renaissance of the LNP solely to the decision to phase out proceedings. In the late 1990s, electronic publishing (as well as first ideas about open access) had started to emerge, though initially mostly for scientific journals. Some e-only journals were launched at that time such as *New Journal of Physics* or *Journal of High-Energy Physics*. Much ahead of everyone else at Springer, Wolf Beiglböck, also the architect of the EPJ collaboration with the Italian and French Physical Societies—whereby *Zeitschrift für Physik*, *Il Nuovo Cimento*, and *Journal de Physique* merged in 1997/1998 to form the *European Physical Journal* series (epj.org)—immediately saw the need to test the waters and the e-only journal *EPJ direct* was launched. Though this journal proved ephemeral in the end—EPJ eventually developed altogether different approaches to e-only and open access—the lesson was well taken.

Thus, when the chance came to test electronic editions of books at Springer in the early 2000s, we immediately volunteered the full LNP series: not only the new releases, but also including a full retro-digitization of the series back to volume 1 (and, of course, including the 73 volumes of the LNP monograph subseries published between 1991 and 2003). As if this were not enough, we decided to further bet on the then equally new technique of print-on-demand and subsequently managed to reduce recurrent issues with the large warehouse stock of LNP volumes while also keeping older volumes available and even bringing out-of-print volumes back to life. What is now the normal procedure for all academic book publishing was at that time truly pioneering work. As one of the first series fully available online, the LNP was among the publisher's most downloaded book series in the hard sciences (as per annual statistics of chapter downloads), though eventually being overtaken by the much larger *Lecture Notes in Computer Science* and *Lecture Notes in Mathematics* series.

Another important initiative around that time centered on indexing. Publishing in journals with impact factors became an increasing concern for researchers seeking tenure or applying for grants. This left them less time to devote to writing books or book chapters. We therefore introduced a journal-type style for referencing LNP books and chapters (this was shown at the bottom of each chapter title page) in order to encourage the citation of either the whole book or of individual chapters with a simple and well-defined bibliographical reference similar to that of a journal article. We applied for the series to be indexed in the Web of Science's (WoS) journal database. Indeed, a few book series were indexed in that database and had obtained an impact factor. Unfortunately, this was the time when Thomson Reuters, the then owner of WoS, must already have decided not only not to index further book series but also to actually remove existing book series from the journal database and to build up a new database specifically for books, which indeed came on the market sometime later. So we had to live with the fact that journals increasingly represented competition for well-prepared edited article collections, and in particular for tutorial reviews. One of the consequences was that we concentrated even more on short introductory monographs and, in 2007, on the basis of the former *Journal de Physique IV*, we launched the journal *EPJ Special Topics*, which was quickly indexed in WoS and able from then on to accommodate such edited lecture-notes-style article collections for those volume editors who preferred this to the book format.

These years devoted to the relaunch of LNP were also very interesting times in terms of the topical coverage of the series. Traditionally, mathematical and theoretical physics as well as astrophysics were well covered, reflecting of course the research interests of the founding editors, as were specific fields like high-energy, fluid, or statistical physics. Yet, many mainstream topics, e.g., atomic, molecular, or condensed matter physics, were hardly covered at all. With varying success, attempts were made to reach out more systematically to these core physics communities but also to altogether new communities that were hardly present in the series until then, such as biological or chemical physics.

For a while, we had also toyed with the idea of launching a separate lecture notes series on astronomy/astrophysics—indeed, it is well known that, when it comes to research article publishing, the physics and astronomy communities do not overlap very much in terms of jointly used journals. Eventually, we decided to keep both these fields within LNP but to share the corresponding handling of the series. Thus, as of then, my Springer colleague Ramon Khanna has taken responsibility as publishing editor for all astronomy/astrophysics titles, also appointing appropriate experts to the editorial board and with the specific aim of publishing more volumes on topics at the growing interfaces between physics and astronomy (e.g., astroparticle physics, cosmology, . . .).

Particularly interesting were some lively discussions we had with the expanding the editorial board about what ramifications should still be considered core physics and thus be either covered or excluded. Controversial examples were the emerging fields of socio- and econophysics. Eventually we decided to create a new book publishing program, called Springer Complexity, and to merge these topics with other initiatives in the field of complex systems. For this, we had already the well-known book series *Synergetics* edited by Prof. Hermann Haken, and on the initiative of new colleagues at Springer in applied sciences, a lecture-notes-type series *Understanding Complex Systems* was added, with many physicists who had moved to such interdisciplinary research topics joining the newly constituted editorial board of the Springer Complexity program.

Last but not least, since we received many good proposals that were pitched at the undergraduate rather than graduate level, the publisher eventually took the decision in 2012 to launch a new series *Undergraduate Lecture Notes in Physics* which has turned out to be a very successful series in its own right.

In 2007, I had the pleasure of appointing Wolf Beiglböck, Jürgen Ehlers, Klaus Hepp, and Hans Weidemüller as Founding Editors of the series with the right to act as editors for as long as they please. I would like to take this opportunity to thank them personally (Jürgen Ehlers, alas, passed away in 2008) and all of the series editors I had the privilege of closely working with M. Bartelmann, R. Beig, W. Domcke, B.-G. Englert, U. Frisch, F. Guinea, P Hänggi, G. Hasinger, W. Hillebrandt, M. Hjorth-Jensen, D. Imboden, R.L. Jaffe, W. Janke, M. Lewenstein, R. Lipowsky, H.v. Löhneysen, M. Mangano, I. Ojima, J.F. Pinton, J.-M. Raimond, A. Rubio, M. Salmhofer, W. Schleich, D. Sornette, S. Theisen, D. Vollhardt, W. Weise, J. Wess, J.D. Wells, G. Zank and J. Zittartz.

It is truly rewarding to now see the milestone publication of LNP 1000 in a book series that has been able to re-invent and rejuvenate itself at regular intervals to remain one of the trusted and highly regarded sources of such tutorial state-of-the-art volumes.

Executive Publishing Editor, Physics   Christian Caron
Managing Editor of the *Lecture Notes in Physics* 2003–2019

Preface

It is a great honor for the editors of *Lecture Notes in Physics* to contribute to the 1000th volume of one of the most important and influential book series in physics. Like every physicist, we have fond memories of finding a volume within LNP that we read in our formation as researchers and scholars, or when we ventured into a new subfield. They have taught us the fundamentals and key developments of the new research field; they have elucidated our understanding of fields we currently work on; and they have been a forum for the most up-to-date ideas that have been recently solidified within physics, thus enabling a synthesis on which further research can build. The work of the LNP series has been of particular value to early career researchers.

The contributions within this volume are written by current editors of the LNP series. Our contributions are intended to celebrate the many valuable roles LNP has played over the years by contributing chapters that illustrate the goals listed above for the series. Some contributions illustrate the role of synthesizing key recent progress. Others illustrate the value of speaking the often unspoken, but valuable, ideas of researchers about the foundations of their field. All are meant to edify and to be of particular interest to young researchers working to obtain a deeper understanding beyond the standard textbook material.

In Chap. 1, Dante Kennes and Angel Rubio discuss the fascinating process of controlling quantum materials in their contribution, "A New Era of Quantum Materials Mastery and Quantum Simulators In and Out of Equilibrium." Quantum materials are materials whose identities cannot be expressed without explicit quantum mechanical language. For example, they may be systems that have emergent or topological collective phenomena made possible by quantum mechanics. Quantum materials are playing a significant role in the development of modern material science. A key desire in the quantum materials realm is to be able to control them. By control, one means having the ability to apply some external driving imposition, such as pressure or photons, on the material to change one or more of its properties in a desirable way. Non-linear interactions yielding non-equilibrium quasi-stable states are of particular interest in studying these materials. The applications of this field range widely from engineering advances, such as twistronics, to advances in understanding fundamental physics, such as simulating the behavior of fermions in flat and curved space or with changing (effective) mass.

In Chap. 2, I discuss the long-standing issue of naturalness in the contribution, "Evaluation and utility of Wilsonian Naturalness." For many decades, physicists have recognized that standard formulations of quantum field theory render the cosmological constant unnatural compared to the scales of fundamental physics around us, such as the electron mass, Planck's constant, etc. An elementary scalar Higgs boson with mass much less than the Planck scale also suffers from this same problem. Perhaps it is not a problem, some might say; it may be a philosophical misconception. However, there is some historical precedence for naturalness being useful, and I argue that there is a rigorous technique to assess naturalness by evaluating fine-tunings across mass thresholds that match low and high-energy quantum field theories. Every naturalness worry in particle physics can be articulated precisely with that technique, and some implications for new physics are discussed.

In Chap. 3, Roberta Citro and Ofelia Durante review the importance of geometric phases in cyclical evolution in their contribution, "The Geometric Phase: Consequences of Classical and Quantum Physics." In physics, we have always been interested in the cyclical evolution of a system. Most systems are not cyclical in every possible way, but perhaps only in one or a few ways. There can be residual differences—a geometric phase—that arise at the completion of a cycle with slowly changing parameters. The most celebrated phase in the quantum world has been the residual phase driving the Aharonov-Bohm effect. But there are many others, including a classical geometric phase in the Foucault pendulum as well as the quantum mechanical geometric phase in the polarization of light in an optical fiber. Citro and Durante give several other interesting examples that demonstrate the unifying principles that important physical phases arise through the systems' geometric and topological configurations. These have important physical implications, which in turn can lead to interesting applications to technology.

In Chap. 4, Joan Fraxanet, Tymoteusz Salamon, and Maciej Lewenstein review a central aspect of quantum computing in their contribution, "The coming decades of quantum simulation." Quantum mechanics has been around for more than a century, and we have made many computations of quantum systems, such as the hydrogen atom, since its early days. We have even simulated quantum mechanical systems on a computer, but these simulations have in a sense all been utilizing classical computers which rely on definite states of 0 or 1 in their computations. But quantum computing enables truly quantum states that are not as restricted. They, therefore, have the prospect of achieving a much more efficient and much better simulation of a truly quantum mechanical system without having to "classicalize it" on a standard computer architecture. Fraxanet, Salamon, and Lewenstein explain why we should be excited for this application of quantum computing to rise dramatically over the coming decades, and what that can mean to us in our pursuit of basic scientific knowledge and quantum engineering applications.

In Chap. 5, Wolfgang Schleich, Iva Tkáčová, and Lucas Happ give us significantly deeper appreciation of physical systems described by complex functions in their contribution, "Insights into complex functions." Complex-valued functions are a staple in the tool box of physicists. On day one of any quantum mechanics course we learn that the state of the system is described by a complex-valued function, called the wave function. But as physicists we rarely expend much effort in learning what insights we can gain on the system through deeper understanding of its mathematical description of a complex-valued function. This contribution gives deeper insight through the introduction of Newton flows and the Cauchy-Riemann differential equations. These concepts help us visualize better, and therefore understand better, what are otherwise difficult, higher-dimensional complex spaces. It also gives greater insights into solving important problems, such as finding the zeros of a function.

In Chap. 6, Gary Zank, Lingling Zhao, Laxman Adhikari, Daniele Telloni, Justin C. Kasper, and Stuart D Bale report on the intriguing and complex physics of the solar corona in their contribution, "Exploring the hottest atmosphere with the Parker Solar Probe." The solar corona is extremely hot, and it is what drives the solar wind in the solar system. One of the mysteries of the Sun is precisely how its corona is heated. One of the key ideas is that it is accomplished via dissipation of low-frequency magnetohydrodynamics turbulence. Simulations of this heating idea are compared to the observations made by the Parker Solar Probe, which was launched in July 2018 and is continuing to take data. Zank et al. show that there is good correspondence of the magnetohydrodynamics turbulence theory explanation of heating with the detailed observational patterns measured by the Parker Solar Probe, thus going a long way toward resolving this interesting mystery.

In the last chapter, Michael Tschaffon, Iva Tkáčová, Helmut Maier, and Wolfgang Schleich give us a primer on the Riemann hypothesis and Dirichlet L-functions. A Dirichlet L-series is a special type of infinite series for a restricted complex number (real part greater than 1). The analytic continuation of the Dirichlet L-series over the full complex plane is a Dirichlet L-function. Dirichlet L-functions have an illustrious history in number theory. An interesting feature of these functions is that they can be shown to be equivalent to an Euler product over all prime numbers. The famous Riemann hypothesis states that the zeroes of the Riemann zeta function, a special case of the Dirichlet L-functions, are only at negative integers and complex numbers whose real part is 1/2. In addition to these fascinating properties of the Dirichlet L-functions studied by mathematicians, it turns out that there are physics applications of Dirichlet L-functions. In particular, the Riemann zeta function plays a prominent role in string theory, including for example the ability to reformulate Veneziano amplitudes in terms of ratios of Riemann zeta functions. More generally, these functions have utility in black hole physics, symmetry determinations in string theory, and other applications. In this chapter, a thorough description is given of Dirichlet L-functions with an audience of physicists in mind.

We hope you enjoy this volume, and the many future volumes to come within the *Lecture Notes in Physics* series.

Ann Arbor, MI, USA   James D. Wells

Contents

Contributors

Laxman Adhikari, Center for Space Plasma and Aeronomic Research (CSPAR) and Department of Space Science, University of Alabama in Huntsville, Huntsville, AL, USA

Laxman Adhikari received his Ph.D. in Physics from the University of Alabama in Huntsville (USA) in 2015. He is a research scientist in the Department of Space Science (SPA) at the University of Alabama in Huntsville. He has more than 7 years of experience in heliospheric plasma research, specializing in MHD turbulence transport modeling and spacecraft data analysis. He has compared the theoretical results of the turbulence transport model equations with the measurements of the Parker Solar Probe (PSP), Solar Orbiter (SolO)/Metis, Helios 2, Ulysses, Voyager 2, Pioneer 10, WIND, ACE, and New Horizons Solar Wind Around Pluto (NH SWAP), ranging from 3.5 Solar radii to 75 astronomical unit (au). As first author or co-author, he has published more than 40 papers in peer-reviewed journals.

Stuart D. Bale, Physics Department, University of California, Berkeley, CA, USA

Stuart D. Bale is a Professor of Physics at the University of California, Berkeley. He is the former Director of the Berkeley Space Sciences Laboratory (SSL) there. He received his PhD in Physics from the University of Minnesota in 1994. After three years of postdoctoral work at Queen Mary College, University of London, he joined UC Berkeley. He has held visiting positions at the Observatoire de Paris, Meudon (Univ. Paris VII), the University of Sydney, and Imperial College, London. He is the recipient of the 2003 US Presidential Early Career Award for Scientists and Engineers (PECASE) and a Fellow of the American Physical Society (APS). Bale is interested in experimental plasma astrophysics and particle energization in astrophysical and heliospheric plasmas. He is the NASA Principal Investigator for experiments on the Wind, STEREO, and Parker Solar Probe missions and for the NASA Lunar Surface Electromagnetics (LuSEE) program, which will put a low frequency radio astronomy telescope on the lunar surface in 2025.

Roberta Citro, Department of Physics "E.R. Caianiello", University of Salerno, Fisciano, Italy

Roberta Citro is Full Professor of Theoretical Matter Physics at the Department of Physics of the University of Salerno (Italy). In recent years, she has matured professional experience in many-body techniques of low-dimensional systems and in quantum transport in nanostructures. She has recently established a research team/activity on quantum transport and superconductivity in low-dimensional systems. Prof. Citro completed her PhD in Physics at the University of Salerno (Italy) in 1998 defending a thesis on high-temperature superconductors. After her graduation, she was, first, a Post Doc Fulbright fellow at the Physics Department of Rutgers University (New Jersey, USA) where she collaborated with the Condensed Matter Theory Group, completing original works on the puzzling phase diagram of coupled spin chains by means of bosonizations and renormalization group methods. In 2007, she was a Marie Curie fellow under the EU's Sixth Framework Programme-Mobility Action at the Laboratoire de Physique et Modélisation des Milieux Condensés (LPMMC, CNRS) in Grenoble (FR) where she collaborated with Dr. Anna Minguzzi and Prof. Frank Hekking. Here, she acquired knowledge in the field of atomic and molecular systems, working on the problem of non-equilibrium dynamics of a Bose gas subjected to a time-dependent perturbation. She has an active synergetic activity (member of the organizing committees of several international conferences in non-equilibrium physics and quantum gases and a national meeting in superconductivity, coordinator of the Doctorate in Physics and Innovation Technology, referee of the APS, IOP, and Nature Group journals, PI of various national/EU research projects).

Ofelia Durante, Department of Physics "E.R. Caianiello", University of Salerno, Fisciano, Italy

Ofelia Durante is a researcher at the Physics Department "E. Caianiello" of the University of Salerno (Italy). In 2017, she obtained the master degree in physics, focusing on superconductivity at the University of Salerno in collaboration with the National Enterprise for Nanoscience and Nanotechnology (NEST) Laboratory of Scuola Normale Superiore, Pisa (Italy), where she acquired expertise in nano-fabrication and electric/magnetic characterizations on Al-based Josephson devices formed by InAs nanowires. In 2021, she earned PhD in material science at the University of Salerno. During her PhD program, she focused on the study of morphological and structural properties of the metamaterials formed by dielectric oxides (e.g., SiO_2, TiO_2, Ta_2O_5, Al_2O_3, ZrO_2) used to reduce the thermal noise in the mirror prototypes for gravitational waves detection, acquiring in-depth expertise in atomic force microscopy, X-Ray diffraction, and Raman spectroscopy. In this regard, she collaborated with Laboratoire des Matériaux Avancés (LMA), and she joined the International collaboration LIGO-Virgo-KAGRA. She co-authored ca. 40 papers in international peer-reviewed journals.

Joana Fraxanet, ICFO—Institut de Ciencies Fotoniques, The Barcelona Institute of Science and Technology, Barcelona, Spain

Joana Fraxanet studied a double degree in physics and mathematics at the University of Barcelona and obtained her Master's degree at the University of Leiden in 2017. Since then, she has been doing her doctoral thesis at ICFO in the Quantum Optics Theory Group, under the supervision of Maciej Lewenstein. She is currently researching the properties of quantum many-body systems, focusing on the study of exotic phases with topological properties in the context of cold atoms for quantum simulation. She is also interested in the development of quantum algorithms for the study of strongly correlated matter.

Lucas Happ, "Few-body Systems in Physics" Laboratory, RIKEN Nishina Center for Accelerator-Based Science, 2-1 Hirosawa, Wakō, Saitama, Japan, and Institut für Quantenphysik and Center for Integrated Quantum Science and Technology (IQST), Universität Ulm, Ulm, Germany

Lucas Happ is a special postdoctoral researcher at the Nishina Center for Accelerator-Based Science at RIKEN, Japan. He started his studies in physics in 2010, and from 2015 to 2021, he did his Master thesis and his PhD under the guidance of Maxim A. Efremov at Ulm University. After earning his PhD, he became postdoctoral researcher at the Institute of Quantum Physics, before moving in 2022 to the RIKEN Nishina Center for Accelerator-Based Science in Japan for his current postdoctoral position. He obtained main scientific achievements in the field of quantum few-body physics, where he established universal properties, whose applications range from subatomic particles to astrophysics. Further areas of research include theoretical quantum optics and quantum information theory. He is the author of several published articles in peer-reviewed journals.

Justin C. Kasper, BWX Technologies, Inc., Washington DC, USA and Department of Climate and Space Sciences and Engineering, University of Michigan, Ann Arbor, MI, USA

Justin C. Kasper is the Deputy Chief Technology Officer for BWX Technologies (BWXT). He maintains a fractional appointment as professor in the University of Michigan Department of Climate and Space Sciences and Engineering and was previously a Smithsonian Institution civil servant. Dr. Kasper is an experimental physicist by training, with experience developing electromagnetic and particle sensors and systems for the exploration of space and operation in extreme environments. He is the principal investigator of the SWEAP Investigation on Parker Solar Probe and of the six-spacecraft SunRISE Explorer mission. He has received the Presidential Early Career Award for Scientists and Engineers, the Henry Russel Award, and numerous NASA awards and is a member of the American Geophysical Union and the American Physical Society. He received a PhD in Physics from the Massachusetts Institute of Technology.

Dante M. Kennes, Institut für Theorie der Statistischen Physik, RWTH Aachen University and JARA-Fundamentals of Future Information Technology, Aachen, Germany, and Max Planck Institute for the Structure and Dynamics of Matter, Center for Free-Electron Laser Science (CFEL), Hamburg, Germany

Dante Kennes is a professor at the RWTH Aachen University and visiting scientist at the Max Planck Institute for the Structure and Dynamics of Matter. His research focus lies on the theoretical study of ultrafast quantum materials design, non-equilibrium quantum many-body dynamics, and twisted van der Waals heterostructures. He has put forward a number of important developments to push the boundaries of renormalization group- or tensor network-based methods applicable to the out-of-equilibrium realm of strongly correlated systems. Recently, he has also made important contributions to the quickly emerging field of designing two-dimensional quantum materials. These advances allow to realize quantum materials with specific tailored properties enabling novel quantum functionalities.

Maciej Lewenstein, ICFO—Institut de Ciencies Fotoniques, The Barcelona Institute of Science and Technology, Barcelona, Spain and ICREA, Barcelona, Spain

Maciej Lewenstein is ICREA Research Professor and leads the quantum optics theory group at ICFO (the Institute of Photonic Sciences) in Castelldefels— Barcelona, Catalonia. He graduated from Warsaw University in 1978 and joined the Centre for Theoretical Physics of the Polish Academy of Sciences in Warsaw, where he remained for 15 years, becoming a professor in 1993. He finished his PhD in Essen in 1983 and habilitated in 1986 in Warsaw. He has spent several long-term visits at the University of Essen in Germany, at Harvard University with Roy J. Glauber (Nobel 2005), at the Saclay Nuclear Research Centre (CEA) near Paris, and at the Joint Institute for Laboratory Astrophysics in Boulder, Colorado. He was on the faculty of Centre CEA in Saclay during the period 1995–1998, and of Leibniz University in Hannover over the period 1998–2005. In 2005 he moved to Catalonia. His research interests include quantum optics, quantum physics, quantum information, attosecond science, and statistical physics. His other passion is jazz and avant-garde music and is an acclaimed jazz writer and critic.

Helmut Maier, Institut für Reine Mathematik, Universität Ulm, Ulm, Germany

Helmut Maier is Senior Professor of Mathematics at Ulm University, Germany. He graduated with a Diploma in Mathematics from the University of Ulm in 1976, under the supervision of Professor H.-E. Richert. After a period as a scientific employee at the University of Ulm, he started his doctoral studies at the University of Michigan in 1979 and received his PhD in 1981 under the supervision of Professor J. Ian Richards. After postdoctoral positions at the University of Michigan and the Institute for Advanced Study, Princeton, Maier took a position at the

University of Georgia, first as an assistant professor, later on as associate and full professor. Since 1993, Maier is a professor at the University of Ulm, Germany.

One of his most important scientific achievements is the invention of Maier's Matrix Method that led him and others to the discovery of unexpected irregularities in the distribution of prime numbers. Collaborators of Maier include I. Bezděková, P. Erdős, C. Feiler, J. Friedlander, A. Granville, D. Haase, A. J. Hildebrand, S. Konyagin, M. L. Lapidus, H. L. Montgomery, J. Neuberger, C. Pomerance, M. Rassias, W. Schleich, G. Tenenbaum, and E. Wirsing.

Ángel Rubio, Max Planck Institute for the Structure and Dynamics of Matter, Center for Free-Electron Laser Science (CFEL), Hamburg, Germany, Center for Computational Quantum Physics, Simons Foundation Flatiron Institute, New York, NYUSA, and Nano-Bio Spectroscopy Group and ETSF, Universidad del País Vasco UPV/EHU- 20018, San Sebastían, Spain

Ángel Rubio is the director of the theory department of the Max Planck Institute for the Structure and Dynamics of Matter and distinguished research scientist at the Simons Foundation's Flatiron Institute (NY, USA). His research interests are rooted in the modeling and theory of electronic and structural properties of condensed matter as well as in the development of new theoretical tools to investigate the electronic response of materials and molecules and to characterize and predict new non-equilibrium states of matter. He is acknowledged as pioneer and leader in the area of computational materials physics and one of the founders of modern *theoretical spectroscopy*. In the last years, he has pioneered the development of the theoretical framework of *quantum electrodynamical density functional theory (QEDFT)* that enables the ab-initio modeling of strong light-matter interaction phenomena in materials, nanostructures, and molecules.

Tymoteusz Salamon, ICFO—Institut de Ciencies Fotoniques, The Barcelona Institute of Science and Technology, Barcelona, Spain

Tymoteusz Salamon is a PhD student in the Quantum Optics Theory Group lead by Prof. Maciej Lewenstein at ICFO, Barcelona (Spain). He has obtained his BSc degree from Wroclaw University of Technology (Poland) and the MSc degree from Ulm University, (Germany). His research focuses on the quantum simulation of twisted bilayer material with cold atoms in optical lattices.

Wolfgang P. Schleich, Institut für Quantenphysik and Center for Integrated Quantum Science and Technology (IQ^{ST}) Universität Ulm, Ulm, Germany, Hagler Institute for Advanced Study, Institute for Quantum Science and Engineering (IQSE), and Texas A&M AgriLife Research, Texas A&M University, College Station, TX, USA

Wolfgang P. Schleich is Professor of Theoretical Physics and Director of the Institute of Quantum Physics at Ulm University, Germany. From 1980 to 1984, he did his diploma thesis and his PhD under the guidance of Marlan O. Scully at the Ludwig-Maximilians-Universität München, with an intermediate research visit (1982/83) at the Institute of Modern Optics, Albuquerque, USA. From 1984 to 1986, he worked as a postdoctoral fellow with John Archibald Wheeler at the Center for Theoretical Physics in Austin, Texas, USA, and then as a research scientist at the Max Planck Institute of Quantum Optics in Garching, Germany, under Herbert Walther. In 1991, he moved to his current position at Ulm University. He is the author of several books, including Quantum Optics in Phase Space and Elements of Quantum Information. His areas of research include theoretical quantum optics, physics of cold atoms and analogies to solid-state physics, fundamental questions of quantum mechanics, general relativity, number theory, statistical physics, and non-linear dynamics. Some of his most important scientific achievements are related to the role of quantum phase space in quantum optics, for which he has received numerous national and international awards, most recently the Herbert Walther Award of the German Physical Society and the Optical Society of America.

Daniele Telloni, INAF—Astrophysical Observatory of Torino, Pino Torinese, Italy

Daniele Telloni is staff researcher astronomer at INAF-Astrophysical Observatory of Turin. He has a long-term experience in the analysis of remote-sensing observations of the solar corona and of in-situ measurements of solar wind properties. His scientific interests include coronal plasma heating, solar wind acceleration, MHD turbulence, dissipative processes in space plasma, evolution of solar transients, such as Coronal Mass Ejections, and Space Weather. He is Co-I of SWA and Metis aboard Solar Orbiter, of ASPIICS aboard PROBA-3, of SCORE-2 aboard HERSCHEL, and of the MidEx NASA Solaris solar mission. He has published more than hundred peer-reviewed papers and 20 conference proceedings.

Iva Tkáčová, Department of Physics, Faculty of Electrical Engineering and Computer Science, VSB-Technical University of Ostrava, Ostrava – Poruba, Czech Republic

Iva Tkáčová (**née Bezděková**) is Assistant Professor at the Department of Physics, Faculty of Electrical Engineering and Computer Science, VSB—Technical University of Ostrava, Czech Republic. In 2018, she obtained her PhD in Mathematical Physics from the Faculty of Nuclear Sciences and Physical Engineering, Czech Technical University in Prague under the guidance of Igor Jex and Martin Štefaňák. Her areas of interest are quantum physics, especially quantum walks, number theory, and classical optics.

Michael E. N. Tschaffon, Institut für Quantenphysik, Universität Ulm, Ulm, Germany

Michael E. N. Tschaffon is a PhD student at the Institute of Quantum Physics, Ulm University, Germany. From 2018 to 2021, he was working on his BSc and MSc thesis under the guidance of Prof. Dr. Matthias Freyberger and Prof. Dr. Maxim A. Efremov, accordingly. His research interest is focused on foundations of quantum mechanics, quantum information, and quantum electrodynamics.

James D. Wells, Leinweber Center for Theoretical Physics, University of Michigan, Ann Arbor, MI, USA

James D. Wells is a professor of Physics at the University of Michigan (USA). As a theoretical physicist, his research explores ideas designed to solve outstanding "origins" problems in fundamental physics: the origin of gauge symmetries, dark matter, flavor violations, CP violation, and mass. Professor Wells is a fellow of the American Physical Society, a recipient of an Outstanding Junior Investigator (OJI) Award from the US Department of Energy, and a Sloan Fellowship from the Alfred P. Sloan Foundation.

Gary P. Zank, Center for Space Plasma and Aeronomic Research (CSPAR) and Department of Space Science, University of Alabama in Huntsville, Huntsville, AL, USA

Gary P. Zank received his PhD in Applied Mathematics from the University of Natal in South Africa in 1987. Gary is an eminent scholar and distinguished professor, director of Center for Space Plasma and Aeronomic Research (CSPAR), and chair of the Department of Space Science (SPA) at the University of Alabama in Huntsville. Gary has been recognized in his field through the receipt of numerous honors and awards throughout his career. In 2017, he was named the University of Alabama Board of Trustees Trustee Professor, the first and only University of Alabama System faculty member to achieve this position. In part, this was in recognition of Dr Zank being elected in 2016 as a member of the US National Academy of Sciences, the only person in AL to be a member of this august body. He was recognized internationally in 2015 with the AOGS Axford Medal, the highest honor given by the Asia Oceania Geosciences Society (AOGS). Other awards include his being a fellow of the American Geophysical Union, the American Physical Society, and the American Association for the Advancement of Science. In 2017, he was also elected an AOGS Honorary Member and was chosen by the International Space Science Institute (ISSI) to be the 2017 Johannes Geiss Fellow. One of his publications has been recognized as one of the twelve "classic papers" ever published in the Journal of Plasma Physics. Gary is dedicated to his research, which is clearly represented in his achievements over the years and categorizes him as a cutting-edge leader in the world of space physics.

Lingling Zhao, Center for Space Plasma and Aeronomic Research (CSPAR) and Department of Space Science, University of Alabama in Huntsville, Huntsville, AL, USA

Lingling Zhao received her PhD in Space Physics from the University of Chinese Academy of Sciences in China in 2013, July. She is an assistant professor of the Department of Space Science (SPA) at the University of Alabama in Huntsville (UAH). Before joining UAH, she has been working on the propagation and acceleration of energetic particles since her PhD. She has accumulated rich experience in modeling the transport of galactic cosmic rays, anomalous cosmic rays, and interstellar pickup ions. Currently, she is working on the solar wind turbulence and energetic particles observed by Parker Solar Probe and Solar Orbiter. She developed the graduate course *Analysis of Spacecraft Data* for the Department of Space Science at UAH and received the UAH CSPAR Science Achievement Award in 2019 and UAH Researcher of the Year Award in 2020.

A New Era of Quantum Materials Mastery and Quantum Simulators In and Out of Equilibrium

Dante M. Kennes and Angel Rubio

Contents

D. M. Kennes (✉)
Institut für Theorie der Statistischen Physik, RWTH Aachen University and JARA-Fundamentals of Future Information Technology, Aachen, Germany

Max Planck Institute for the Structure and Dynamics of Matter, Center for Free-Electron Laser Science (CFEL), Hamburg, Germany
e-mail: dante.kennes@rwth-aachen.de

A. Rubio
Max Planck Institute for the Structure and Dynamics of Matter, Center for Free-Electron Laser Science (CFEL), Hamburg, Germany

Center for Computational Quantum Physics, Simons Foundation Flatiron Institute, New York, NY, USA

Nano-Bio Spectroscopy Group and ETSF, Universidad del País Vasco UPV/EHU, San Sebastián, Spain
e-mail: angel.rubio@mpsd.mpg.de

© The Author(s), under exclusive license to Springer Nature Switzerland AG 2023
R. Citro et al. (eds.), *Sketches of Physics*, Lecture Notes in Physics 1000,
https://doi.org/10.1007/978-3-031-32469-7_1

1

Abstract

We provide a perspective on the burgeoning field of controlling quantum
materials at will and its potential for quantum simulations in and out equilibrium.
After briefly outlining a selection of key recent advances in controlling materials
using novel high fluence lasers as well as in innovative approaches for novel
quantum materials synthesis (especially in the field of twisted two-dimensional
solids), we provide a vision for the future of the field. By merging state of
the art developments we believe it is possible to enter a new era of quantum
materials mastery, in which exotic and for the most part evasive collective as well
as topological phenomena can be controlled in a versatile manner. This could
unlock functionalities of unprecedented capabilities, which in turn can enable
many novel quantum technologies in the future.

1   Preamble

Out of equilibrium phenomena are ubiquitous in many fields of science from
materials to chemistry to bioscience. They dictate how a system reacts to specific
stimuli and how the nonlinear interactions between system and driving can realize
effects that cannot be achieved otherwise. In solids entirely novel states of matter
may be unveiled under driving conditions. However, those states tend to have
finite lifetimes determined by many different parameters. In fact, much like the
phenomena of biological life itself, a nonequilibrium quasi-stationary state in a solid
requires constant "feeding" by driving to evolve and survive.

Over the last decades the authors, coworkers and many other experimental and
theoretical groups worldwide have devoted their research attention towards unravel-
ing the microscopical mechanism behind this fascinating world of driven materials
with properties which can be switched on demand by external stimuli. In these
materials, many different competing orders and degrees of freedom do play a crucial
role necessitating a close consideration of correlations between many particles with
different statistics and interaction strengths (electrons, ions, photons)—and alike
all quasiparticle hybrid-states, that can form (such as polaron, polariton, magnon,
plasmaron, plexciton, ...) and which can lead to new components out of which to
build materials properties in a lego-like fashion [1–10].

This essay will provide a personal and clearly biased view of the authors on how
science has progressed through this scientific journey (we apologize in advance if
some contributions are not properly recognized, we want to provide our personal
view of the field and its evolution and not to present an exhaustive review of the
whole field). We outline the abundance of new challenges and exciting emergent

new physics that came along with the journey so far and that we are just beginning to explore the exciting road ahead. We highlight some of our own humble contributions and what we think the future will bring focusing on—as the title of this essay suggests– "mastering materials". The theater stage of this essay, if you will, is thus one of a quantum theater. In this the main actors for us are electrons and ions (neutrons and protons). The directors here, which advise the actors on their actions, are light fields (photons) or other forms of external stimuli such as pressure or static electric fields. The audience are novel ultrafast time-resolved spectroscopic tools, which probe the changes to the materials' properties. However, when we bring together directors (e.g. photons) and actors (e.g. electrons) in the quantum realm we can realize fascinating matter-light-hybrid states giving rise to emerging, highly tunable polaritonic quantum matter both in their ground and excited quantum state.

This essay starts with a brief introduction and recap of some aspects of the field of research of nonequilibrium material phenomena. This summary is very contextual and far from exhaustive. It is simply supposed to set the right frame, following the famous quote

> If I have seen further it is by standing on the shoulders of giants.   –Isaac Newton

from a letter to Robert Hooke in 1675, introducing some background on which to rest the discussion.

2   Introduction

Controlling nature has always been one of the pragmatic driving forces of physics and human kind in general. As such, striving for control proceeds by far the emergence of modern quantum mechanics and its application in terms of quantum information science, which is raising tremendous attention at the moment for it could revolutionize many technological sectors. This complements the fundamental goals of science which are to foster our basic understanding of nature and elevates them also to a pragmatic level. However, it should not be forgotten, that it is this basic understanding that allows us to exert control over physical systems, explore their novel functionalities and finally in turn realize innovational technologies after all. In this sense it is frequently true that

> insight must precede application.   –Max Planck

Along these lines, obtaining novel forms of control over *quantum many-body systems* and therefore a deepened understanding of these systems is currently a major research objective as it enables rising quantum technologies with unprecedented capabilities. First applications of quantum technologies have already let to revolutionary new developments in the past and are expected to yield even more profound advancements in the future [11]. Here the *objective of control* is central as quantum physics by its very nature tends to have a tendency to escape strong control paradigms. Obtaining a firm grasp on truly quantum properties such as entanglement, collective emergent behavior as well as collective quantum coherence is a—as Feynman has put it—"golly [...] wonderful problem, because it doesn't look so easy" [12].

In solids quantum control relies in large parts on so-called *quantum materials*, which are systems hosting a wide range of *emergent collective and/or topological phenomena* and which are integral in realizing such next-generation technologies [2, 6, 13] including quantum computing [14]. The term "quantum material" might seem like a misnomer as in reality all materials rely on quantum mechanics. However, the phrase was introduced [6] to differentiate materials whose specific properties are rooted on the fundamental laws and quantization of quantum mechanics evading a quasi-classical interpretation from those materials whose properties can still be understood in such a quasi-classical approach. As a consequence quantum materials have recently moved into the center of modern condensed matter and quantum technologies research. Challenging our theoretical understanding, the fascinating phenomena found in quantum materials, ranging from unconventional superconductivity to topologically protected edge modes, emerge from a delicate interplay between spin, charge, lattice, and orbital degrees of freedom, as well as from the geometric aspects of their wave functions [6–8]. Recent technological progress on fabricating quantum materials with particular, desirable qualities has advanced the field of *materials science* into a new era [1–5], where mastery over the quantum properties is at the vanguard of the current attention. The combined major breakthroughs in materials science, in computational physics and in the field of creating light sources have further let to giant leaps in understanding materials, their quantum aspects as well as their interplay with driving forces. At the interface of materials science and the science of light, novel subfields are born at an exciting rate. This has given rise to the ideas of quantum materials engineering using light, controlling properties by heterostructuring of 2d materials and playing with their stacking arrangement (such as including a twist between one layer to the next) as well as cavity controlled chemistry and cavity materials engineering [3,4,9,10,15–21]. Although, still in its infancy these concepts have already brought forth intriguing demonstrations of quantum materials mastery in parts by using external stimuli.

In the same spirit of advancing the quantum-control paradigm, two major trail-blazing research directions have very recently emerged which will be the focus of this essay: (i) (twisted) van der Waals materials (also called moiré materials) as novel platforms of quantum materials engineering and (ii) nonequilbrium driving, most prominently by light-matter coupling in vacuum and in and out of equilibrium in a cavity environments– to access the physics beyond the restrictive linear-response, quasi-equilibrium regime. In particular combining the two, as will be discussed, will unlock light-matter hybrids in two-dimensional twisted heterostructures (with 2 or more layers [22–24]).

2.1   The Rise of (Twisted) van der Waals Heterostructures

It is not by accident that early archaeological periods (Stone Age, Copper Age, Bronze Age, Iron Age) are often named after the materials used in state-of-the-art technologies at the time. It emphasizes the role material discoveries have always played in the ways and fails of civilizational development; in the early times resting mainly on metallurgy. This is a tradition that, although reaching back as far as the beginning of civilization, continues to this day. Modern technological breakthroughs are still enabled by innovational materials platforms and there is no reason to assume the dawning of the quantum information age will be any different. In fact, recently the potential dawning of the topological age, in which functional topological quantum materials take center stage, has been put forward [25]. Our rapidly growing understanding of the fundamental building blocks of matter has led to a novel way of materials engineering: A bottom-up approach in which properties of solids can be engineered at will by growing crystals in a predefined manner. This hope fuels the fields of solid state physics, chemical physics as well as materials science working hand in hand on a deepened understanding of the properties to expect for a given chemical composition of a solid as well as their experimental realization. One line of thought along this direction which emerged are atomically-thin two-dimensional materials. Sparked by the landmark discovery of graphene (awarded the Nobel prize of physics in 2010), two-dimensional materials, which were prior to this discovery believed not to be able to exist,[1] have quickly become a crucial scientific pillar of materials science. Building a three dimensional crystal out of specific two-dimensional layers have raised the hope that the crystal properties can be controlled in a fully novel way. The building blocks, that are atomically-thin van der Waals materials, are rather versatile, with graphene, boron-nitride, different transition metal dichalcogenides (e.g. WSe_2, MoS_2, . . .), phosphorenes (SnS, GeS,. . .) or group-III chalcogenides (GaS, InSe,. . .) being just some marked examples [27]. The idea is often compared to playing with legos where the building block of crystals can be engineered from different two-dimensional layers by a controlled stacking sequence [28].

Very recently another impressive arena within the realm of stacked two-dimensional van der Waals materials received tremendous attention [29–35]. In contrast to lego blocks, which have to be stacked neatly atop each other to hold together and for that purpose feature the same lattice constant on top of each block, van der Waals materials might be staked at a twist [10, 29, 36–38] or different materials might feature different lattice constants, both of which would lead to a (classical geometric) effect known as a *moiré pattern*. Dreaded by anyone who

---

[1] The Mermin-Wagner theorem [26] forbids spontaneous breaking of a continuous symmetry at finite temperature. Therefore, crystalization (breaking the continuous spatial translation symmetry) is forbidden in two-dimensions and indeed free-standing two-dimensional materials will buckle. However, being supported on a substrate the world of two-dimensional materials becomes accessible.

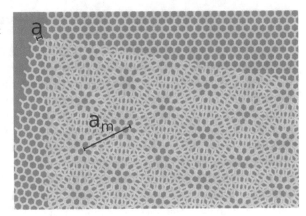

Fig. 1 Moiré pattern of two honeycomb lattices stacked at 10° angle. The lattice constant of one honeycomb lattice a is increased to a_m. For smaller twist angle the factor of lattice constant increase becomes larger

works in front of a camera and has a liking for narrow patterned cloths this effect produces a large scale interference pattern which arises when two similar patterns are stacked at a slight twist or when two patterns deviate only slightly in their periodicity (or which arises when the finite resolution of a camera tries to record a very finely structured textile or other material). When the lattice constant mismatch or the twist is small the periodicity of the interference pattern can be large. This means that the overall lattice constant can be engineered by, e.g, the twist angle between adjacent two-dimensional materials. For small twist angles or lattice constant mismatches the lattice constant of the combined moiré material is tuned form the one of the individual chemical bonds a (typically on the Angstrom (10^{-10} m) scale) to the one of the moiré pattern a_m (typically on the 100 nm (10^{-7} m) scale) constituting an impressive 2–3 orders of magnitude enhancement. This is shown for the example of a hexagonal lattice (such as found in sheets of graphene) in Fig. 1. Put more generally, this provides an unprecedented control knob on material properties which is usually not available in conventional solids where the lattice constant is given by chemical properties. In fact, moiré physics has been omnipresent in surface science for many decades [39, 40] (and references therein) but just now took off with the emergence of two-dimensional heterostructures. This is because the tunable kinetic energy scales of the moiré crystals allow to tune the interplay of kinetic with interaction, interlayer or spin-orbit energy scales. This can be utilized to control competing correlated orders and topology in a novel way [10].

2.2   Towards Nonequilibrium Quantum Materials Design

In materials science much progress has been made in the last few decades in controlling solids routinely by growth and chemical means (such as introducing dopents, substitutions or vacancies or using different substrates to grow the materials on). Another route is the usage of stacking two-dimensional materials as building blocks of crystals [28, 41]. While all of these chemical means over the years have led to many fascinating insights and a certain degree of control over interesting material properties, it is still limited in its flexibility, reversibility as well as time dependent controllability. This has recently triggered the quest for control beyond the chemical possibilities of materials using external stimuli (e.g. light, pressure, strain, electric fields,. . .), which is a field that has made major experimental leaps in the last few years. On a theoretical level, this requires the description of many-body quantum systems out of equilibrium, which in general poses a formidable challenge. In particular, many materials exhibiting interesting emergent behavior or topological properties show an unprecedentedly strong response to external perturbations, such as pressure or light fields. This on the one hand opens up the route towards efficient control paradigms, but on the other also challenges our fundamental theoretical understanding with many problems in the field being unsolved to this day, triggering a strong research interest in the subject matter. All of this has given rise to the dream of *quantum materials on demand* [1, 2] which subsumes the broad goal of programming and controlling quantum properties at will by external stimuli, traversing the more conventional idea of chemical control.

One of the triggers of this new research direction has been the development of light sources with unprecedented brilliance, e.g., in free electron lasers building on the technology developed for synchrotrons, which themselves originally played their most prominent role in the completely different field of high energy physics. Although synchrotrons and cyclotrons continue to be celebrated in the field of high energy physics, it has now been realized that the tunable way of creating high energy photonic radiation (light) by accelerated charges has a usage in its own rights: to probe and excite solids beyond the standard paradigm of optics. In a nutshell this new technology relies on synchrotron radiation being generated as an ensemble of electrons is accelerated through a so-called undulator, a magnetic structure sketched in Fig. 2. For a free electron laser the radiation created is amplified as it bounces through and forth the electron ensemble leading to coherent emission of light; a quantum effect which exponentially increases the brilliance of the laser. With this and other key technologies now at hand, it is time to study the pressing questions of

Fig. 2 Schematic of a free electron laser. A particle accelerator is used to create a beam of high kientic energy electrons which are sourced into a magnetic field which will force them on a periodic path. Due to the bremsstrahlung high fluency lasing is achieved

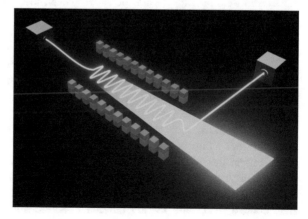

how light and matter interact, how light can control materials on ultrafast time scales (femto- to picoseconds) and what exotic effects may lie beyond the much discussed weak driving regime. These advances were matched with impressive developments in the field of ultrafast time resolved electron diffraction to measure changes on very fast time scales [42, 43].

3   Manipulating Materials with Light

When one aims to *probe* the properties of a solid one usually tries to apply a small perturbation and measures how the solid will react. Within linear response, i.e. for perturbations that are very weak, one can relate this response of the solid (described by so-called response functions) to its thermal static or dynamic properties [44, 45]. That is to say that the weak perturbation is a probe in the true sense of being able to generate insights into the equilibrium properties of the solid (without the perturbation present), but does not alter (to any significance) these properties while probing them. For decades the linear response paradigm has served condensed matter physicists extremely well and it finds application in almost any interpretation of equilibrium measurements [46]. However, insightful as this may be, recently, an alternative vantage point has moved into the center of attention, where the probe now is being strongly intensified to such degree that it generates a surgical alteration to the solid [1,2,5,42,47–49]. In these cases the former probe now acts as a *pulse* to the system and properties are being engineered which lie firmly beyond the limitations set by thermal equilibrium (i.e. we are dealing with the non-perturbative response of the system and the previous linear-response paradigm based on perturbation theory in the applied field does not hold). To rephrase, while in thermal equilibrium the control knobs available to scientists are given basically by chemical compositions (synthesis, growth, doping, . . .), pressure, static electric/magnetic fields, strain and temperature, the nonequilibrium realm offers the tantalizing opportunity of time-dependent control. This upgrades the flexibility of the state of the system from being a thermal one (described by in principle one parameter, temperature) to any state achievable by a nonequilibrium pulse. However, this fascinating paradigm comes with immense experimental and theoretical challenges.

On the experimental side it is not straightforward to generate such strong perturbing pulses in a controlled fashion. Recently, advances in light sources received a tremendous experimental push with key developments being made in ultrafast spectroscopies, as well as within creating tailored strong laser sources that can *specifically drive desired excitations in materials in an almost surgical way* [5, 50–62]. As a consequence structural light is one of the leading experimental vehicles to achieve such nonequilibrium control over solids, not least because it comes in many flavors: We can differentiate between using light in the classical as well as in the deeply quantum limit to achieve control each with and without carrying intrinsic angular momentum; the latter opening up the blossoming field of chiral light sources. The quantum nature of light can be relevant in the solid state context when considering tailored environments (e.g in cavities and other environments) connecting to the fields of circuit QED and polaritonic matter [1–6, 8–10, 15, 63, 64].

On the flip side, simulation capabilities with respect to nonequilibrium setups have developed significantly now enabling the prediction of novel nonthermal phenomena. Such recent theoretical developments allow us to provide guidance to pump-probe experiments, even beyond specific limits such as slowly or quickly evolving pump fields [65–75]. Key proposals for control include: (1) exciting the system into a thermally close-by *metastable state* with different properties [76–79], (2) the controlled excitation of specific vibrational degrees of freedom, dubbed *non-linear phononics*, see e.g. [50, 65, 73], or (3) the direct coupling of electronic degrees of freedom to a driving light field, called *Floquet engineering*, see e.g. [52, 53, 61, 80–84] and will be discussed in more detail below. The latter is based on the tenet that the dressed quantum states of the material under continuously applied, oscillating perturbation are stationary, endowing the driven material with different properties. The current frontier is to define, classify, and potentially realize strongly interacting Floquet phases and control the strength and form of the interaction. One of the key recent advances in the field of Floquet engineering concerns cavity engineering to boost the (prototypically small) light-matter coupling by the presence of a mirror cavity. This approach is sometimes refered to as *cavitronics* [85–92]. In the latter context, current research efforts address the questions of describing light-matter coupled materials in a cavity from an ab-initio perspective [19]. In the last years a fundamental step towards that goal has been made by merging quantum electrodynamics and density functional theory. This quantum electrodynamical density functional theory (QEDFT) [19, 93] provides a unique framework to explore, predict and control light-matter hybrid states which can be created both by driving and by coupling to cavity quantum fluctuations. QEDFT has matured to a valuable tool to address, e.g. the modification of chemical landscapes for molecules embedded in cavities [94–96], but describing general materials in cavities coupled to a strong light field still requires better descriptions of the exchange-and-correlation terms in the theory. In particular, the different intertwined competing energy scales of light, matter and collective degrees of freedom poses challenges to the numerical description, e.g., because they define time-scales which might vary by orders of magnitude or require the accurate modeling of coupling between lattice and electronic degrees of freedom beyond standard frozen phonon accounts [18, 97]. Furthermore, the present theoretical ab-initio understanding of quantum cavity dressing is mainly based on the assumption that the confined photon-field can be described within the dipole approximation, with only very few exceptions even in the simpler realm of model Hamiltonians addressing the issue of relaxing this approximation [64].

3.1 Shooting at the Crystal Lattice

In a series of experiments [50, 51, 54–56], it was shown that the fragile balance between competing states of matter can be tuned by exciting the vibrational degrees

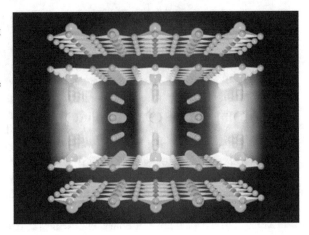

Fig. 3 Illustration of nonlinear phononics. Exciting non-linearly coupled phonon modes can be used to change material properties in an ultrafast, dynamic way (figure from ©Joerg Harms, MPSD Hamburg)

in a lattice (phonons) using light [98, 99]; see Fig. 3. In the experiments, an infrared active phonon mode of a crystal is resonantly addressed directly via light in the THz regime. Due to non-linear phonon-phonon coupling between the infrared active mode (coordinate Q_{IR}) to a Raman active mode (coordinate Q_R) of the type

$$H_{coup} = \alpha Q_{IR}^2 Q_R + \beta Q_{IR}^2 Q_R^2, \tag{1}$$

the lattice is on average stretched along the Raman phonon mode, if we average over the square of the quickly oscillating infrared mode. This mechanism relies on the non-linear coupling of phonons and was consequently dubbed *non-linear phononics*. It was demonstrated that due to this deformation of the lattice the charge density wave order in cuprates was suppressed which in turn unsheaths the competing superconducting order. Theoretical progress has been made in understanding this interplay using density functional theory [65] as well as the functional renormalization group [66]. Ref. [66] focuses on the theoretical description of competing orders and how their interplay is tuned. It relies on quasi-equilibrium arguments, which allows for a qualitative account of the experiments, but neglects the dynamic aspect of the problem. In Ref. [65] a static approach within density functional theory is used to map out the free energy landscape under deformation along the Raman modes (Q_R) from first principles. Along these lines the coupling between the phonon modes Eq. (1) can be determined and treating the modes classically the dynamics resulting from a stimulation of the infrared modes is deduced. A dynamical treatment of a similar system (albeit for a model Hamiltonian), neglecting the quantum nature of the infrared mode, but keeping the quantum nature of the Raman mode can be found in Refs. [67–69]. Phrased loosely, the authors find amplification of superconducting order by parametrically amplified electron-phonon coupling (between the Raman mode and the electrons). Enhancing superconductivity by light (or any other means for that matter) is of course very intriguing with the ultimate technological goal being room temperature superconductivity.

The arguably most puzzling and intriguing report of light-induced superconductivity, however, was given in another experiment, where THz light was shone on the organic superconductor K_3C_{60} [57, 58]. Here, there is no competing charge density wave order. However, measuring the optical properties in a time-resolved fashion after the pump has been applied the authors report signatures akin to a superconducter at temperatures far above the equilibrium $T_c \approx 20\,\mathrm{K}$ on picosecond time scales. At such high temperatures the equilibrium system is metallic. Experimentally, superconducting signatures in the optical conductivity persist clearly up to temperatures as high as $100\,\mathrm{K}$ suggesting a striking factor of five enhancement of T_c by nonequilibrium control. In further experiments, recently it was shown that light can induce superconducting-like properties in K_3C_{60} for much longer times scales [60] and a similar phenomenology was reported in organic salts [59] To date the mechanism of this light-induced phenomena is still subject of heated debate: Whether phonons play a role or not or whether the state is indeed a superconducting one (and what the meaning of transient superconductivity really is) remain open (e.g see [100–102]).

3.2   Shooting at Electrons

On even larger frequency scales then the ones discussed above it is possible to address the electronic degrees of freedom directly. The light field acts as a periodic driving field to the electrons so no additional transduction mechanism like in the above discussed case of non-linear phononics (via lattice vibrations) is needed. This *Floquet engineering* has emerged as a viral field of physics and embodies studies about how many-body systems can be geared by a periodic drive; see Fig. 4. Floquet engineering has already been successfully demonstrated in the field of ultracold gases [103], but solid state experiments are quickly catching up. E.g., in Refs. [52, 53] the authors demonstrate experimentally that Floquet replicas of Dirac cones can be observed in the band structure using time-resolved

Fig. 4 Illustration of floquet engineering of electrons. Addressing electrons in a lattice (blue periodic potential) by light can be used for Floquet engineering. The electronic Hamiltonian can be altered effectively using the periodic perturbation by the light field

APRES studies of the topological insulator Bi_2Se_3. In the pioneering experiment of [61], graphene illuminated by a circularly polarized mid-IR pulse was shown to exhibit an anomalous Hall effect. This is arguably the first experimental solid state demonstration of topological Floquet states [61], which contribute to the underlying transport phenomena [80, 104–106]. However, although the emergent Floquet–Berry curvature also has a non-zero contribution, one finds that the light-induced Hall effect dominantly originates from the imbalance of photo carrier distribution in momentum space [105] in the strong field regime. This finding indicates that intrinsic transport properties of materials can be overwritten by external driving and this may open a way to ultrafast optical-control of transport properties of materials. Another important factor is the dissipation in the material: Due to dissipation Floquet states cannot be formed in the weak field regime but only once the field strength increases, the photon-dressing effect becomes more significant and overcomes the dissipation effects. In the strong field regime, dissipation acts as to stabilize the Floquet states [105].

From a theoretical point of view Floquet systems are under heavy investigation (for a recent reviews see [5, 107, 108]). Especially the question of runaway heating in continuously driven interacting systems is of paramount interest. One promising route is to use optimal control theory to put the system into a (close approximant of a) Floquet eigenstate, which does not evolve under drive [109]. If runaway heating of a generic interacting system by the external drive cannot be avoided (on the relevant time scales), then in principle the state after long time will be a featureless infinite-temperature-like one [110–118]. Beyond many-body localized systems [119] at least two regimes where runaway heating can be suppressed were previously identified [82, 120, 121]: (i) the "Magnus regime" of high frequency as well as (ii) frequencies well within the energy gap of the undriven system (if the system is gapped). In the former, an effective language where the Hamiltonian is replaced by its time average can be used (or one can include higher orders in this Magnus expansion to go beyond this [122]). This allows one to steer some of the electronic properties in interesting ways. Even more flexible control can be obtained in the latter regime, where it was shown that for a Hubbard type system the effective magnetic interaction can be reversed in sign [123] or that charge density wave order can dynamically be stabilized or destabilized at will [82]. In the same frequency regime, where electrons can be addressed directly, an arguably even more intriguing way of obtaining quantum control is to directly address the electronic wave function to tune topological properties (note that the experimental examples [52, 53] discussed above were performed on the topological insulator Bi_2Se_3 already). When the light matter coupling becomes sufficiently large the systems exhibit strongly-correlated electron-photon eigenstates [3], which can be entangled and disentangled by the light matter interplay with intriguing topological applications ahead.

3.3   Amplifying the Light

In the past few decades tremendous efforts have been spend in increasing the brilliance of light sources. One goal is to achieve light strengths for which materials can be controlled using external laser sources. This is a formidable challenge as light-matter coupling, governed by the fine-structure constant $\alpha \approx 1/137$, is quite small. Complementary, to this route of increasing the light field strength, very recently an alternative avenue emerged in which one tries to amplify the light-matter coupling itself instead. In the field of cavity-QED [4] this secondary route has led to the idea of cavity quantum materials engineering [4,9], for which material properties are altered by the presence of cavities confining the light field and in turn enhancing the light-matter coupling geometrically; see Fig. 5. This modification can be obtained either by coupling to the vacuum fluctuations of the light field in such a cavity or by driving the cavity modes with now much weaker laser strengths as the light matter coupling is enhanced.

In this sense, here, we shift the focus on how to realize nonequilibrium states of matter by coupling them instead of to an external drive to the quantized electromagnetic modes in a cavity; the hope here being to realize a novel groundstate (which as such is longlived by construction and no heating problem occurs!) of the coupled cavity-matter system. Nonequilibrium offers the tantalizing opportunity to prepare long-lived metastable states, inaccessible by thermal pathways which can be viewed as (non-global) minima in the free energy of the system. Without a drive the system might never reach these states, as the free energy is minimized globally. However, using specific drive protocols, one might prepare one of these states as a long-lived meta-stable state, which is hidden from a purely equilibrium protocol. Alternatively, with cavity engineering, the first hope is to stabilize these states with the help of strong coupling to a light field. To give a specific example, it was shown both theoretically [80, 104, 105] and experimentally [61] that graphene can experience a light-induced anomalous Hall effect upon irradiation by circularly polarized light

Fig. 5 Illustrations of the concepts of polaronic chemistry and cavitronics. Left: in polaronic chemistry transitions in molecules are effected and investigated by the cavity mode fluctuations. Center: coupling a material (a two dimensional transition metal dichalcogenide in the illustration) to the vaccum fluctuations of a cavity can alter the material's properties. Right: engineering the cavity properties, the modes' polarization that couples to molecules or solids can be made circular. This allows to study the influence of chiral light on matter. (left and right figure from ©Enrico Ronca, Joerg Harms, MPSD Hamburg)

(which corresponds to the design of the celebrated Haldane Hamiltonian [80, 104]). Dissipation, decoherence and heating of the material play a fundamental role in this nonequilibrium QAH state [105]. However, using a cavity as a mediator of the QAH effect would remove those unwanted contributions turning the QAH-state into the ground state of the material embedded in a chiral cavity, for example. In other words, this Floquet-to-cavity-QED mapping allows to translate the striking properties of the Floquet phases induced by (chiral) light into properties of an equilibrium material in a (chiral) cavity [15]. Furthermore, by taking the best of both worlds, i.e., to have a (chiral) driving as well as a (chiral) cavity we gain an enormous toolkit to manipulate material properties at will [15, 124]. The cross-talking that is induced due to the coupled nature of the cavity-matter system between the external driving of the different subsystems has not been explored extensively so far. It promises novel spectroscopic means to generate and control the quantumness of the induced response of the coupled system and to harness the full potential of the nature of the light and/or the matter system [15].

Currently the overwhelming majority of theoretical studies in the context of polaritonic chemistry and material sciences starts from the dipole and single-mode approximation. The coupling to non-standard light (with angular momentum, breaking of time-reversal symmetry etc.) is then commonly described by adapting the dipole approximation in one way or the other. We are currently pushing the highly ambitious and complementary approach, where we use first-principles QED simulations in full minimal coupling to investigate the effect of cavities on a combined light-matter system. Already for the standard linearly-coupled case there are strong differences in observables when comparing the dipole approximation with such a minimal coupling ansatz [19, 125, 126]. While going beyond the dipole approximations for finite systems is in principle straightforward, doing so for extended systems leads to several fundamental issues. Full minimal-coupling is not compatible with the periodicity of the Bloch ansatz that is the basis of virtually all solid-state descriptions. Only in the dipole approximation in velocity (Coulomb) gauge the periodicity is respected and already for the formally equivalent length gauge picture the usual periodicity of the solid is changed and a novel polaritonic unit cell appears [127–129] leading to a Polaritonic Hofstadter butterfly and cavity control of the quantized Hall conductance [130]. Recently effects arising from this have raised tremendous attention and led to the demonstration of a modification of the quantized integer Hall conductance in a high-mobility two-dimensional electron gas embedded in a subwavelength split-ring resonator [131]. In a nutshell the formation of Landau polaritons in the cavity result in a modifcation of the quantized Hall conductance $\sigma_{x}y = \frac{e^2}{h}\nu$ with ν being an integer to a non-integer $\sigma_{xy} = \frac{e^2}{h}\nu\frac{\omega_c^2}{\omega_c^2+\omega_p^2}$ with $\omega_p = \sqrt{\frac{e^2 n_e}{m_e \epsilon_0}}$ and $\omega_c - \frac{eB}{m_e}$ being the diamagnetic shift and cyclotron frequency, respectively. Both the diamagnetic shift and cyclotron frequency depend only on fundamental constants (electron charge e, electron mass m_e, dielectric constant ϵ_0) as well the magnetic field B and the electron density n_e. This observed change in the integer quantum Hall conductivity can only be explained by a cavity-mediated electron hopping via the nonlocal nature of the cavity vacuum fields [131].

Experimental and theoretical research on the control of materials by cavity vacuum fields is destined to considerably accelerate and expand. Controlling the nature of the quantum vacuum surrounding a material inside a resonating cavity can alter its properties imprinting the symmetry of the cavity photons on the matter and design novel quantum materials and phenomena [1, 15]. From the theoretical point of view, future possible developments encompass the generalization to spatially non-homogeneous photon modes, multimode and lossy cavities [3, 4, 9, 15, 19, 21]. New predictions for quantum phenomena mediated by cavity photon modes include the enhancement of superconductivity and establishing photon-mediated super-conductivity, photon-induced magnetism, ferroelectricity, cavity control of many body interactions in materials and photon-driven topological phenomena in two-dimensional heterostructures and beyond [3, 4, 10, 132]. Those are among few of the new lines of research that to be encompassed in the near future under the emerging field of *cavity materials engineering*.

4   Novel Materials and Avenues of Time-Resolved Control

As we are developing more intricate quantum technologies the platforms on which to run these technologies also increasingly move into the center of attention. Recently, many breakthroughs in materials science have given rise to the exper-imental realization of a plethora of different exotic phenomena, many of which build inherently on the principles of quantum mechanics [133]. In this section we complement the brief excerpt summarizing current light-based control schemes of the previous section by outlining a selective assortment of topics on such materials-based control, their intrinsic collective behavior and how they interplay with driving.

4.1   Fabricating New Quantum Materials: A Novel Twist

Stacking layers of van der Waals (vdW) materials has become a flexible avenue towards electronic band structure engineering and marked the rapid rise of vdW heterostructures [41, 134] in condensed matter physics and materials science. Today, vdW heterostructures are celebrated for their potential in nano-electronics, quantum information and the basic sciences, as the control of band structures allows to tune transport, topological and correlated properties with a high degree of flexibility. For example, the celebrated Dirac cone at the Fermi level of undoped monolayer graphene grants access to the physics of linearly dispersing electrons. Turning this around, graphene can function as a condensed matter platform to study the physics of ultra-relativistic particles usually described by this kind of dispersion. This builds a bridge between condensed matter and high-energy physics/cosmology allowing to study hallmark questions of high energy physics such as the Klein paradox [135], or by choosing other two-dimensional materials axyon fields [136] or Weyl magnetic monopoles [137], in a solid state laboratory.

To go even further, some of these analogies between high-energy/cosmology and solid state physics even find interesting potential applications such as the Klein paradox enhancing transport properties in graphene [138]. When the stacked layers have a slight lattice constants mismatch [139], or are slightly rotated with respect to each other [10, 29, 36, 37, 140, 141], a long-wavelength periodic modulation is found. This is referred to as a "moiré superlattice". A moiré superlattice modifies the electronic bandstructure [142] and can, in selected cases, result in the formation of low-energy sub-bands [36, 141]. Following such a route, isolated flat bands have been realized in a wide range of graphene-based structures including twisted bilayer graphene (TBG) [34,35,143–147], double bilayer graphene [148–152], ABC trilayer graphene/BN [153–155] and twisted trilayer graphene [156,157]. On the one hand, this engineering of electronic bands via the moiré superlattice can be utilized to steer topological properties [158–162]. On the other hand, flat band engineering provides a handle to tune the kinetic energy scales to the ones of electronic interactions which can facilitate control of emergent electronic phases such as correlated insulators, superconductors and quantum magnets [10, 34, 35, 37, 156]. Additionally, the wave functions of the flat bands are topologically tied to the dispersive bands in twisted bilayer graphene which was also found to harbor intriguing effects: For example this topological properties give rise to competing Chern insulating states [161] as well as unconventional light-matter couplings [163]. This flexibility also allows to realize strongly correlated topological phases [164, 165], which are highly relevant for quantum information [2].

As briefly stated above correlated phases of matter and topological properties are at the vanguard of condensed matter research for their high potential in key technological applications [1, 2], such as high-temperature superconductivity, high speed and huge capacity memory devices, impeccable security application (in quantum cryptography) or novel information technologies (e.g. in quantum computing). These highly sought after functionalities explains the tremendous interest in twisted vdW materials as candidate systems with highly controllable properties [10] (fit for our increasingly information based society and its need for impeccable security).

In the most studied case of TBG, the flat bands appear only in certain narrow ranges around specific twist angles, so-called magic angles, the largest of which is $1.1 \pm 0.1°$ [34, 36, 143]. This selective appearance arises from a delicate interplay between the layer hybridization energy and twist-determined band displacements in momentum space [34, 36, 143]. The sharp magic angle structure for TBG leads to difficulties in terms of materials design, as slight uncertainties in twist angle result in widely varying band widths and therefore physics. Recently, studies of twisted vdW heterostructures that combine moiré patterns with using semiconducting transition metal dichalcogenides (TMDs) [164, 164, 166, 166–170], twisted bilayer boron nitride [171] or graphitic systems where a band gap is induced by electric fields [148–155], revealed that in these materials the bandwidth varies continuously and smoothly with the twist angle between layers. The absence of sharp magic angles makes these systems less sensitive to small angle variations, which is advantageous for experiments and technological applications in quantum devices. In trilayer graphene on boron nitrite and twisted double bilayer graphene this phenomenon was also demonstrated by using a transverse displacement field to induce a semiconductor bandgap [148–151, 153]. In comparison to these graphetic systems, intrinsic semiconducting transition metal dichalcogenides (TMDs) provide several potential advantages being less restricted in accessible moiré wavelengths and requiring no additional displacement field [164, 164, 166, 166–169]. In this case the bandwidth varies continuously with twist angle, without the need for additional displacement fields, and therefore is widely tunable.

By exploiting the large choice of available TMD materials, properties not present in graphene, such as strong spin-orbit coupling, can be accessed as well. This leads to an additional degree of control to modify electronic properties of twisted materials, and brings several interesting (correlated) states of matter within experimental reach, such as Mott states, Wigner crystals, quantum anamolous Hall states among others [166, 172–183]. In twisted TMD heterobilayers, such as $MoSe_2$ on WSe_2 additionally control over excitons—bound electron-hole states– of unprecedented pristine nature as well as their condensation was demonstrated [178, 184, 185]. Going beyond TMDs, alternative materials have been proposed to yield additional control knobs to tune various electronic and structural properties. For example, we have recently shown that twisted bilayers of GeSe can be used to achieve one-dimensional flat bands in a controlled fashion [186]. Alternatively, twisted bilayer $MnBi_2Te_4$ was proposed to realize flat Chern-insulators with intriguing topological properties [187]. Even for correlated oxides moiré engineering has been demonstrated to alter electronic properties [188]. All of this demonstrates a fascinating opportunity: The catalog of vdW materials is huge with a tremendous potential for control.

This short discussion exemplifies something profound: While much focus has been put on studying graphitic systems [189], with TBG still at the uncontested center of attention, there is a huge combinatoric space of available chemical compositions to exploit which will unlock some of the game changing promises made by quantum materials design; see Fig. 6.

4.2   Exotic Collective Phenomena and Their Control

Two thrusts are currently at the vanguard of condensed matter physics: topological and collective phases of matter.

The former is a relatively young field and relies on the fact that geometric properties of the wave functions can give rise to very robust novel boundary physics in solids. This robustness –sometimes refereed to as topological protection—is hailed as an important ingredient in future quantum technologies. The arguably earliest discovery of this paradigm is the quantum Hall effect dominated by robust counter-propagating edge modes carrying currents. This effect was later generalized to the anomalous case for which the modes are polarized in one direction as well

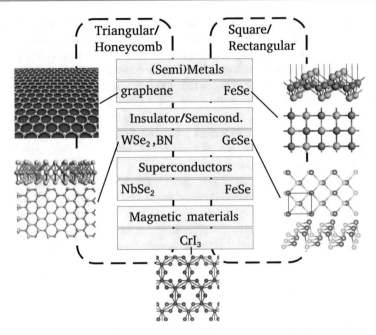

Fig. 6 (Incomplete) characterization of twisted vdW heterostructures via their mono-layer lattice symmetry as well as the phase of matter which they display in equilibrium. Even this very incomplete list hints at the vast combinatoric possibilities offered by twisting these structures

as to the quantum spin Hall effect for which spin-polarization occurs. In general Chern insulators exhibit the general phenomenology of robust edge modes linked to non-trivial geometric aspects of the wave function. Following this line of thought it is nowadays understood that at an interface to a topologically trivial system (such as the vacuum) a closing of the gap in Chern insulators and a corresponding edge mode is imperative. Soon after the concept of such topological gaps protecting edge modes in topological insulators and superconductors was introduced, multiple classes of systems were identified hosting gapless bulk modes but yet topological edge states are found. One of these classes of systems are Weyl semimetals [207–209], which host an even number of gap closing points with opposite chirality of the wave functions at these closing points, which are connected by topological edge modes at interfaces to a vacuum. Thus, such a system exhibits features "in between" of those of a topological insulator (edge modes) and a metallic system (ungapped). The gap closing points, the so-called Weyl nodes, which determine the edge mode properties were demonstrated to be highly susceptible to chiral light fields [210–213]. To generalize these concepts further, higher-order topological insulators were introduced [203, 204]. In higher-order topological system the edge modes are not $D - 1$ but $D - n$ with $n > 1$ dimensional, where D is the dimension of the system. E.g., here robust *corner* (zero dimensional) states at the surface of two-dimensional systems in contrast to the (one-dimensional) edge states discussed above are found.

All combinations of dimensionality of the system and the surface states can be discussed and give rise to a rich playground to control and modify surface properties by engineering the bulk material. Nowadays, many of these topological phenomena are understood in terms of topological invariants, the most clear example of which is the integer quantum hall effect being characterized by the integer-valued Chern number.

In contrast, collective phases of matter refers to the physics emerging in systems where particles are (strongly) correlated among each other. This emergence can give rise to entirely novel physics and is the foundation of phenomena such as quantum magnetism and superconductivity, which find widespread applications already. However, the catalog of emergent phenomena extends far beyond these two well-studied cases. They, e.g., include more exotic types of insulators, such as Mott insulators, for which the insulating behavior is driven by many-body interactions, Wigner crystals, for which interaction lead to a spontaneous breaking of the spatial symmetry (crystallization) or charge density wave formations which freeze charge movement in place. It can also give rise to fundamentally important concepts such as Luttinger Liquids in quantum wires, which exhibit power-law dominated transport properties firmly beyond an effective single particle picture or the prediction of excitonic condensates, which are analog to superconductors, but with the paired electrons (cooper pairs) being replaced by condensed particle-hole pairs instead. For the study of these highly complex and intuition-defying phenomena new simulation techniques are brought to life. Complementing the traditional route of characterizing these phenomena in and out of equilibrium using techniques from quantum many-body theory, nowadays there is a strong push towards quantum simulations [10, 214, 215]. In quantum simulations, in contrast to the traditional computing approach, a highly controlled quantum system is realized physically which ought to mimic the original Hamiltonian under scrutiny. This approach again (see above) relates back to key ideas of R. Feynman [12]:

> Nature isn't classical, dammit, and if you want to make a simulation of nature, you'd better make it quantum mechanical. –Richard Feynman

In a sense this provides a full loop in our approach as we want to understand the quantum nature of materials, we try to abstract the relevant parts in a model Hamiltonian, which we then again solve using a quantum mechanical simulator [214]. Intriguingly, as mastery over solids increases the basis of such in and out of equilibrium quantum simulators might move from, e.g., ultracold gases of atoms confined in optical lattices [215] towards a material platform in certain cases, as we have recently proposed in Ref. [10] in the context of "a quantum moiré simulators".

At the interface of topology and electron correlations—as usual in science—an even more rich catalog of physics is discovered. Strong correlations can give rise to fractionalization of charges, e.g. in the fractional quantum Hall effect or fractional Chern insulators. These systems could harbor intriguing anyonic quasiparticles with braiding statistics going beyond the simpler fermionic (-1) or bosonic $(+1)$ cases. These particles are believed to be important in quantum information sciences as

Table 1 Phases of matter being driven either by topology, correlations or both. We indicate a non exhaustive list of (twisted) van der Waals materials related references where the corresponding phases of matter were discussed. This clearly highlights the versatility of van der Waals materials engineering to realize these exotic condensed matter phenomena

Electronic Topology	← Both →	Electronic Correlations
Quantum Hall Effect (QHE) [190]	Fractional QHE [139]	
Anomalous QHE [182, 191]		
Quantum Spin Hall Effect [164]		
	Quantum Spin Liquids [192]	Quantum Magnetism [155]
	Topological Mott Insulators [165]	Mott Insulators [181]
		Wigner Crystals [183, 193]
Chern Insulators [161]	Correlated Chern Insulators [155, 194]	
	Fractional Chern Insulators [191, 195]	
	Topo. Superconductors (TSC) [196, 197]	(Un)conventional SC [35, 198, 199]
	Topological CDW [200]	Charge Density Waves (CDW) [201]
		Nematic States [152]
	Topological Luttinger Liquids	Luttinger Liquids [186, 202]
Higher-Order Topological Insulators [203]	Higher-Order TSC [204]	
Weyl Semimetals [205]		
	Topo. Excitonic Insulators [206]	Excitonic Insulators [185]

they could allow topologically protected and universal quantum gate operations. In the context of quantum magnetism, spin liquid phases, phases which avert magnetic ordering and are often characterized by intriguing topological properties are also raising tremendous attention at the moment [192] just as well as topological superconductors which might host evasive Majorana edge modes. A short summary of the intriguing behavior in topological and/or correlated systems is presented in Table 1.

With this exciting plethora of phenomena at hand (of which we have barely scratched the surface!), which have only partially been discovered experimentally let alone exploited in technologies, it comes to no surprise that scientists are pushing hard on the boundaries of what can be engineered in artificial systems. With every new material or other avenue of control such as stacking or driving in our tool belt, we come closer to the goal of quantum materials mastery and therefore the realization of the fascinating physics hinted at above. This plausibilizes the exponential growth in the field of materials science that we witnessed in the last decades. It also explains why twistronics and ultrafast materials science is on the rise. In the end we want to obtain materials mastery on the level of the quantum degrees of freedom and be able to modify topological and collective phases at will, which in the best case will enable new technologies. Twistronics is still very much in its infancy, but initial experiments show the difficulties in obtaining large, clean areas that are unstrained and have a homogenous twist for which the flat band physics discussed above dominates. In ultrafast materials science the large driving fields required to modify materials properties substantially, come with heating; a major challenge to be overcome in the future. Therefore, new ideas and concepts are required to pave the road ahead.

5   The Road Ahead: The Future of 2D Materials Is Bright

The future of two-dimensional materials is bright. As hinted at by the references in Table 1 many of the fascinating topological and many-body phenomena prototypical to condensed matter research have already been proposed and/or realized in such material platforms. Yet, there is more! On the one hand, there are many additional known phases of matter and phenomena going beyond the short exposition of Table 1 in particular and this brief overview in general which we expect to be realized in two-dimensional materials in the future. On the other hand, the journey of exploring the consequences of topology, correlations and other unifying themes in condensed matter research are far from complete. Exciting surprises are expected to be unveiled and discoveries to be made in the future, not least building on the extremely versatile platform that is two-dimensional materials.

We believe that there is also another meaning in which the future of two-dimensional materials is bright: the very promising route of interacting with heterostructures using light in the different versions that we have described in the previous pages. This provides a unique platform in which to explore many inter-disciplinary concepts from fields ranging from quantum thermodynamics, materials science, quantum optics, particle physics, cavity-QED, dissipative (non-hermitian quantum) systems, fluid dynamics and more. Many new phenomena at the interface of those disciplines could be realized using this condensed matter platform and would enable a better understanding of the physics (as it becomes relatively easily measurable!). It would help in the development of our understanding of new states of matter that build on those cross-disciplinary concept, as for example realizations of novel quasiparticle condenses and polaritonic quantum matter (based on light-matter hybrids) become available. In the previous sections we have presented an overview of the impressive achievements in the field of control of nonequilibrium phenomena in materials and the realization of new phenomena and functionalities. We have also highlighted the immense possibilities offered by novel 2D heterostructure platform combined with optical cavities (quantum light) or interacting with a laser field (pump probe studies). We have focused here mainly on the impact on materials but the present findings can have applications to chemistry (control of chemical reactions and energy transfer phenomena) as well as biophysics.

5.1   Polaritonic Chemistry

Seminal experimental results have uncovered that by engineering the vacuum of the electromagnetic field via, e.g. optical cavities, chemical reactions can be changed. Since this can happen at standard ambient conditions in solvation, such a novel control technique is very promising for future quantum technological applications. This field of polaritonic chemistry has become a rapidly developing one over the last years. For example, the hybridisation of light-matter (quantum) states within a cavity, i.e. the formation of polaritons, has lead to astonishing experimental discoveries. It was observed that vibrational strong coupling can inhibit [216], steer [217] and even catalyze [218] a chemical process. Moreover, seminal measurements were published on the control of photo-chemical reactions [219], energy transfer [220], the realization of single molecular strong coupling [221] and even evidence for the increase of the critical temperature in superconductors was reported [222]. polaritonic chemistry is a notoriously hard problem to solve from a fundamental theoretical perspective [223]. This originates from the strong hybridization of light and matter, which a priori requires to treat not only the matter part with "chemical" accuracy, but also the involved electro-magnetic fields. The theoretical understanding of such cavity-mediated chemical reactions is very challenging. Besides the well-known complexity of large chemical systems we have also to consider the strong coupling to cavity modes and collective effects. Very recently, first ab-initio simulation methods emerged, which solve the Pauli-Fierz Hamiltonian numerically. Computational implementations involve quantum electrodynamical density functional theory (QEDFT) [224] and coupled cluster theory for molecular polaritons [94]. Early applications of these novel ab-initio methods indeed suggest that the widespread phenomenological models cannot capture the entire story of polaritonic chemistry and important effects are absent. Linear response QEDFT has demonstrated that strong local modifications of the electronic structure emerge in the vicinity of impurities, embedded within a collectively coupled environment. By construction, collective phenomenological models did not include this effect. The observation of this complex interplay has started a paradigmatic shift in the understanding of polaritonic chemistry, since it re-introduces the principle of locality for polaritons [223], which is key for the understanding of chemical processes (e.g. reactions). Furthermore, QEDFT Ehrenfest dynamics reveal that the cavity can correlate multiple vibrational degrees of freedom, which redistributes energy from a specific bond to other degrees of freedom and eventually suppresses the bond breaking [225]. Still there is plenty of experimental and theoretical work to be done to unravel all miscrocopical and collective processes behind the control of chemical phenomena in cavities as well as their reliable prediction.

5.2   High Energy Physics (and Beyond) in a Condensed Matter Lab

The interdisciplinary character of the research discussed so far is also evident when connecting condensed matter to high energy physics. To provide further examples of highly intriguing directions of future research (beyond the above), we name type-I and -II Dirac Fermions in condensed matter systems as a direct link between the low-energy excitation of a Dirac-material in solid-state physics and that of a quantum theory in flat (type-I Dirac fermions) or curved (type-II Dirac fermions) spaces from general relativity. Another example briefly touched upon in the main text is provided by the chiral anomaly in astrophysics and cosmology. As mentioned analyzing the behavior of Weyl nodes in materials allows for a solid state inroad into this phenomena. In certain solid state systems and conditions, it is possible to tune the effective fermion mass experimentally [226]. Thus, one can study the effectiveness of the chiral magnetic effect as a function of the effective fermion mass. With suitable rescaling of the relevant quantities, one could transfer such measurements in condensed matter to astrophysical environments. Therefore, this route might allow us to address open questions about how effective the chiral magnetic effect can be in astrophysical environments from a very controlled condensed matter viewpoint.

Strongly correlated phenomena and the emergence of hydrodynamic flow in materials (exotic metals) is an alternative, interesting avenue of research that connects to such a multidisciplinary approach. The tools and advances briefly highlighted in the present essay would enable the control and realization of these states in a real material.

5.3   (Quantum) Floquet Materials Engineering

In Floquet materials engineering, beyond a variety of Floquet topological phases, such as Floquet topological insulators and Floquet Weyl semimetals for example, which replicate existing equilibrium analogues, there is a treasure trove of possible Floquet states, which have no equilibrium counterpart. The driven Floquet state, however, is hard to realize experimentally in materials because processes such as dissipation, decoherence, impurity scattering, excited state lifetimes and heating strongly limit the survival of the Floquet states. Furthermore, the actual population of Floquet states is specific to the driving protocol. As we discussed in the subsection 3.2, especially the question of runaway heating in continuously driven interacting systems is of fundamental interest as it, if it cannot be avoided, dictates a featureless infinite-temperature-like state at long-times. One direction to avoid this, which we believe will be of major relevance in the future is the Floquet-to-cavity-QED mapping [15]. This generalizes the concepts of polaritonic chemistry towards the solid state realm. The major theme here being that controlling the nature of the quantum vacuum acting on a material inside a resonating cavity allows to steer some of the electronic properties in interesting ways. Despite the considerable theoretical difficulties in describing the driven-dissipative nature of these hybrid light-matter correlated systems, the developments have fueled theoretical work to not only explain the experimental observations, but to lay out a set of proposals and to define an agenda for strongly-correlated electron-phonon-photon systems (which can led to new light-matter hybrids in interacting polaritons).

Furthermore, cavity-modes can both mediate materials' correlations and be an additional knob to measure and control electronic systems that are interacting from the outset. Such photon mediated interactions can be engineered to be drastically different and induce long-ranged interactions and non-local entanglement. When matter with intrinsic strong electron correlations is embedded in a cavity, qualitatively new phenomena are expected to emerge as a whole zoo of polaritonic quasiparticles come to life. Just like electron-hole pairs (excitons) can be hybridized with light, various collective modes (e.g. Goldstone modes, Higgs particles, Bogoliubov quasiparticles, etc.) can be mixed with photons leading to non-local composite polaritons [16]. Driving the cavity can induce condensation of this new form of light-matter hybrids with strong and tuneable interactions, paving the way to quantum control of coherent phases.

5.4   Twistronics for Ultrafast Quantum Materials Design

In the previous sections, we have introduced the concept of moiré van der Waals heterostructures as robust solid-state based quantum simulation platform to study strongly correlated physics and topology such as Mott and fractional Chern insulator, ferromagnetic and correlated QAH insulator, quantum chiral spin liquid, superconductivity (d-wave, triplet pairing, in a Kitaev lattices,...), Majorana fermions, and many more [10]. A high-impact direction of future research is to use this novel materials platform to enhance the capabilities of ultrafast quantum materials design. We believe this to be a particularly fruitful direction as the electronic band structures of moiré crystals can be engineered in a particularly flexible manner. As mentioned, currently one of the main limitations of the field are heating effects, which are detrimental to control. This is where moiré engineering of materials can intervene, holding the timely promise that the properties of these twisted vdW materials can be controlled with high flexibility, allowing for a unique bidirectional optimization, see Fig. 7, of the driving and of the material towards each other (the current state-of-the-art is mainly to engineer the driving towards the material due to lack of material control in conventional solids). Furthermore, in moiré materials different control mechanisms are moved to comparable energy scales utilizing the twist-induced reduction of the kinetic energy; see top panel of Fig. 8. This allows us to address and exploit them in a synergistic fashion. The methodologically ambitious goal of combining the fields of twisted van der Waals materials with the one of ultrafast quantum materials design is thus particularly timely and synergistic in nature. It advances nonequilibrium quantum materials design into a new era, beyond the current limits set by the lack of device flexibility and by heating effects.

Fig. 7 Bidirectional optimization approach for driven materials' phenomena enabled by innovative quantum materials design such as twistronics. In a synergistic fashion the optimization of twisted vdW materials' properties to the drive and vice versa will advance nonequilibrium control of solids beyond the current limitations set by detrimental heating effects and the limited flexibility of conventional solid state based platforms

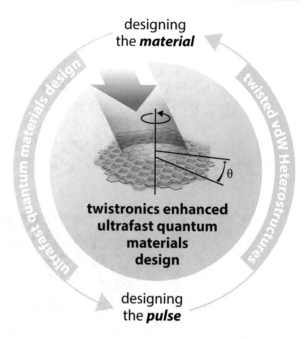

Finally, we propose "Cavity Twistronics", i.e. to combine cavity quantum electrodynamics (QED), quantum reservoir engineering and 2D-twisted van der Waals heterostructures to build a novel and unique platform that enables the seamless realization and control of a plethora of interacting quantum phenomena, including exotic, elusive and not-yet envision correlated and topological phases of matter. An illustration is shown in the bottom panel of Fig. 8. This research direction will open up exciting new opportunities for the study of the interaction of quantized light with quantum many body systems with a number of intriguing observations and few exotic predictions ranging from enhancing superconductivity [88, 222] to driving paraelectric-ferroelectric transition in perovskites [15, 227] to control of excitons in 2D materials [228] to strongly correlated Bose–Fermi mixtures of degenerate electrons and dipolar excitons [229] to emergent topological phases. Cavity control of many-body interactions in strength and form, realized in these highly tunable platforms is an particularly interesting direction of Cavity-QED. Complementary, cavity enhancement of electron-phonon coupling [88] could strengthen interactions and provide unique means to control properties of moiré bands. This facilitates the microscopic understanding of superconductivity and other strongly correlated phenomena inside a cavity. Moving towards 3D material's embedding we can envision stackings of 2D materials to be a promising route. E.g. stacks at alternating twist meaning that the twist angle alternates between two values, zero and α, could be particularly promising [24]. If α is small, in-plane localized sites will emerge by the moiré interference, and these sites will lie directly atop each other in the out-of-plane direction We expect this to give rise to a 3D quantum Hall state. Utilizing the

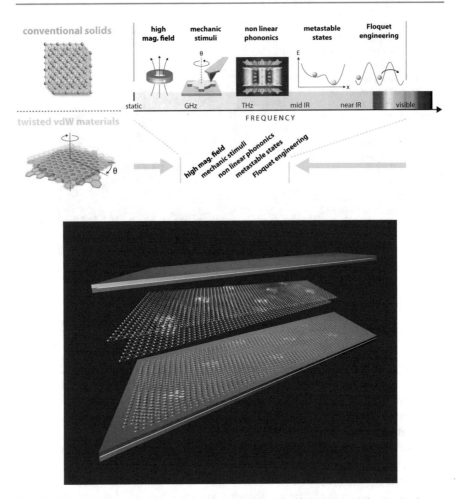

Fig. 8 Mechanisms of control in novel quantum materials. Top: frequency scale illustrating different nonequilibrium control routes in conventional solids (top and images in the center) as well as in the case of twisted vdW materials (bottom). For the latter, control mechanisms are squeezed to similar energy scales. Bottom: illustration of cavitwistronics. By utilizing the advantages of both cavitronics and twistronics novel physical regimes and control forms should be accessible (lower panel from ©Joerg Harms, MPSD Hamburg)

cavity modes as synthetic dimensions we could realize a 4D-quantum Hall effect [230] and even go to effectively higher dimensions.

Looping back to the connection to other fields of physics, highly engineered moiré materials hold further promises to extend concepts of high-energy physics. For example, entirely new fermionic quasiparticles beyond those known from high-energy physics could be engineered in cavities. In quantum field theory, three types of fermions play a fundamental role in our understanding of nature: Majorana, Dirac, and Weyl. They are constrained by the Poincare symmetry in high-energy physics. In contrast, electrons and quasi-particles in materials obey the space group symmetries, which are less constrained. Therefore, beyond Dirac, Weyl and Majorana fermions, new fermions with non-trivial topological properties could be observed [231]. Inspired by this idea, and using the cavity modes as additional synthetic dimensions, we propose to extend the theoretical group symmetry classification to the larger light-matter hybrid space to identify conditions under which spin-1 and spin-3/2 fermions as well as other (fractal) fermions can be created.

6   Closing

In closing, the field of manipulating quantum materials with light is, in our humble opinion, destined to yield many exciting surprises in the coming years. We hope that this has become apparent from the collection of topic we chose to discuss above. However, the reader should be advised that the list of phenomena and routes of future research we provided is not complete. Quite the opposite is true, we have barely penetrated but the most superficial of surface in this quickly emerging field. New discoveries are to be made with some high impact advances sketched above which will be complemented, like is usually in science, by unforeseen discoveries in the near future. We are very excited to be part of this collective and intriguing endeavor and agree with E. Teller when he said that

> the science of today is the technology of tomorrow.. –Edward Teller.

With that in mind we look forward to the future of the field of quantum materials control with the highest expectations.

Acknowledgments We thank Jörg Harms for helping with some of the figures. DMK acknowledges the Deutsche Forschungsgemeinschaft (DFG, German Research Foundation) for support through RTG 1995, within the Priority Program SPP 2244 "2DMP"—443273985 and under Germany's Excellence Strategy—Cluster of Excellence Matter and Light for Quantum Computing (ML4Q) EXC 2004/1-390534769. AR acknowledges financial support from the Cluster of Excellence 'CUI: Advanced Imaging of Matter'- EXC 2056—project ID 390715994 and SFB-925 "Light induced dynamics and control of correlated quantum systems"—project 170620586 of the Deutsche Forschungsgemeinschaft (DFG), Grupos Consolidados (IT1249-19). We acknowledge support from the Max Planck-New York City Center for Non-Equilibrium Quantum Phenomena. The Flatiron Institute is a division of the Simons Foundation.

This is a preprint of the following chapter: D.M. Kennes, A. Rubio, "A New Era of Quantum Materials Mastery and Quantum Simulators In and Out of Equilibrium" published in Sketches of Physics: The Celebration Collection, edited by Roberta Citro, Morten Hjorth-Jensen, Maciej Lewenstein, Angel Rubio, Wolfgang P. Schleich, James D. Wells and Gary P. Zank, 2022, Springer reproduced with permission of Springer. The final authenticated version is available online at: http://dx.doi.org/[DOI shall be updated once given]

References

1. D.N. Basov, R.D. Averitt, D. Hsieh, Towards properties on demand in quantum materials. Nat. Mater. **16**, 1077 (2017)
2. Y. Tokura, M. Kawasaki, N. Nagaosa, Emergent functions of quantum materials. Nat. Phys. **13**, 1056 (2017)
3. J. Bloch, A. Cavalleri, V. Galitski, M. Hafezi, A. Rubio, Strongly-correlated electron-photon systems. Nature **606**, 41–48 (2022)
4. D.N. Basov, A. Asenjo-Garcia, P.J. Schuck, X. Zhu, A. Rubio, Polariton panorama. Nanophotonics **10**, 549–577 (2021)
5. A. de la Torre, D.M. Kennes, M. Claassen, S. Gerber, J.W. McIver, M.A. Sentef, Colloquium: nonthermal pathways to ultrafast control in quantum materials. Rev. Mod. Phys. **93**, 041002 (2021)
6. B. Keimer, J.E. Moore, The physics of quantum materials. Nat. Phys. **13**, 1045 (2017)
7. P. Narang, C.A.C. Garcia, C. Felser, The topology of electronic band structures. Nat. Mater. **20**, 293 (2021)
8. Y. Tokura, K. Yasuda, A. Tsukazaki, Magnetic topological insulators. Nat. Rev. Phys. **1**, 126 (2019)
9. F. Schlawin, D.M. Kennes, M.A., Sentef, Cavity quantum materials. arXiv:2112.15018 (submitted to Applied Physics Reviews) (2022)
10. D.M. Kennes, M. Claassen, L. Xian, A. Georges, A.J. Millis, J. Hone, C.R. Dean, D.N. Basov, A.N. Pasupathy, A. Rubio, Moiré heterostructures as a condensed-matter quantum simulator. Nat. Phys. **17**, 155 (2021)
11. J. Preskill, Quantum computing in the NISQ era and beyond. Quantum **2**, 79 (2018)
12. R.P. Feynman, Simulating physics with computers. Int. J. Theor. Phys. **21**, 467 (1982)
13. J. Orenstein, J. Moore, T. Morimoto, D. Torchinsky, J. Harter, D. Hsieh, Topology and symmetry of quantum materials via nonlinear optical responses. Annu. Rev. Condens. Matter Phys. **12**, 247 (2021)
14. J.M. Smith, Introducing materials for quantum technology. Mater. Quant. Technol. **1**, 010201 (2020)
15. H. Hübener, U. De Giovannini, C. Schäfer, J. Andberger, M. Ruggenthaler, J. Faist, A. Rubio, Engineering quantum materials with chiral optical cavities. Nat. Mater. **20**, 438 (2021)
16. D.N. Basov, A. Asenjo-Garcia, P.J. Schuck, X. Zhu, A. Rubio, Polariton panorama. Nanophotonics **10**, 549 (2021)
17. Y. Ashida, A. Imamoglu, E. Demler, Cavity quantum electrodynamics at arbitrary light-matter coupling strengths. Phys. Rev. Lett. **126**,153603 (2021)
18. S. Latini, D. Shin, S.A. Sato, C. Schäfer, U. De Giovannini, H. Hübener, A. Rubio, The ferroelectric photo ground state of SrTiO3: cavity materials engineering. Proc. Natl. Acad. Sci. **118**, e2105618118 (2021). https://doi.org/10.1073/pnas.2105618118. https://arxiv.org/abs/https://www.pnas.org/content/118/31/e2105618118.full.pdf
19. M. Ruggenthaler, N. Tancogne-Dejean, J. Flick, H. Appel, A. Rubio, From a quantum-electrodynamical light–matter description to novel spectroscopies. Nat. Rev. Chem. **2**, 0118 (2018)
20. F.J. Garcia-Vidal, C. Ciuti, T.W. Ebbesen, Manipulating matter by strong coupling to vacuum fields. Science **373**, eabd0336 (2021). https://arxiv.org/abs/https://www.science.org/doi/pdf/10.1126/science.abd0336
21. C. Genet, J. Faist, T.W. Ebbesen, Inducing new material properties with hybrid light–matter states. Phys. Today **74**, 42 (2021). https://arxiv.org/abs/https://doi.org/10.1063/PT.3.4749
22. F. Liu, W. Wu, Y. Bai, S.H. Chae, Q. Li, J. Wang, J. Hone, X.Y. Zhu, Disassembling 2d van der waals crystals into macroscopic monolayers and reassembling into artificial lattices. Science **367**, 903 (2020). https://arxiv.org/abs/https://www.science.org/doi/pdf/10.1126/science.aba1416

23. K. Yao, N.R. Finney, J. Zhang, S.L. Moore, L. Xian, N. Tancogne-Dejean, F. Liu, J. Ardelean, X. Xu, D. Halbertal, K. Watanabe, T. Taniguchi, H. Ochoa, A. Asenjo-Garcia, X. Zhu, D.N. Basov, A. Rubio, C.R. Dean, J. Hone, P.J. Schuck, Enhanced tunable second harmonic generation from twistable interfaces and vertical superlattices in boron nitride homostructures. Sci. Adv. **7**, eabe8691 (2021). https://arxiv.org/abs/https://www.science.org/doi/pdf/10.1126/sciadv.abe8691

24. L. Xian, A. Fischer, M. Claassen, J. Zhang, A. Rubio, D.M. Kennes, Engineering three-dimensional moiré flat bands. Nano Lett. **21**, 7519 (2021)

25. A.P. Ramirez, B. Skinner, Dawn of the topological age? Phys. Today **73**, 30 (2020). https://doi.org/10.1063/PT.3.4567

26. N.D. Mermin, H. Wagner, Absence of ferromagnetism or antiferromagnetism in one- or two-dimensional isotropic heisenberg models. Phys. Rev. Lett. **17**, 1133 (1966)

27. K.S. Novoselov, A. Mishchenko, A. Carvalho, A.H.C. Neto, 2d materials and van der waals heterostructures. Science **353**, aac9439 (2016). https://arxiv.org/abs/https://www.science.org/doi/pdf/10.1126/science.aac9439

28. A.K. Geim, I.V. Grigorieva, Van der waals heterostructures. Nature **499**, 419 (2013)

29. G. Li, A. Luican, J.M.B. Lopes dos Santos, A.H. Castro Neto, A. Reina, J. Kong, E.Y. Andrei, Observation of van hove singularities in twisted graphene layers. Nat. Phys. **6**, 109 (2010)

30. D. Wong, Y. Wang, J. Jung, S. Pezzini, A.M. DaSilva, H.Z. Tsai, H.S. Jung, R. Khajeh, Y. Kim, J. Lee, S. Kahn, S. Tollabimazraehno, H. Rasool, K. Watanabe, T. Taniguchi, A. Zettl, S. Adam, A.H. MacDonald, M.F. Crommie, Local spectroscopy of moiré-induced electronic structure in gate-tunable twisted bilayer graphene. Phys. Rev. B **92**, 155409 (2015)

31. K. Kim, A. DaSilva, S. Huang, B. Fallahazad, S. Larentis, T. Taniguchi, K. Watanabe, B.J. LeRoy, A.H. MacDonald, E. Tutuc, Tunable moiré bands and strong correlations in small-twist-angle bilayer graphene. Proc. Natl. Acad. Sci. U.S.A. **114**, 3364 (2017)

32. N.R. Finney, M. Yankowitz, L. Muraleetharan, K. Watanabe, T. Taniguchi, C.R. Dean, J. Hone, Tunable crystal symmetry in graphene–boron nitride heterostructures with coexisting moiré superlattices. Nat. Nanotechnol. **14**, 1029 (2019)

33. D. Edelberg, H. Kumar, V. Shenoy, H. Ochoa, A.N. Pasupathy, Tunable strain soliton networks confine electrons in van der waals materials. Nat. Phys. **16**, 1097 (2020)

34. Y. Cao, V. Fatemi, A. Demir, S. Fang, S.L. Tomarken, J.Y. Luo, J.D. Sanchez-Yamagishi, K.Watanabe, T. Taniguchi, E. Kaxiras, R.C. Ashoori, P. Jarillo-Herrero, Correlated insulator behaviour at half-filling in magic-angle graphene superlattices. Nature **556**, 80 (2018)

35. Y. Cao, V. Fatemi, S. Fang, K. Watanabe, T. Taniguchi, E. Kaxiras, P. Jarillo-Herrero, Unconventional superconductivity in magic-angle graphene superlattices. Nature **556**, 43 (2018)

36. R. Bistritzer, A.H. MacDonald, Moiré bands in twisted double-layer graphene. Proc. Natl. Acad. Sci. **108**, 12233 (2011)

37. L. Balents, C.R. Dean, D.K. Efetov, A.F. Young, Superconductivity and strong correlations in moiré flat bands. Nat. Phys. **16**, 725 (2020)

38. E.Y. Andrei, D.K. Efetov, P. Jarillo-Herrero, A.H. MacDonald, K.F. Mak, T. Senthil, E. Tutuc, A. Yazdani, A.F. Young, The marvels of moiré materials. Nat. Rev. Mat. **6**, 201 (2021)

39. M. Rocca, T.S. Rahman, L. Vattuone, *Springer Handbook of Surface Science*, 1st edn. (Springer, Cham, 2020)

40. I. Matsuda, Introduction for the special issue "moiré patterns and surface science". Vacuum Surface Sci. **61**, 704 (2018)

41. C.R. Dean, A.F. Young, I. Meric, C. Lee, L. Wang, S. Sorgenfrei, K. Watanabe, T. Taniguchi, P. Kim, K.L. Shepard, J. Hone, Boron nitride substrates for high-quality graphene electronics. Nat. Nanotechnol. **5**, 722 (2010)

42. J.A. Sobota, Y. He, Z.X. Shen, Angle-resolved photoemission studies of quantum materials. Rev. Mod. Phys. **93**, 025006 (2021)

43. E.J. Sie, C.M. Nyby, C.D. Pemmaraju, S.J. Park, X. Shen, J. Yang, M.C. Hoffmann, B.K. Ofori-Okai, R. Li, A.H. Reid, S. Weathersby, E. Mannebach, N. Finney, D. Rhodes, D. Chenet, A. Antony, L. Balicas, J. Hone, T.P. Devereaux, T.F. Heinz, X. Wang, A.M. Lindenberg, An ultrafast symmetry switch in a Weyl semimetal. Nature **565**, 61 (2019)

44. G. Onida, L. Reining, A. Rubio, Electronic excitations: density-functional versus many-body green's-function approaches. Rev. Mod. Phys. **74**, 601 (2002)
45. M. Cardona, G. Güntherodt, *Light Scattering in Solids II* (Springer, Berlin, 1982)
46. R. Kubo, Statistical-mechanical theory of irreversibleProcesses. I. Generaltheory and simple applications to magnetic and conduction problems. J. Phys. Soc. Jpn. **12**, 570 (1957)
47. G. Wang, A. Chernikov, M.M. Glazov, T.F. Heinz, X. Marie, T. Amand, B. Urbaszek, Colloquium: excitons in atomically thin transition metal dichalcogenides. Rev. Mod. Phys. **90**, 021001 (2018)
48. M. Buzzi, M. Först, A. Cavalleri, Measuring non-equilibrium dynamics in complex solids with ultrashort x-ray pulses. Philos. Trans. R. Soc. A Math. Phys. Eng. Sci. **377**, 20170478 (2019). https://arxiv.org/abs/https://royalsocietypublishing.org/doi/pdf/10.1098/rsta.2017.0478
49. F.H.L. Koppens, T. Mueller, P. Avouris, A.C. Ferrari, M.S. Vitiello, M. Polini, Photodetectors based on graphene, other two-dimensional materials and hybrid systems. Nat. Nanotechnol. **9**, 780 (2014)
50. M. Först, C. Manzoni, S. Kaiser, Y. Tomioka, Y. Tokura, R. Merlin, A. Cavalleri, Nonlinear phononics as an ultrafast route to lattice control. Nat. Phys. **7**, 854 (2011). arXiv:1101.1878
51. M. Först, R. Mankowsky, H. Bromberger, D.M. Fritz, H. Lemke, D. Zhu, M. Chollet, Y. Tomioka, Y. Tokura, R. Merlin, J.P. Hill, S.L. Johnson, A. Cavalleri, Displacive lattice excitation through nonlinear phononics viewed by femtosecond X-ray diffraction. Solid State Commun. **169**, 24 (2013)
52. Y.H. Wang, H. Steinberg, P. Jarillo-Herrero, N. Gedik, Observation of Floquet-Bloch states on the surface of a topological insulator. Science **342**, 453 (2013)
53. F. Mahmood, C.K. Chan, Z. Alpichshev, D. Gardner, Y. Lee, P.A. Lee, N. Gedik, Selective scattering between Floquet-Bloch and Volkov states in a topological insulator. Nat. Phys. **12**, 306 (2016). arXiv:1512.05714
54. W. Hu, S. Kaiser, D. Nicoletti, C.R. Hunt, I. Gierz, M.C. Hoffmann, M. Le Tacon, T. Loew, B. Keimer, A. Cavalleri, Optically enhanced coherent transport in $YBa_2Cu_3O_{6.5}$ by ultrafast redistribution of interlayer coupling. Nat. Mater. **13**, 705 (2014)
55. R. Mankowsky, A. Subedi, M. Först, S.O. Mariager, M. Chollet, H.T. Lemke, J.S. Robinson, J.M. Glownia, M.P. Minitti, A. Frano, M. Fechner, N.A. Spaldin, T. Loew, B. Keimer, A. Georges, A. Cavalleri, Nonlinear lattice dynamics as a basis for enhanced superconductivity in YBa2Cu3O6.5. Nature **516**, 71 (2014). arXiv:1405.2266
56. S. Kaiser, C.R. Hunt, D. Nicoletti, W. Hu, I. Gierz, H.Y. Liu, M. Le Tacon, T. Loew, D. Haug, B. Keimer, A. Cavalleri, Optically induced coherent transport far above Tc in underdoped $YBa_2Cu_3\,O_{6+\delta}$. Phys. Rev. B **89**, 184516 (2014). arXiv:1205.4661
57. M. Mitrano, A. Cantaluppi, D. Nicoletti, S. Kaiser, A. Perucchi, S. Lupi, P. Di Pietro, D. Pontiroli, M. Riccò, S.R. Clark, D. Jaksch, A. Cavalleri, Possible light-induced superconductivity in K3 C60 at high temperature. Nature **530**, 461 (2016)
58. T.F. Nova, A. Cartella, A. Cantaluppi, M. Först, D. Bossini, R.V. Mikhaylovskiy, A.V. Kimel, R. Merlin, A. Cavalleri, An effective magnetic field from optically driven phonons. Nat. Phys. **13**, 132 (2017). arXiv:1512.06351
59. M. Buzzi, D. Nicoletti, M. Fechner, N. Tancogne-Dejean, M.A. Sentef, A. Georges, T. Biesner, E. Uykur, M. Dressel, A. Henderson, T. Siegrist, J.A. Schlueter, K. Miyagawa, K. Kanoda, M.-S. Nam, A. Ardavan, J. Coulthard, J. Tindall, F. Schlawin, D. Jaksch, A. Cavalleri, Photomolecular high-temperature superconductivity. Phys. Rev. X **10**, 031028 (2020)

60. M. Budden, T. Gebert, M. Buzzi, G. Jotzu, E. Wang, T. Matsuyama, G. Meier, Y. Laplace, D. Pontiroli, M. Riccò, F. Schlawin, D. Jaksch, A. Cavalleri, Evidence for metastable photo-induced superconductivity in K3C60. Nat. Phys. **17**, 611–618 (2021). https://doi.org/10.1038/s41567-020-01148-1

61. J.W. McIver, B. Schulte, F.-U. Stein, T. Matsuyama, G. Jotzu, G. Meier, A. Cavalleri, Light-induced anomalous hall effect in graphene. Nat. Phys. **16**, 38 (2020)

62. K.R. Beyerlein, A.S. Disa, M. Först, M. Henstridge, T. Gebert, T. Forrest, A. Fitzpatrick, C. Dominguez, J. Fowlie, M. Gibert, J.-M. Triscone, S.S. Dhesi, A. Cavalleri, Probing photoinduced rearrangements in the NdNiO$_3$ magnetic spiral with polarization-sensitive ultrafast resonant soft x-ray scattering. Phys. Rev. B **102**, 014311 (2020)

63. J.B. Curtis, A. Grankin, N.R. Poniatowski, V.M. Galitski, P. Narang, E. Demler, Cavity magnon-polaritons in cuprate parent compounds (2021). arXiv:2106.07828

64. Y. Ashida, A. Imamoglu, E. Demler, Nonperturbative waveguide quantum electrodynamics (2021). arXiv:2105.08833

65. A. Subedi, A. Cavalleri, A. Georges, Theory of nonlinear phononics for coherent light control of solids. Phys. Rev. B **89**, 220301(R) (2014). arXiv:1311.0544

66. A.A. Patel, A. Eberlein, Light-induced enhancement of superconductivity via melting of competing bond-density wave order in underdoped cuprates. Phys. Rev. B **93**, 195139 (2016). arXiv:1602.05964

67. M. Knap, M. Babadi, G. Refael, I. Martin, E. Demler, Dynamical Cooper pairing in nonequilibrium electron-phonon systems. Phys. Rev. B **94**, 214504 (2016)

68. A. Komnik, M. Thorwart, BCS theory of driven superconductivity. Eur. Phys. J. B **89**, 244 (2016). arXiv:1607.03858

69. M. Babadi, M. Knap, I. Martin, G. Refael, E. Demler, Theory of parametrically amplified electron-phonon superconductivity. Phys. Rev. B **96**, 014512 (2017). arXiv:1702.02531

70. Y. Murakami, N. Tsuji, M. Eckstein, P. Werner, Nonequilibrium steady states and transient dynamics of conventional superconductors under phonon driving. Phys. Rev. B **96**, 045125 (2017)

71. M.A. Sentef, Light-enhanced electron-phonon coupling from nonlinear electron-phonon coupling. Phys. Rev. B **95**, 6 (2017). arXiv:1702.00952

72. N. Bittner, T. Tohyama, S. Kaiser, D. Manske, Possible light-induced superconductivity in a strongly correlated electron system. J. Phys. Soc. Jpn. **88**, 044704 (2019). https://doi.org/10.7566/JPSJ.88.044704

73. D.M. Kennes, E.Y. Wilner, D.R. Reichman,, A.J. Millis, Transient superconductivity from electronic squeezing of optically pumped phonons. Nat. Phys. **13**, 479 (2017). arXiv:1609.03802v1

74. S. Beaulieu, S. Dong, N. Tancogne-Dejean, M. Dendzik, T. Pincelli, J. Maklar, R.P. Xian, M.A. Sentef, M. Wolf, A. Rubio, L. Rettig, R. Ernstorfer, Ultrafast dynamical lifshitz transition. Sci. Adv. **7**, eabd9275 (2021). https://www.science.org/doi/pdf/10.1126/sciadv.abd9275

75. S. Dong, M. Puppin, T. Pincelli, S. Beaulieu, D. Christiansen, H. Hübener, C.W. Nicholson, R.P. Xian, M. Dendzik, Y. Deng, Y.W. Windsor, M. Selig, E. Malic, A. Rubio, A. Knorr, M. Wolf, L. Rettig, R. Ernstorfer, Direct measurement of key exciton properties: energy, dynamics, and spatial distribution of the wave function. Nat. Sci. **1**, e10010 (2021). https://onlinelibrary.wiley.com/doi/pdf/10.1002/ntls.10010

76. L. Stojchevska, I. Vaskivskyi, T. Mertelj, P. Kusar, D. Svetin, S. Brazovskii, D. Mihailovic, Ultrafast switching to a stable hidden quantum state in an electronic crystal. Science **344**, 177 (2014). arXiv:1401.6786v3

77. J. Zhang, X. Tan, M. Liu, S.W. Teitelbaum, K.W. Post, F. Jin, K.A. Nelson, D.N. Basov, W. Wu, R.D. Averitt, Cooperative photoinduced metastable phase control in strained manganite films. Nat. Mater. **15**, 956 (2016). arXiv:1512.00436

78. S.W. Teitelbaum, T. Shin, J.W. Wolfson, Y.H. Cheng, I.J. Porter, M. Kandyla, K.A. Nelson, Real-time observation of a coherent lattice transformation into a high-symmetry phase. Phys. Rev. X **8**, 31081 (2018)

79. D. Mihailovic, The importance of topological defects in photoexcited phase transitions including memory applications. Appl. Sci. **9**, 890 (2019)
80. T. Oka, H. Aoki, Photovoltaic Hall effect in graphene. Phys. Rev. B **79**, 081406 (2009)
81. M. Bukov, L. D'Alessio, A. Polkovnikov, Universal high-frequency behavior of periodically driven systems: from dynamical stabilization to Floquet engineering. Adv. Phys. **64**, 139 (2015)
82. D.M. Kennes, A. de la Torre, A. Ron, D. Hsieh, A.J. Millis, Floquet engineering in quantum chains. Phys. Rev. Lett. **120**, 127601 (2018)
83. N. Dasari, M. Eckstein, Transient Floquet engineering of superconductivity. Phys. Rev. B **98**, 235149 (2018)
84. M.S. Rudner, N.H. Lindner, Band structure engineering and non-equilibrium dynamics in floquet topological insulators. Nat. Rev. Phys. **2**, 229 (2020)
85. D. Hagenmüller, J. Schachenmayer, S. Schütz, C. Genes, G. Pupillo, Cavity-enhanced transport of charge. Phys. Rev. Lett. **119**, 223601 (2017)
86. F. Schlawin, A. Cavalleri, D. Jaksch, Cavity-mediated electron-photon superconductivity. Phys. Rev. Lett. **122**, 133602 (2019)
87. J.B. Curtis, Z.M. Raines, A.A. Allocca, M. Hafezi, V.M. Galitski, Cavity quantum eliashberg enhancement of superconductivity. Phys. Rev. Lett. **122**, 167002 (2019)
88. M.A. Sentef, M. Ruggenthaler, A. Rubio, Cavity quantum-electrodynamical polaritonically enhanced electron-phonon coupling and its influence on superconductivity. Sci. Adv. **4**, (2018). https://doi.org/10.1126/sciadv.aau6969. https://advances.sciencemag.org/content/4/11/eaau6969.full.pdf
89. M. Kiffner, J.R. Coulthard, F. Schlawin, A. Ardavan, D. Jaksch, Manipulating quantum materials with quantum light. Phys. Rev. B **99**, 085116 (2019)
90. M. Bello, G. Platero, J.I. Cirac, A. González-Tudela, Unconventional quantum optics in topological waveguide QED. Sci. Adv. **5**, (2019). https://doi.org/10.1126/sciadv.aaw0297. https://arxiv.org/abs/https://advances.sciencemag.org/content/5/7/eaaw0297.full.pdf
91. J. Kuttruff, D. Garoli, J. Allerbeck, R. Krahne, A. De Luca, D. Brida, V. Caligiuri, N. Maccaferri, Ultrafast all-optical switching enabled by epsilon-near-zero-tailored absorption in metal-insulator nanocavities. Commun. Phys. **3**, 114 (2020)
92. L. Huang, L. Xu, M. Rahmani, D.N. Neshev, A.E. Miroshnichenko, Pushing the limit of high-Q mode of a single dielectric nanocavity. Adv. Photon. **3**, 1 (2021)
93. J. Flick, M. Ruggenthaler, H. Appel, A. Rubio, Atoms and molecules in cavities, from weak to strong coupling in quantum-electrodynamics (QED) chemistry. Proc. Natl. Acad. Sci. **114**, 3026 (2017). https://www.pnas.org/content/114/12/3026.full.pdf
94. T.S. Haugland, E. Ronca, E.F. Kjønstad, A. Rubio, H. Koch, Coupled cluster theory for molecular polaritons: changing ground and excited states. Phys. Rev. X **10**, 041043 (2020)
95. J. Flick, C. Schäfer, M. Ruggenthaler, H. Appel, A. Rubio, Ab initio optimized effective potentials for real molecules in optical cavities: photon contributions to the molecular ground state. ACS Photon. **5**, 992 (2018)
96. C. Schäfer, M. Ruggenthaler, H. Appel, A. Rubio, Modification of excitation and charge transfer in cavity quantum-electrodynamical chemistry. Proc. Natl. Acad. Sci. **116**, 4883 (2019). https://www.pnas.org/content/116/11/4883.full.pdf
97. S. Pisana, M. Lazzeri, C. Casiraghi, K.S. Novoselov, A.K. Geim, A.C. Ferrari, F. Mauri, Breakdown of the adiabatic born–oppenheimer approximation in graphene. Nat. Mater. **6**, 198 (2007)
98. D. Shin, H. Hübener, U. De Giovannini, H. Jin, A. Rubio, N. Park, Phonon-driven spin-floquet magneto-valleytronics in mos2. Nat. Commun. **9**, 638 (2018)
99. H. Hübener, U. De Giovannini, A. Rubio, Phonon driven floquet matter. Nano Lett. **18**, 1535 (2018)
100. D.M. Kennes, E.Y. Wilner, D.R. Reichman, A.J. Millis, Transient superconductivity from electronic squeezing of optically pumped phonons. Nat. Phys. **13**, 479 (2017)
101. A. Nava, C. Giannetti, A. Georges, E. Tosatti, M. Fabrizio, Cooling quasiparticles in A3C60 fullerides by excitonic mid-infrared absorption. Nat. Phys. **14**, 154 (2018). arXiv:1704.05613

102. P.E. Dolgirev, A. Zong, M.H. Michael, J.B. Curtis, D. Podolsky, A. Cavalleri, E. Demler, Periodic dynamics in superconductors induced by an impulsive optical quench (2021). arXiv:2104.07181
103. A. Eckardt, Colloquium: atomic quantum gases in periodically driven optical lattices. Rev. Mod. Phys. **89**, 011004 (2017). arXiv:1606.08041
104. T. Kitagawa, T. Oka, A. Brataas, L. Fu, E. Demler, Transport properties of nonequilibrium systems under the application of light: photoinduced quantum Hall insulators without Landau levels. Phys. Rev. B **84**, 235108 (2011)
105. S.A. Sato, J.W. McIver, M. Nuske, P. Tang, G. Jotzu, B. Schulte, H. Hübener, U. De Giovannini, L. Mathey, M.A. Sentef, A. Cavalleri, A. Rubio, Microscopic theory for the light-induced anomalous Hall effect in graphene. Phys. Rev. B **99**, 214302 (2019)
106. M. Nuske, L. Broers, B. Schulte, G. Jotzu, S.A. Sato, A. Cavalleri, A. Rubio, J.W. McIver, L. Mathey, Floquet dynamics in light-driven solids. Phys. Rev. Res. **2**, 043408 (2020)
107. T. Oka, S. Kitamura, Floquet engineering of quantum materials. Annu. Rev. Condens. Matter Phys. **10**, 387 (2019). https://doi.org/10.1146/annurev-conmatphys-031218-013423
108. U.D. Giovannini, H. Hübener, Floquet analysis of excitations in materials. J. Phys. Mater. **3**, 012001 (2019)
109. A. Castro, S.A. Sato, U.D. Giovannini, H. Hübener, A. Rubio, Floquet engineering the band structure of materials with optimal control theory. Phys. Rev. Res. **4**, 033213 (2022)
110. L. D'Alessio, M. Rigol, Long-time behavior of isolated periodically driven interacting lattice systems. Phys. Rev. X **4**, 041048 (2014)
111. D.A. Abanin, W. De Roeck, F. Huveneers, Exponentially slow heating in periodically driven many-body systems. Phys. Rev. Lett. **115**, 256803 (2015)
112. T. Mori, T. Kuwahara, K. Saito, Rigorous bound on energy absorption and generic relaxation in periodically driven quantum systems. Phys. Rev. Lett. **116**, 120401 (2016)
113. T. Kuwahara, T. Mori, K. Saito, Floquet-Magnus theory and generic transient dynamics in periodically driven many-body quantum systems. Ann. Phys. **367**, 96 (2016)
114. W.W. Ho, I. Protopopov, D.A. Abanin, Bounds on energy absorption and prethermalization in quantum systems with long-range interactions. Phys. Rev. Lett. **120**, 200601 (2018)
115. D. Abanin, W.D. Roeck, W.W. Ho, F. Huveneers, A rigorous theory of many-body prethermalization for periodically driven and closed quantum systems. Commun. Math. Phys. **354**, 809 (2017)
116. D.A. Abanin, W. De Roeck, W.W. Ho, F. Huveneers, Effective Hamiltonians, prethermalization, and slow energy absorption in periodically driven many-body systems. Phys. Rev. B **95**, 014112 (2017)
117. T. Mori, T.N. Ikeda, E. Kaminishi, M. Ueda, Thermalization and prethermalization in isolated quantum systems: a theoretical overview. J. Phys. B At. Mol. Opt. Phys. **51**, 112001 (2018)
118. M. Claassen, Flow renormalization and emergent prethermal regimes of periodically-driven quantum systems (2021). arXiv:2103.07485
119. S.A. Weidinger, M. Knap, Floquet prethermalization and regimes of heating in a periodically driven, interacting quantum system. Sci. Rep. **7**, 45382 (2017). arXiv:1609.09089
120. M. Bukov, M. Kolodrubetz, A. Polkovnikov, Schrieffer-Wolff transformation for periodically driven systems: strongly correlated systems with artificial gauge fields. Phys. Rev. Lett. **116**, 125301 (2016)
121. N. Walldorf, D.M. Kennes, J. Paaske, A.J. Millis, The antiferromagnetic phase of the Floquet-driven Hubbard model. Phys. Rev. B **100**, 121110 (2019)
122. W. Magnus, On the exponential solution of differential equations for a linear operator. Commun. Pure Appl. Math. **7**, 649 (1954)
123. J.H. Mentink, K. Balzer, M. Eckstein, Ultrafast and reversible control of the exchange interaction in Mott insulators. Nat. Commun. **6**, 6708 (2015). arXiv:1407.4761v1
124. M.A. Sentef, J. Li, F. Künzel, M. Eckstein, Quantum to classical crossover of floquet engineering in correlated quantum systems. Phys. Rev. Res. **2**, 033033 (2020)
125. J. Flick, D.M. Welakuh, M. Ruggenthaler, H. Appel, A. Rubio, Light–matter response in nonrelativistic quantum electrodynamics. ACS Photon. **6**, 2757 (2019)

126. R. Jestädt, M. Ruggenthaler, M.J.T. Oliveira, A. Rubio, H. Appel, Light-matter interactions within the ehrenfest–maxwell–pauli–kohn–sham framework: fundamentals, implementation, and nano-optical applications. Adv. Phys. **68**, 225 (2019). https://doi.org/10.1080/00018732. 2019.1695875

127. V. Rokaj, M. Penz, M.A. Sentef, M. Ruggenthaler, A. Rubio, Quantum electrodynamical bloch theory with homogeneous magnetic fields. Phys. Rev. Lett. **123**, 047202 (2019)

128. V. Rokaj, M. Ruggenthaler, F.G. Eich, A. Rubio, Free electron gas in cavity quantum electrodynamics. Phys. Rev. Research **4**, 013012 (2022)

129. C.J. Eckhardt, G. Passetti, M. Othman, C. Karrasch, F. Cavaliere, M.A. Sentef, D.M. Kennes, Quantum floquet engineering with an exactly solvable tight-binding chain in a cavity (2021). arXiv:2107.12236

130. V. Rokaj, M. Penz, M.A. Sentef, M. Ruggenthaler, A. Rubio, Polaritonic hofstadter butterfly and cavity-control of the quantized hall conductance (2021). arXiv:2109.15075

131. F. Appugliese, J. Enkner, G.L. Paravicini-Bagliani, M. Beck, C. Reichl, W. Wegscheider, G. Scalari, C. Ciuti, J. Faist, Breakdown of the topological protection by cavity vacuum fields in the integer quantum hall effect. Science **375**(6584), 1030–1034 (2021)

132. C. Bao, P. Tang, D. Sun, S. Zhou, Light-induced emergent phenomena in 2d materials and topological materials. Nat. Rev. Phys. **4**, 33 (2022)

133. L. Jaeger, *The Second Quantum Revolution* (Springer, Basel, 2018)

134. L. Wang, I. Meric, P.Y. Huang, Q. Gao, Y. Gao, H. Tran, T. Taniguchi, K. Watanabe, L.M. Campos, D.A. Muller, J. Guo, P. Kim, J. Hone, K.L. Shepard, C.R. Dean, One-dimensional electrical contact to a two-dimensional material. Science **342**, 614 (2013)

135. O. Klein, Die reflexion von elektronen an einem potentialsprung nach der relativistischen dynamik von dirac. Z. Phys. **53**, 157 (1929)

136. J. Gooth, B. Bradlyn, S. Honnali, C. Schindler, N. Kumar, J. Noky, Y. Qi, C. Shekhar, Y. Sun, Z. Wang, B.A. Bernevig, C. Felser, Axionic charge-density wave in the weyl semimetal (tase4)2i. Nature **575**, 315 (2019)

137. J.-Z. Ma, Q.S. Wu, M. Song, S.-N. Zhang, E.B. Guedes, S.A. Ekahana, M. Krivenkov, M.Y. Yao, S.Y. Gao, W.H. Fan, T. Qian, H. Ding, N.C. Plumb, M. Radovic, J.H. Dil, Y.M. Xiong, K. Manna, C. Felser, O.V. Yazyev, M. Shi, Observation of a singular weyl point surrounded by charged nodal walls in PTGA. Nat. Commun. **12**, 3994 (2021)

138. M.I. Katsnelson, K.S. Novoselov, A.K. Geim, Chiral tunnelling and the klein paradox in graphene. Nat. Phys. **2**, 620 (2006)

139. C.R. Dean, L. Wang, P. Maher, C. Forsythe, F. Ghahari, Y. Gao, J. Katoch, M. Ishigami, P. Moon, M. Koshino, T. Taniguchi, K. Watanabe, K.L. Shepard, J. Hone, P. Kim, Hofstadter's butterfly and the fractal quantum Hall effect in moiré superlattices. Nature **497**, 598 (2013)

140. J.M.B. Lopes dos Santos, N.M.R. Peres, A.H. Castro Neto, Graphene bilayer with a twist: electronic structure. Phys. Rev. Lett. **99**, 256802 (2007)

141. E. Suárez Morell, J.D. Correa, P. Vargas, M. Pacheco, Z. Barticevic, Flat bands in slightly twisted bilayer graphene: tight-binding calculations. Phys. Rev. B **82**, 121407 (2010)

142. C.-H. Park, L.I. Yang, Y.-W. Son, M.L. Cohen, S.G. Louie, Anisotropic behaviours of massless Dirac fermions in graphene under periodic potentials. Nat. Phys. **4**, 213 (2008)

143. M. Yankowitz, S. Chen, H. Polshyn, Y. Zhang, K. Watanabe, T. Taniguchi, D. Graf, A.F. Young, C.R. Dean, Tuning superconductivity in twisted bilayer graphene. Science **363**, 1059 (2019)

144. A. Kerelsky, L.J. McGilly, D.M. Kennes, L. Xian, M. Yankowitz, S. Chen, K. Watanabe, T. Taniguchi, J. Hone, C. Dean, A. Rubio, A.N. Pasupathy, Maximized electron interactions at the magic angle in twisted bilayer graphene. Nature **572**, 95 (2019)

145. A.L. Sharpe, E.J. Fox, A.W. Barnard, J. Finney, K. Watanabe, T. Taniguchi, M.A. Kastner, D. Goldhaber-Gordon, Emergent ferromagnetism near three-quarters filling in twisted bilayer graphene. Science **365**, 605 (2019)

146. X. Lu, P. Stepanov, W. Yang, M. Xie, M.A. Aamir, I. Das, C. Urgell, K. Watanabe, T. Taniguchi, G. Zhang, A. Bachtold, A.H. MacDonald, D.K. Efetov, Superconductors , orbital magnets and correlated states in magic-angle bilayer graphene. Nature **574**, 20 (2019)

147. M. Serlin, C.L. Tschirhart, H. Polshyn, Y. Zhang, J. Zhu, K. Watanabe, T. Taniguchi, L. Balents, A.F. Young, Intrinsic quantized anomalous Hall effect in a moiré heterostructure. Science **367**, 900 (2020)
148. X. Liu, Z. Hao, E. Khalaf, J.Y. Lee, Y. Ronen, H. Yoo, D. Haei Najafabadi, K. Watanabe, T. Taniguchi, A. Vishwanath, P. Kim, Tunable spin-polarized correlated states in twisted double bilayer graphene. Nature **583**, 221 (2020)
149. Y. Cao, D. Rodan-Legrain, O. Rubies-Bigorda, J.M. Park, K. Watanabe, T. Taniguchi, P. Jarillo-Herrero, Tunable correlated states and spin-polarized phases in twisted bilayer–bilayer graphene. Nature **583**, 215 (2020)
150. C. Shen, Y. Chu, Q. Wu, N. Li, S. Wang, Y. Zhao, J. Tang, J. Liu, J. Tian, K. Watanabe, T. Taniguchi, R. Yang, Z.Y. Meng, D. Shi, O.V. Yazyev, G. Zhang, Correlated states in twisted double bilayer graphene. Nat. Phys. **16**, 520 (2020)
151. G.W. Burg, J. Zhu, T. Taniguchi, K. Watanabe, A.H. MacDonald, E. Tutuc, Correlated insulating states in twisted double bilayer graphene. Phys. Rev. Lett. **123**, 197702 (2019)
152. C. Rubio-Verdú, S. Turkel, L. Song, L. Klebl, R. Samajdar, M.S. Scheurer, J.W.F. Venderbos, K. Watanabe, T. Taniguchi, H. Ochoa, L. Xian, D.M. Kennes, R.M. Fernandes, Á. Rubio, A.N. Pasupathy, Moiré nematic phase in twisted double bilayer graphene. Nat. Phys. **18**, 196–202 (2022). https://doi.org/10.1038/s41567-021-01438-2
153. G. Chen, L. Jiang, S. Wu, B. Lyu, H. Li, B.L. Chittari, K. Watanabe, T. Taniguchi, Z. Shi, J. Jung, Y. Zhang, F. Wang, Evidence of a gate-tunable Mott insulator in a trilayer graphene moiré superlattice. Nat. Phys. **15**, 237 (2019)
154. G. Chen, A.L. Sharpe, P. Gallagher, I.T. Rosen, E.J. Fox, L. Jiang, B. Lyu, H. Li, K. Watanabe, T. Taniguchi, J. Jung, Z. Shi, D. Goldhaber-Gordon, Y. Zhang, F. Wang, Signatures of tunable superconductivity in a trilayer graphene moiré superlattice. Nature **572**, 215 (2019)
155. G. Chen, A.L. Sharpe, E.J. Fox, Y.-H. Zhang, S. Wang, L. Jiang, B. Lyu, H. Li, K. Watanabe, T. Taniguchi, Z. Shi, T. Senthil, D. Goldhaber-Gordon, Y. Zhang, F. Wang, Tunable correlated chern insulator and ferromagnetism in a moiré superlattice. Nature **579**, 56 (2020)
156. J.M. Park, Y. Cao, K. Watanabe, T. Taniguchi, P. Jarillo-Herrero, Tunable strongly coupled superconductivity in magic-angle twisted trilayer graphene. Nature **590**, 249 (2021)
157. Z. Hao, A.M. Zimmerman, P. Ledwith, E. Khalaf, D.H. Najafabadi, K. Watanabe, T. Taniguchi, A. Vishwanath, P. Kim, Electric field tunable superconductivity in alternating twist magic-angle trilayer graphene. Science **371**(6534), 1133–1138 (2021). https://doi.org/10.1126/science.abg0399
158. S. Chen, M. He, Y.-H. Zhang, V. Hsieh, Z. Fei, K. Watanabe, T. Taniguchi, D.H. Cobden, X. Xu, C.R. Dean, M. Yankowitz, Electrically tunable correlated and topological states in twisted monolayer–bilayer graphene. Nat. Phys. **17**, 374 (2021)
159. H. Polshyn, J. Zhu, M.A. Kumar, Y. Zhang, F. Yang, C.L. Tschirhart, M. Serlin, K. Watanabe, T. Taniguchi, A.H. MacDonald, A.F. Young, Electrical switching of magnetic order in an orbital chern insulator. Nature **588**, 66 (2020)
160. B. Lian, Z.D. Song, N. Regnault, D.K. Efetov, A. Yazdani, B.A. Bernevig, Twisted bilayer graphene. iv. exact insulator ground states and phase diagram. Phys. Rev. B **103**, 205414 (2021)
161. P. Stepanov, M. Xie, T. Taniguchi, K. Watanabe, X. Lu, A.H. MacDonald, B.A. Bernevig, D.K. Efetov, Competing zero-field chern insulators in superconducting twisted bilayer graphene. Phys. Rev. Lett. **127**, 197701 (2021)
162. Y. Choi, H. Kim, Y. Peng, A. Thomson, C. Lewandowski, R. Polski, Y. Zhang, H.S. Arora, K. Watanabe, T. Taniguchi, J. Alicea, S. Nadj-Perge, Correlation-driven topological phases in magic-angle twisted bilayer graphene. Nature **589**, 536 (2021)
163. G.E. Topp, C.J. Eckhardt, D.M. Kennes, M.A. Sentef, P. Törmä, Light-matter coupling and quantum geometry in moiré materials. Phys. Rev. B **104**, 064306 (2021)
164. F. Wu, T. Lovorn, E. Tutuc, I. Martin, A.H. Macdonald, Topological Insulators in twisted transition metal dichalcogenide homobilayers. Phys. Rev. Lett. **122**, 86402 (2019)

165. B.B. Chen, Y.D. Liao, Z. Chen, O. Vafek, J. Kang, W. Li, Z.Y. Meng, Realization of topological mott insulator in a twisted bilayer graphene lattice model. Nat. Commun. **12**, 5480 (2021)
166. F. Wu, T. Lovorn, E. Tutuc, A.H. Macdonald, Hubbard model physics in transition metal dichalcogenide moiré bands. Phys. Rev. Lett. **121**, 26402 (2018)
167. M.H. Naik, M. Jain, Ultraflatbands and shear solitons in moiré patterns of twisted bilayer transition metal dichalcogenides. Phys. Rev. Lett. **121**, 266401 (2018)
168. D.A. Ruiz-Tijerina, V.I. Fal'Ko, Interlayer hybridization and moiré superlattice minibands for electrons and excitons in heterobilayers of transition-metal dichalcogenides. Phys. Rev. B **99**, 30 (2019)
169. C. Schrade, L. Fu, Spin-valley density wave in moiré materials. Phys. Rev. B **100**, 035413 (2019)
170. L. Wang, E.M. Shih, A. Ghiotto, L. Xian, D.A. Rhodes, C. Tan, M. Claassen, D.M. Kennes, Y. Bai, B. Kim, K. Watanabe, T. Taniguchi, X. Zhu, J. Hone, A. Rubio, A.N. Pasupathy, C.R. Dean, Correlated electronic phases in twisted bilayer transition metal dichalcogenides. Nat. Mater. **19**, 861 (2020)
171. L. Xian, D.M. Kennes, N. Tancogne-Dejean, M. Altarelli, A. Rubio, Multiflat bands and strong correlations in twisted bilayer boron nitride: doping-induced correlated insulator and superconductor. Nano Lett. **19**, 4934 (2019)
172. E. Dagotto, Correlated electrons in high-temperature superconductors. Rev. Mod. Phys. **66**, 763 (1994)
173. D.J. Scalapino, A common thread: the pairing interaction for unconventional superconductors. Rev. Mod. Phys. **84**, 1383 (2012)
174. J.P.F. LeBlanc, A.E. Antipov, F. Becca, I.W. Bulik, G.K.L. Chan, C.M. Chung, Y. Deng, M. Ferrero, T.M. Henderson, C.A. Jiménez-Hoyos, E. Kozik, X.W. Liu, A.J. Millis, N.V. Prokof'ev, M. Qin, G.E. Scuseria, H. Shi, B.V. Svistunov, L.F. Tocchio, I.S. Tupitsyn, S.R. White, S. Zhang, B.X. Zheng, Z. Zhu, E. Gull, Solutions of the two-dimensional hubbard model: benchmarks and results from a wide range of numerical algorithms. Phys. Rev. X **5**, 041041 (2015)
175. A. Mazurenko, C.S. Chiu, G. Ji, M.F. Parsons, M. Kanász-Nagy, R. Schmidt, F. Grusdt, E. Demler, D. Greif, M. Greiner, A cold-atom fermi–hubbard antiferromagnet. Nature **545**, 462 (2017)
176. J. Kang, O. Vafek, Strong coupling phases of partially filled twisted bilayer graphene narrow bands. Phys. Rev. Lett. **122**, 246401 (2019)
177. K. Seo, V.N. Kotov, B. Uchoa, Ferromagnetic mott state in twisted graphene bilayers at the magic angle. Phys. Rev. Lett. **122**, 246402 (2019)
178. Z. Wang, D.A. Rhodes, K. Watanabe, T. Taniguchi, J.C. Hone, J. Shan, K.F. Mak, Evidence of high-temperature exciton condensation in two-dimensional atomic double layers. Nature **574**, 76 (2019)
179. Y. Cao, D. Rodan-Legrain, J.M. Park, N.F.Q. Yuan, K. Watanabe, T. Taniguchi, R.M. Fernandes, L. Fu, P. Jarillo-Herrero, Nematicity and competing orders in superconducting magic-angle graphene. Science **372**, 264 (2021). https://arxiv.org/abs/https://www.science.org/doi/pdf/10.1126/science.abc2836
180. M. Christos, S. Sachdev, M.S. Scheurer, Superconductivity, correlated insulators, and wess–zumino–witten terms in twisted bilayer graphene. Proc. Natl. Acad. Sci. **117**, 29543 (2020). https://arxiv.org/abs/https://www.pnas.org/content/117/47/29543.full.pdf
181. T. Li, S. Jiang, L. Li, Y. Zhang, K. Kang, J. Zhu, K. Watanabe, T. Taniguchi, D. Chowdhury, L. Fu, J. Shan, K.F. Mak, Continuous mott transition in semiconductor moiré superlattices. Nature **597**, 350 (2021)
182. T. Li, S. Jiang, B. Shen, Y. Zhang, L. Li, Z. Tao, T. Devakul, K. Watanabe, T. Taniguchi, L. Fu, J. Shan, K.F. Mak, Quantum anomalous hall effect from intertwined moiré bands. Nature **600**, 641 (2021)

183. H. Li, S. Li, E.C. Regan, D. Wang, W. Zhao, S. Kahn, K. Yumigeta, M. Blei, T. Taniguchi, K. Watanabe, S. Tongay, A. Zettl, M.F. Crommie, F. Wang, Imaging two-dimensional generalized wigner crystals. Nature **597**, 650 (2021)
184. C. Jin, E.C. Regan, A. Yan, M. Iqbal Bakti Utama, D. Wang, S. Zhao, Y. Qin, S. Yang, Z. Zheng, S. Shi, K. Watanabe, T. Taniguchi, S. Tongay, A. Zettl, F. Wang, Observation of moiré excitons in wse$_2$/ws$_2$ heterostructure superlattices. Nature **567**, 76 (2019)
185. Y. Shimazaki, I. Schwartz, K. Watanabe, T. Taniguchi, M. Kroner, A. Imamoglu, Strongly correlated electrons and hybrid excitons in a moiré heterostructure. Nature **580**, 472 (2020)
186. D.M. Kennes, L. Xian, M. Claassen, A. Rubio, One-dimensional flat bands in twisted bilayer germanium selenide. Nat. Commun. **11**, 1124 (2020)
187. B. Lian, Z. Liu, Y. Zhang, J. Wang, Flat chern band from twisted bilayer MnBi$_2$Te$_4$. Phys. Rev. Lett. **124**, 126402 (2020)
188. X. Chen, X. Fan, L. Li, N. Zhang, Z. Niu, T. Guo, S. Xu, H. Xu, D. Wang, H. Zhang, A.S. McLeod, Z. Luo, Q. Lu, A.J. Millis, D.N. Basov, M. Liu, C. Zeng, Moiré engineering of electronic phenomena in correlated oxides. Nat. Phys. **16**, 631 (2020)
189. G.A. Tritsaris, S. Carr, Z. Zhu, Y. Xie, S.B. Torrisi, J. Tang, M. Mattheakis, D.T. Larson, E. Kaxiras, Electronic structure calculations of twisted multi-layer graphene superlattices. 2D Mater. **7**, 035028 (2020)
190. Y. Zhang, Y.W. Tan, H.L. Stormer, P. Kim, Experimental observation of the quantum hall effect and berry's phase in graphene. Nature **438**, 201 (2005)
191. M. Claassen, L. Xian, D.M. Kennes, A. Rubiou, Ultra-strong spin-orbit coupling and topological moiré engineering in twisted zrs2 bilayers. arXiv:2110.13370 (2021)
192. G. Semeghini, H. Levine, A. Keesling, S. Ebadi, T.T. Wang, D. Bluvstein, R. Verresen, H. Pichler, M. Kalinowski, R. Samajdar, A. Omran, S. Sachdev, A. Vishwanath, M. Greiner, V. Vuletić, M.D. Lukin, Probing topological spin liquids on a programmable quantum simulator. Science **374**, 1242 (2021). https://www.science.org/doi/pdf/10.1126/science.abi8794
193. T. Smoleński, P.E. Dolgirev, C. Kuhlenkamp, A. Popert, Y. Shimazaki, P. Back, X. Lu, M. Kroner, K. Watanabe, T. Taniguchi, I. Esterlis, E. Demler, A. Imamoglu, Signatures of wigner crystal of electrons in a monolayer semiconductor. Nature **595**, 53 (2021)
194. K.P. Nuckolls, M. Oh, D. Wong, B. Lian, K. Watanabe, T. Taniguchi, B.A. Bernevig, A. Yazdani, Strongly correlated chern insulators in magic-angle twisted bilayer graphene. Nature **588**, 610 (2020)
195. Y. Xie, A.T. Pierce, J.M. Park, D.E. Parker, E. Khalaf, P. Ledwith, Y. Cao, S.H. Lee, S. Chen, P.R. Forrester, K. Watanabe, T. Taniguchi, A. Vishwanath, P. Jarillo-Herrero, A. Yacoby, Fractional chern insulators in magic-angle twisted bilayer graphene. Nature **600**, 439 (2021)
196. D.M. Kennes, M. Claassen, M.A. Sentef, C. Karrasch, Light-induced d-wave superconductivity through floquet-engineered fermi surfaces in cuprates. Phys. Rev. B **100**, 075115 (2019)
197. Y. Wang, J. Kang, R.M. Fernandes, Topological and nematic superconductivity mediated by ferro-su(4) fluctuations in twisted bilayer graphene. Phys. Rev. B **103**, 024506 (2021)
198. M. Oh, K.P. Nuckolls, D. Wong, R.L. Lee, X. Liu, K. Watanabe, T. Taniguchi, A. Yazdani, Evidence for unconventional superconductivity in twisted bilayer graphene. Nature **600**, 240 (2021)
199. Y. Cao, J.M. Park, K. Watanabe, T. Taniguchi, P. Jarillo-Herrero, Pauli-limit violation and re-entrant superconductivity in moiré graphene. Nature **595**, 526 (2021)
200. H. Polshyn, Y. Zhang, M.A. Kumar, T. Soejima, P. Ledwith, K. Watanabe, T. Taniguchi, A. Vishwanath, M.P. Zaletel, A.F. Young, Topological charge density waves at half-integer filling of a moiré superlattice. Nat. Phys. **18**, 42–47 (2022). https://doi.org/10.1038/s41567-021-01418-6
201. W.M. Zhao, L. Zhu, Z. Nie, Q.Y. Li, Q.W. Wang, L.G. Dou, J.G. Hu, L. Xian, S. Meng, S.C. Li, Moiré enhanced charge density wave state in twisted 1t-tite2/1t-tise2 heterostructures. Nat. Mater. **21**, 284–289 (2022). https://doi.org/10.1038/s41563-021-01167-0

202. P. Wang, G. Yu, Y.H. Kwan, Y. Jia, S. Lei, S. Klemenz, F.A. Cevallos, R. Singha, T. Devakul, K. Watanabe, T. Taniguchi, S.L. Sondhi, R.J. Cava, L.M. Schoop, S.A. Parameswaran, S. Wu, One-dimensional luttinger liquids in a two-dimensional moiré lattice (2021). arXiv:2109.04637

203. M.J. Park, Y. Kim, G.Y. Cho, S. Lee, Higher-order topological insulator in twisted bilayer graphene. Phys. Rev. Lett. **123**, 216803 (2019)

204. A. Chew, Y. Wang, B.A. Bernevig, Z.-D. Song, Higher-order topological superconductivity in twisted bilayer graphene (2021). arXiv:2108.05373

205. J. Liu, H. Wang, C. Fang, L. Fu, X. Qian, van der waals stacking-induced topological phase transition in layered ternary transition metal chalcogenides. Nano Lett. **17**, 467 (2017)

206. D. Varsano, M. Palummo, E. Molinari, M. Rontani, A monolayer transition-metal dichalcogenide as a topological excitonic insulator. Nat. Nanotechnol. **15**, 367 (2020)

207. N.P. Armitage, E.J. Mele, A. Vishwanath, Weyl and dirac semimetals in three-dimensional solids. Rev. Mod. Phys. **90**, 015001 (2018)

208. A. Burkov, Weyl metals. Annu. Rev. Condens. Matter Phys. **9**, 359 (2018). https://arxiv.org/abs/https://doi.org/10.1146/annurev-conmatphys-033117-054129

209. B. Yan, C. Felser, Topological materials: weyl semimetals. Annu. Rev. Condens. Matter Phys. **8**, 337 (2017). https://arxiv.org/abs/https://doi.org/10.1146/annurev-conmatphys-031016-025458

210. S. Ebihara, K. Fukushima, T. Oka, Chiral pumping effect induced by rotating electric fields. Phys. Rev. B **93**, 155107 (2016)

211. H. Hübener, M.A. Sentef, U. De Giovannini, A.F. Kemper, A. Rubio, Creating stable Floquet-Weyl semimetals by laser-driving of 3D Dirac materials. Nat. Commun. **8**, 13940 (2017). arXiv:1604.03399

212. L. Bucciantini, S. Roy, S. Kitamura, T. Oka, Emergent Weyl nodes and Fermi arcs in a Floquet Weyl semimetal. Phys. Rev. B **96**, 041126(R) (2017). arXiv:1612.01541

213. O. Deb, D. Sen, Generating surface states in a Weyl semimetal by applying electromagnetic radiation. Phys. Rev. B **95**, 144311 (2017). arXiv:1701.03661

214. A. Trabesinger, Quantum simulation. Nat. Phys. **8**, 263 (2012)

215. I. Bloch, J. Dalibard, S. Nascimbene, Quantum simulations with ultracold quantum gases. Nat. Phys. **8**, 267 (2012)

216. A. Thomas, J. George, A. Shalabney, M. Dryzhakov, S.J. Varma, J. Moran, T. Chervy, X. Zhong, E. Devaux, C. Genet, J.A. Hutchison, T.W. Ebbesen, Ground-state chemical reactivity under vibrational coupling to the vacuum electromagnetic field. Angewandte Chemie Int. Edn. **55**, 11462 (2016). https://arxiv.org/abs/https://onlinelibrary.wiley.com/doi/pdf/10.1002/anie.201605504

217. A. Thomas, L. Lethuillier-Karl, K. Nagarajan, R.M.A. Vergauwe, J. George, T. Chervy, A. Shalabney, E. Devaux, C. Genet, J. Moran, T.W. Ebbesen, Tilting a ground-state reactivity landscape by vibrational strong coupling. Science **363**, 615 (2019). https://arxiv.org/abs/https://www.science.org/doi/pdf/10.1126/science.aau7742

218. H. Hiura, A. Shalabney, Vacuum-field catalysis: accelerated reactions by vibrational ultra strong coupling (2021). ChemRxiv. https://doi.org/10.26434/chemrxiv.7234721.v5

219. B. Munkhbat, M. Wersäll, D.G. Baranov, T.J. Antosiewicz, T. Shegai, Suppression of photo-oxidation of organic chromophores by strong coupling to plasmonic nanoantennas. Sci. Adv. **4**, eaas9552 (2018). https://arxiv.org/abs/https://www.science.org/doi/pdf/10.1126/sciadv.aas9552

220. D.M. Coles, N. Somaschi, P. Michetti, C. Clark, P.G. Lagoudakis, P.G. Savvidis, D.G. Lidzey, Polariton-mediated energy transfer between organic dyes in a strongly coupled optical microcavity. Nat. Mater. **13**, 712 (2014)

221. D. Wang, H. Kelkar, D. Martin-Cano, T. Utikal, S. Götzinger, V. Sandoghdar, Coherent coupling of a single molecule to a scanning fabry-perot microcavity. Phys. Rev. X **7**, 021014 (2017)

222. A. Thomas, E. Devaux, K. Nagarajan, T. Chervy, M. Seidel, D. Hagenmüller, S. Schütz, J. Schachenmayer, C. Genet, G. Pupillo, T.W. Ebbesen, Exploring superconductivity under strong coupling with the vacuum electromagnetic field (2019). arXiv:1911.01459

223. D. Sidler, M. Ruggenthaler, H. Appel, A. Rubio, Chemistry in quantum cavities: exact results, the impact of thermal velocities, and modified dissociation. J. Phys. Chem. Lett. **11**, 7525 (2020)

224. M. Ruggenthaler, J. Flick, C. Pellegrini, H. Appel, I.V. Tokatly, A. Rubio, Quantum-electrodynamical density-functional theory: bridging quantum optics and electronic-structure theory. Phys. Rev. A **90**, 012508 (2014)

225. C. Schäfer, J. Flick, E. Ronca, P. Narang, A. Rubio, Shining light on the microscopic resonant mechanism responsible for cavity-mediated chemical reactivity (2021). arXiv:2104.12429

226. J.J. Cha, J.R. Williams, D. Kong, S. Meister, H. Peng, A.J. Bestwick, P. Gallagher, D. Goldhaber-Gordon, Y. Cui, Magnetic doping and kondo effect in bi2se3 nanoribbons. Nano Lett. **10**, 1076 (2010)

227. Y. Ashida, A. Imamoglu, J. Faist, D. Jaksch, A. Cavalleri, E. Demler, Quantum electrodynamic control of matter: cavity-enhanced ferroelectric phase transition. Phys. Rev. X **10**, 041027 (2020)

228. S. Latini, E. Ronca, U.D. Giovannini, H. Hübener, A. Rubio, Cavity control of excitons in two dimensional materials. Nano Lett. **19**, 3473 (2019)

229. G.L. Paravicini-Bagliani, F. Appugliese, E. Richter, F. Valmorra, J. Keller, M. Beck, N. Bartolo, C. Rössler, T. Ihn, K. Ensslin, C. Ciuti, G. Scalari, J. Faist, Magneto-transport controlled by landau polariton states. Nat. Phys. **15**, 186 (2018)

230. S.C. Zhang, J. Hu, A four-dimensional generalization of the quantum hall effect. Science **294**, 823 (2001). https://arxiv.org/abs/https://www.science.org/doi/pdf/10.1126/science.294.5543.823

231. B. Bradlyn, J. Cano, Z. Wang, M.G. Vergniory, C. Felser, R.J. Cava, B.A. Bernevig, Beyond dirac and weyl fermions: unconventional quasiparticles in conventional crystals. Science **353**, aaf5037 (2016). https://arxiv.org/abs/https://www.science.org/doi/pdf/10.1126/science.aaf5037

Evaluation and Utility of Wilsonian Naturalness

James D. Wells

Contents

Abstract

We demonstrate that many Naturalness tests of particle theories discussed in the literature can be reformulated as straightforward algorithmic finetuning assessments in the matching of Wilsonian effective theories above and below particle mass thresholds. Implications of this EFT formulation of Wilsonian Naturalness are discussed for several theories, including the Standard Model, heavy singlet scalar theory, supersymmetry, Grand Unified Theories, twin Higgs theories, and theories of extra dimensions. We argue that the Wilsonian Naturalness algorithm presented here constitutes an unambiguous, a priori, and meaningful test that the Standard Model passes and which "the next good theory" of particle physics is very likely to pass.

J. D. Wells (✉)
Leinweber Center for Theoretical Physics, University of Michigan, Ann Arbor, MI, USA
e-mail: jwells@umich.edu

© The Author(s), under exclusive license to Springer Nature Switzerland AG 2023
R. Citro et al. (eds.), *Sketches of Physics*, Lecture Notes in Physics 1000,
https://doi.org/10.1007/978-3-031-32469-7_2

1   Naturalness and Theory Viability

There are an infinite number of theories that are perfectly compatible with all known data, with the Standard Model (SM) being only one version of that, or rather one class of such theories. This is certainly true for all high-energy collider physics data, which is a key observables target for particle physics theories, as the decoupling nature of viable beyond the Standard Model (BSM) theories allows their preservation. Just push up the "new physics" scales of these BSM theories and the theories are safe from experimental incompatibility. This is the case for supersymmetry, composite Higgs theories, extra dimensional theories, grand unified theories, and many more.

This underdetermination problem is of course well-known by researchers. Attempts to make a rank order among good theories by invoking such tools as Ockham's razor, parsimoniousness, "beauty", consilience, etc. are largely unsatisfying to a rigorous mind. Nevertheless, we will try to do exactly that here, but perhaps with a half-rigorous mind by invoking some probability arguments against coincidences among random numbers. The set of ideas along these lines goes under the name of "Naturalness criteria" (see, e.g., [1,2]), which as we will review below is intimately connected to finetuning.

Criteria for passing a Naturalness test have been given in the past, even though little discussion has been given to what exactly are the implications for a theory when it does not pass a Naturalness test. The case for a meaningful connection between finetuning assessments, the engine under the hood of Naturalness, and probability has been made in Ref. [3], which connected low probability with high finetuning of Z with respect to large X and Y in $Z = X - Y$. This is generically the approximate form that gives rise to all finetunings across EFT boundaries. However, there it was stated that a critical feature to legitimacy is the a priori definition of any finetuning test used in the assessment of Naturalness. In other words, the test needs to be in place before the theory, or one risks manipulating the Naturalness test to serve propaganda purposes.

In this paper we define a general, algorithmic, and a priori test for Naturalness that serves all the requirements of [3]. The test is based on computing finetunings across particle mass thresholds of a theory whose expression (fields and operators) and calculational scheme is presented according to an a priori decided-upon convention. We find that once we do that, many of the Naturalness criteria employed in the literature that looked disparate and even ad hoc are very nearly exactly what one finds when employing what we will call the Wilsonian Naturalness algorithm. If there are such possibilities to make good a would-be Unnatural theory, then this would be another example of a new theory principle recognized to solve the Wilsonian Naturalness problem, thus reenforcing the value of articulating Wilsonian Naturalness assessments. The Naturalness tests that come automatically under this larger Wilsonian Naturalness algorithm framework include finetuning of m_Z^2 in supersymmetric theories, doublet-triplet splitting problem in Grand Unified theories, and quadratic sensitivity problem to a heavy scalar singlet. We even automatically get similar conclusions about the would-be Naturalness failure of the SM if the Higgs mass were much lighter (lighter than a few GeV), reminiscent of the "finite Naturalness" claims [4].

In the following sections we will first describe what is meant by a theory here, since the Naturalness status is a condition on the theory itself and not directly on observables, so a clear understanding of what constitutes a theory is a required first step. We then describe an algorithmic way to compute all the threshold finetunings in a particle physics theory. After that we connect threshold finetunings, and levels of finetuning, to the notion of Wilsonian Naturalness. We then apply these considerations to several elementary particle theories: Standard Model, supersymmetry, grand unified theories, added scalar singlet, and theories of extra dimensions. A chief goal is to determine in what circumstances we believe that the Wilsonian Naturalness test may relegate these theories to a lower status, i.e., unlikely to be a good description nature. The final conclusion section summarizes the findings.

2 Indexed Theories of a Theory Class

The question of whether a theory is Natural is a question internal to the theory under consideration. For that reason we need to specify what is meant by a theory in order to articulate this internal analysis.

For us, a theory T is an algorithm that maps a specified domain of state descriptors $\{\xi_i\}$ to a collection of target observables $\{O_k\}$:

$$T : \{\xi_i\} \longrightarrow \{O_k\} \tag{1}$$

By state descriptors we mean contingent variables for the state of the system. One could call it the in-state of the system ψ_{in}. For example, an electron with momentum p_1 and a positron with momentum p_2 are the state descriptors, which are known and interpretable by the theory. The output target observables are, for example, total cross-section into $\mu^+\mu^-$, forward-backward asymmetries, differential angular cross-section of e^+e^-, and many more.

In our language the Standard Model (SM) is not a single theory but a well-defined class of theories. The reason is that in order to do a calculation a set of theory parameters must be set $\mathbf{g} = \{g', g, y_t, \lambda, \theta_{\text{QCD}}, \cdots\}$. Thus, the class of theories we call the SM becomes a theory when the $\mathbf{g}$ are set to specific values. In this sense, the theory parameters index the infinite number of SM theories: $T_{\mathbf{g}} = \text{SM}(\mathbf{g})$. One can then find the parameter space of $\mathbf{g}$ values that are in accord with experiment.

We therefore can speak in general of a class T of theories $T_{\mathbf{g}}$ indexed by $\mathbf{g}$ with a parameter space of $\mathbf{g}$ that is consistent with experiment. For any given choice of parameters $\mathbf{g}$ that has been determined to be in accord with experiment, one then can perform an internal analysis on the theory to ask if it satisfies a quality or suffers a defect that we have argued to be important to identify.

In our case we will do an analysis of theories $T_{\mathbf{g}}$ to try to discover whether it passes the Wilsonian Naturalness test. The answer to this question may appear then to be completely divorced from observables and data, but it is not. The easiest way to understand how experiment is crucial in all of these considerations is to realize that it is quite possible that there is a parameters space domain of $\mathbf{g}$, which we can call $\mathbf{g}_A$, that is inconsistent with experiment yet internal analysis says that $T_{\mathbf{g}_A}$ passes the Wilsonian Naturalness test. On the other hand, for the entire domain of $\mathbf{g}$ that is consistent with experiment, internal analysis says that those $T_{\mathbf{g}}$ theories fail the Wilsonian Naturalness test. Therefore, it is the combination of experimental data (observables) and internal analysis (Wilsonian Naturalness test) that calls into question the entire class of theories $T_{\mathbf{g}}$. Neither one of them alone could do it. In other words, Naturalness claims and its impact on the viability of a full class of theories is just as much about experimental data as it is about internal analysis of the theory.

3   Threshold Finetuning and Naturalness Tests

In this section we first define what is meant by threshold finetuning in a theory and give an algorithm for computing it. We then use the numerical output of the threshold finetuning measure to argue that some theories fail the Wilsonian Naturalness test and should be dismissed. We do this in the context of Wilsonian effective theories [5].

By threshold we mean the mass scale of a particle Φ_H in the spectrum of the theory. In QFT we can define a high-scale theory with lagrangian $\mathcal{L}_H$ to be one where the particle is a freely propagating degree of freedom. This theory is valid for momentum scales above and below the threshold particle's mass; however, as is well known in QFT, computations of observables for energies well below a particle's mass are beset with large logarithms that are calculationally difficult to handle. However, from the Appelquist-Carrazone decoupling theorem we know that the effects of the heavy state should decouple at low energies, and one can write a new low-scale theory with lagrangian $\mathcal{L}_L$ that does not have that state. The lagrangian couplings of the high scale theory, such as Yukawa couplings, gauge couplings, and mass terms, we denote collectively by $\mathbf{g}_H$, and the lagrangian couplings of the low scale theory by $\mathbf{g}_L$. Let's call the heavy state Φ_H and all other light states ϕ_i. What we have then is

$$\mathcal{L}_H(\Phi_H, \varphi_i; \mathbf{g}_H) \text{ high scale theory with } \Phi_H, \text{ and} \tag{2}$$

$$\mathcal{L}_L(\varphi_i; \mathbf{g}_L) \text{ low scale theory without } \Phi_H \text{ (integrated out)} \tag{3}$$

where the Φ_H field has been integrated out. Note, no coupling $\mathbf{g}_L$ in the low-scale theory is the "same" as any coupling $\mathbf{g}_H$ in the high-scale theory despite appearances. There are schemes in which they are the same at the threshold, but in principle they are different couplings with different renormalization group flows, etc.

Let us define the mass of Φ_H to be M, which sets our definition of the threshold scale at the boundary between $\mathcal{L}_H$ and $\mathcal{L}_L$. Upon integrating out the Φ_H state at the scale M one finds matching conditions between the $\mathbf{g}_L$ couplings and $\mathbf{g}_H$ couplings at that scale. Any given low-scale theory coupling g_{Li} is matched to a function of high-scale theory parameters at the scale M:

$$g_{Li}(M) = F_i(\mathbf{g}_H(M)) \tag{4}$$

where the F_i functions are to be determined by computation.

At this point to make notation simpler we drop the argument of the couplings and define $g_{Li} \equiv g_{Li}(M)$ and $g_{Hj} \equiv g_{Hj}(M)$, and we define a series of threshold finetunings [6] associated with every low-scale theory parameter:

$$\mathrm{FT}[g_{Li} \,|\, g_{Hj}] = \left| \frac{g_{Hj}}{g_{Li}} \frac{\partial g_{Li}}{\partial g_{Hj}} \right|_{\mu^2 = M^2} \tag{5}$$

We can define the finetuning associated with low-energy parameter g_{Li} to be

$$\mathrm{FT}[g_{Li}] = \max_k \mathrm{FT}[g_{Li} \,|\, g_{Hk}] \tag{6}$$

where μ^2 is the renormalization group scale at which the parameters and logarithms are evaluated. This equation looks rather complex with many calculations to do, but we will have one primary focus and that is the Higgs boson mass.[1] To be precise we mean the m^2 coupling of the $H^\dagger H$ operator of the Standard Model,

$$\mathcal{L}_{\mathrm{SM}} = -m^2 H^\dagger H + \cdots . \tag{7}$$

Let us call the finetuning across some heavy threshold M this simply as "Higgs finetuning" defined as

$$\text{Higgs finetuning} \equiv \mathrm{FT}[m^2] = \max_k \left| \frac{g_{Hk}}{m^2} \frac{\partial m^2}{\partial g_{Hk}} \right|_{\mu^2 = M^2} \tag{8}$$

where the threshold M is the one that gives the largest threshold finetuning in the theory.

For a threshold finetuning definition to be useful in this circumstance it must be defined a priori, meaning that we cannot find a special scheme for writing the

---

[1] The cosmological constant is another coefficient of worry for finetuning and naturalness [7]. However, it is a dimension-zero operator and deeply connected to issues of quantum gravity, which are beyond our scope. For this reason, we do not consider it further when assessing Naturalness within our particle physics theories.

lagrangian, or a special scheme for renormalization for the sole purpose of trying to minimize the finetuning. Instead, we suggest merely using the most standard approach to calculational scheme, $\overline{MS}$ and $\overline{DR}$ (if supersymmetric), and an a priori straightforward/textbook way of writing the lagrangian operators with their coefficients of g, m^2, λ, y_t, θ_{QCD}, etc., and simply apply the above definitions to find the finetunings across each threshold as one flows from the UV to the IR.[2] What we present here is an algorithm for Wilsonian Naturalness and a theory under that calculation has an unambiguous, fixed status, independent of how important one might think that status is, which is to be addressed next.

The key claim here on the utility of this algorithm is that it is highly unexpected that the finetunings across the thresholds should ever be greater than some very large value, such as 10^4 or so. The intuition for this is set by probability calculations based on flat priors of lagrangian parameters [3]. As explained in [3], as we consider higher and higher finetunings the assumed distributions of the lagrangian parameters become more and more irrelevant to our ability to declare that the relationship between parameters that gave rise to such a large finetuning is unlikely.

An important question is what does one mean by a distribution over parameters. First, as mentioned above, the existence of a reasonably smooth distribution of possible choices of the theory parameters is all that is needed. We do not need to know the precise form of the distribution. Furthermore, the notion of a distribution here is equivalent to calling the parameters contingent. The parameters could be contingent in several senses. In the first sense, we say that they are selected on a landscape of possibilities, which is motivated by string theory considerations of low-energy parameters being one of an extraordinarily large but finite number of sets of possibilities [8]. If work related to anthropic principle has no other value it has certainly shown that we can have qualitatively the same universe and features for different choices of the parameters (i.e., different electron mass, different force couplings, and so on), even if we assume the static properties of the universe are rigidly set (existence of electrons, photons, $SU(3)$ gauge theory, etc.). Thus, there is no incompatibility in conceiving of other choices, and no incompatibility in calling the parameters of the theory contingent,[3] and no incompatibility in conjecturing that they have been selected by some deeper principle or mechanism (landscape solutions, baby universes, multiverse, Zeus, etc.).

---

[2] It must be emphasized that the *arbitrariness* of the chosen elements for the algorithm is a feature, as it samples on the (mental) distribution of an infinite set of other possibilities for how a theory might be constructed or how a calculational procedure might be implemented. Nevertheless, maintaining *consistency* of the pre-defined definition is also key so as to not let creep in case-by-case re-assessments of a finetuned theory.

[3] A theory itself may be contingent, just as its parameters are. There may be an infinite number of theories that equally well describe all of particle physics as we know it but based on very different mathematics and constructions, such that the parameters across those theories have a distribution. In that case, fluctuations in our brains selected out the current SM construction that we learn in school, and the SM parameters would be merely numbers selected on a distribution of theories without any need to invoke multiverse distributions.

Although we believe that it can be justified reasonably that high finetuning signifies that a conjectured theory is unlikely to be fruitful, the real efficacy of finetuning is whether or not it works to separate wheat theories from chaff theories. With the definition given above, we are able to answer that question not only in the future (will any of our highly finetuned conjectured theories pan out?) but also in the present (do any of our fruitful theories, such as the SM, have a high threshold finetuning problem?).

In the case of the SM there are a dozens of nontrivial tests of our conjecture that threshold finetuning should remain low. One can compute the threshold finetunings of every low-energy parameter across the many thresholds starting with m_{top}, and continuing down to m_h, m_Z, m_W, m_b, and so on. As we will show in the section devoted to the SM example, the SM passes those tests and therefore passes the overall Wilsonian Naturalness test. This is exactly what we want from this conjecture that purports to identify dismissable theories. The fruitful and accepted SM theory satisfies a very large number of tests that it could have failed. This is the data that supports the Wilsonian Naturalness test presented here.

Before we describe several examples, let us introduce one additional useful definition to help us quantify finetuning. A theory is "level-X finetuned" if the maximum finetuning across every threshold of the theory is 10^X.

$$FT[m^2] = 10^X \quad \longrightarrow \quad \text{Level- X finetuned theory.} \tag{9}$$

The X value is the numerical datum that comes out of the Wilsonian Naturalness assessment of the theory. Using [3] as a guide, I consider level-6 and higher theories to be Unnatural (i.e., fail the Wilsonian Naturalness test) and should be labeled as such. Level-0,1,2 theories pass the Wilsonian Naturalness test in our view. One can imagine many different views on whether Level-3, Level-4 and Level-5 theories pass the Wilsonian Naturalness test. It is hard to argue that they are hopelessly improbable at that level and so in the spirit of conservativeness they should not be labeled Unnatural, but they should be looked upon with some suspicion and they likely need other good theory qualities (consilience, additional explanatory power compared to competitors, etc.) to remain in the good graces of physicists. The reader may disagree with my assessments for viability of each level, but I hope they would think hard about what level is ok and what is not. One can start by asking if a Level-13 theory of the future is likely, without any extra principle brought in to explain its large finetuning. If you say "no way is a Level-13 theory going to happen," like I do, then ask yourself about Level-12, and keep going until you get nervous about being too rigidly judgmental about the viability of a hypothesized new theory. For me, that is Level-6, as stated above.

Let us summarize our definition of Wilsonian Naturalness:

Endo-Natural theory: An endo-Natural theory is one where the finetunings are not high (all are, say, level-4 or lower[4]) across all its particle thresholds when matching EFTs

---

[4] Upon surveying the literature of physicists intuitions, any choice of level-3 through level-6 has been deemed a defensible upper bound on tolerable finetuning.

above and below the thresholds, according to the *a priori* defined algorithms of assessment discussed above.

Exo-Natural theory: A theory may have large finetuning across threshold(s), but those finetunings are explained in principle and are not accidental. This case has no implication of low probability despite its large finetuning, and we call the corresponding theory exo-Natural.

Wilsonian Natural theory: A Wilsonian Natural theory is one that is either endo-Natural or exo-Natural.

Exo-Natural theories are ones with strong UV/IR correlations embedded in them, which in principle can be discerned. We do not discuss them much in what follows, but let us illustrate here an exo-natural theory by considering the Arkani-Hamed-Harigaya theory [9], which was introduced as a possible explanation of the $g - 2$ anomaly [10–12]. By virtue of finetuned cancellations in the dimension-six operators that give rise to $g - 2$, the model "violates the Wilsonian notion of naturalness" in the words of the authors. In our language, the violation is not a violation of Wilsonian Naturalness but rather a violation of endo-Naturalness which comes at the matching interface of the second highest mass state S, where the EFT has a large coefficient of a $g - 2$ inducing dimension-six operator that is nearly exactly cancelled by integrating out the S field at the matching threshold scale m_S. But it is this violation of endo-Naturalness that points to the correct supposition that this was no accidental finetuned cancellation and the EFT under consideration (just below the heavier L particle mass) is incomplete and there are surely new particle(s) or new principles at play that *are not manifest within the EFT and are yet to be found.* Indeed that is the case. Thus, the theory is an exo-Natural theory, and is therefore Wilsonian Natural. It is for these reasons that it must be emphasized that there is no claim here that endo-Naturalness must always hold in any theory. Rather, the claim is on the implication of when endo-Naturalness does not hold: there are new particle(s) or principle(s) yet to be discovered because the theory is secretly exo-Natural—i.e., always Wilsonian Natural.

In summary, let us formulate the main conjecture of the paper as follows:

Wilsonian Naturalness conjecture: Large accidental finetunings in EFT matching across a particle threshold is highly improbable. Any such finetuning that may occur should be pursued as a sign for the existence of new particle(s) or principle(s) that render the large finetuning as a non-accidental result (i.e., it is secretly an exo-Natural theory). Furthermore, any conjectured theory that relies on large, unexplained accidental finetuning in EFT matching across particle threshold(s) is unlikely to be a good description of nature. Summary: Wilsonian Naturalness is expected to be satisfied by the next useful theory of nature beyond the Standard Model[5].

The subject of the rest of this paper is to illustrate the meaning and value of this conjecture.

---

[5] Of course, one is free to speculate on the possibility that a more radical idea could come along that does not involve even lagrangians or Wilsonian effective field theories in any currently recognizable way, where Wilsonian Naturalness tests would not be directly applicable.

4   The Standard Model

Statements abound in the literature that "the SM suffers from the Naturalness problem", or more or less equivalently, the hierarchy problem and the finetuning problem [13]. We need to find a theory that "cures the SM's naturalness problem" is another common refrain. However, there is no place for such talk when it comes to a highly successful theory like the SM. If some dreamt-up criteria ends up labeling the SM with a Naturalness problem, then I want every other theory I come up with to also have a Naturalness problem just like it.

There is no reasonable sense in which the SM has a naturalness problem or any other problem intrinsic to the theory itself such that it should be relegated to lower status. It is a successful first-class theory, culminating in the discovery of the Higgs boson which was a non-trivial corroboration of one of its peculiar features. However, one could try to expand beyond the surface meaning of "Naturalness problem" and say something that is reasonable: The SM has no principle or mechanism suggested within it that can protect the Higgs boson mass if new states that couple to it are introduced at large scales, and since we have no reason to believe that such new states are forbidden in nature, we realize that there is a problem—a "Naturalness problem of proliferation of new states that we generically expect"—in understanding how such new states could exist while the Higgs boson is so light. We can call this formulation the "proliferation naturalness problem" and it was argued for in [14]. As we will discuss in Sect. 8, this "proliferation problem" is what is implicitly behind declaring a theory to have a "hierarchy problem", where the hierarchy is between the SM states and exotic new states generically expected to exist in nature well above the weak scale.

The above articulation of the "proliferation naturalness problem" of the SM is really about contemplating simple new theories that are concatenations of the SM, and it by no means suggests that the SM itself is unnatural. The SM has done its job, and whatever Naturalness test one implements the SM better pass it to be a worthwhile test. It is up to any new theory, on the other hand, which might include new states at higher mass scales, to make sure it passes its own Naturalness test.

In what sense can we be sure that the SM passes its Naturalness test? In other words, how does one determine that the criteria set forward earlier when applied to the SM does not condemn it inappropriately as an unnatural theory? The answer is by straightforward calculation. At every particle mass threshold (m_t, m_h, m_Z, m_W, m_b, etc.) compute the low-energy (theory below mass threshold) couplings $\mathbf{g}_L$ in terms of the high-energy (theory above mass threshold) couplings $\mathbf{g}_H$ and compute the threshold finetunings using $\overline{MS}$ renormalized parameters. The answer is that there are no large threshold finetunings in the theory at all, and it therefore it passes the Naturalness test. The Wilsonian Naturalness algorithm correctly retains the SM.

It should be noted that the SM had very many opportunities to fail its test, since there was always a possibility that among the many matching conditions at the numerous thresholds one could have been highly finetuned. No large finetuning was found, but it could have been.

Perhaps the best opportunity for the finetuning to have manifest itself is across the top quark threshold. The matching of the m^2 above and below the top mass after electroweak symmetry breaking requires us to inspect the m_h^2 coefficient of $\frac{m_h^2}{2}h^2$ operator above and below the top mass. One finds the leading term to be

$$m_h^2(m_t)_L = m_h^2(m_t)_H + \frac{3m_t^4}{4\pi^2 v^2} + O(y_t^2 m_h^2) \tag{10}$$

where $v \simeq 246\,\text{GeV}$ and y_t is the top-quark Yukawa coupling.

We can compute the finetuning across the m_t threshold and we find

$$\text{FT}[m_h^2|m_t] \simeq \frac{3m_t^4}{\pi^2 v^2 m_h^2} \tag{11}$$

Inserting $m_t = 173\,\text{GeV}$ and $m_h = 125\,\text{GeV}$ into this equation one finds FT = 0.3 which is $O(1)$ as we expect most finetunings to be across thresholds. This is a low finetuning that is consistent with a Natural theory.

One should note that there would be nothing wrong in principle with the SM if the Higgs boson were much lighter. For example if m_h had been found to be less than $\sim 1\,\text{GeV}$ the finetuning value across the top quark threshold would be FT $\gtrsim 10^3$. In that case, the theory would have failed its Wilsonian Naturalness test, which would not have been physically impossible but rather highly improbable. Again, it is to be expected that such a large finetuning did not show up. This is reminiscent of a similar conclusion using a different approach to assess Naturalness—finite Naturalness [4]—applied specifically to the top quark contribution to the Higgs mass.

The summary of this section is that the SM has low finetunings across matchings of EFTs across its mass thresholds and therefore passes its Naturalness test. There are dozens of non-trivial tests that could have come to a different conclusion. This gives confidence that our primary theory at the present (the SM) does not register as a failure in the Naturalness evaluation with which we plan to asses conjectured theories. This is in contrast to illogically charging the SM with a lethal naturalness problem and then finding new theories that do not. The SM is Natural. Or differently said, the SM does not suffer from Unnaturalness.

5 Adding a Heavy Singlet

One of the simplest ways to extend the SM is to add a real singlet scalar σ to the spectrum. One can call this theory SM+σ for short. The lagrangian is

$$\mathcal{L}_{SM+\sigma} = \mathcal{L}_{SM} + \frac{1}{2}(\partial_\mu \sigma)^2 - \frac{1}{2}m_\sigma^2 \sigma^2 - \frac{\eta_\sigma}{2}H^\dagger H \sigma^2 + \frac{\lambda_\sigma}{4}\sigma^4 \tag{12}$$

Let us suppose that the mass of the σ-particle is higher than the masses of the other particles in the spectrum, and let's also call the effective theory that includes the σ particle $\mathcal{L}_{\sigma+} = \mathcal{L}_{SM+\sigma}$. We shall see that if the mass of the σ particle is too high then the matching encounters a finetuning problem, matching the discussion of [15].

Given the high mass of the σ particle we can integrate it out and are left with a low energy lagrangian $\mathcal{L}_{\sigma-}$ below the σ-mass threshold which is the SM lagrangian plus many higher dimensional operators, such as $O_6 = |H|^6$. After some analysis we can see that no operator in $\mathcal{L}_{\sigma-}$ suffers from a finetuning of matching across the m_σ threshold except possibly the coefficient m^2 of the operator $|H|^2$. In that case the matching is

$$m_{(-)}^2 = m_{(+)}^2 - \frac{\eta_\sigma m_\sigma^2}{16\pi^2}\left[1 - \ln\left(\frac{m_\sigma^2}{\mu^2}\right)\right] \tag{13}$$

where for clarity we have defined

$$m_{(\pm)}^2 = m^2 \text{ evaluated at } q^2 = m_\sigma^2(1 \pm \epsilon), \text{ where } \epsilon \ll 1. \tag{14}$$

In other words $m_{(-)}^2$ is the coefficient of $|H|^2$ in the low-energy effective theory just below the m_σ threshold after the σ-particle has been integrated out, and $m_{(+)}^2$ is the coefficient of $|H|^2$ in the high-energy theory above the m_σ threshold that includes the σ particle.

The calculation of the finetuning across the threshold yields:

$$\text{FT}[m^2] = \left|\frac{m_\sigma^2}{m^2}\frac{\partial m^2}{\partial m_\sigma^2}\right| = \left|\frac{\eta_\sigma m_\sigma^2}{8\pi^2 m_h^2}\ln\left(\frac{m_\sigma^2}{\mu^2}\right)\right| \tag{15}$$

where the last term utilizes the leading order result in the low-energy theory that $m^2 = -m_h^2/2$. We come across here the first instance where we think carefully how to deal with the arbitrary matching scale μ.

In the computation of observables the final result cannot depend on the arbitrary scale μ in the loop integrals. In practice, the result does not depend on μ at the order of the calculation, and a substantial residual μ dependence is an indication of the importance of including higher order corrections. On the other hand, the computation of the finetuning through sensitivity to loop effects, such as our example in Eq. 15, depends on μ. Not only that, choosing $\mu = m_\sigma$ in Eq. 15 appears to conclude that there is no finetuning independent of the mass m_σ^2.

Is not then the strong dependence of finetuning on the arbitrary scale choice μ of matching an indication of the worthlessness of finetuning? No. The proper way to think of the scale choice is that it is indeed arbitrary, and any reasonable choice from the point of view of theory calculability can be employed and the result should not be finetuned across the boundary. Reasonable choices the community has long made is $m_\sigma/2 < \mu < 2m_\sigma$, which translates into $\ln(m_\sigma^2/\mu^2)$ ranging from -1.4 to 1.4. Thus, we should vary μ over the entire range and determine what the average finetuning is. That is one precise prescription on how to deal with μ dependence. Another prescription which is much simpler to implement and yields roughly the same results is to replace $\ln(\Delta/\mu^2)$ with both $+1$ and -1 and average over the two resulting values for finetuning. We go with that prescription unless there is a case where replacing with $+1$ and -1 gives artificial cancellations, and then we revert back to the original prescription of averaging finetuning over choices $-1.4 < \ln(\Delta/\mu^2) < 1.4$. Our singlet scalar example here has no artificial cancellations by replacing $\ln(\Delta/\mu^2) = \pm 1$ and this prescription applied to our example then gives

$$\mathrm{FT}[m^2] = \frac{\eta_\sigma m_\sigma^2}{8\pi^2 m_h^2} = 0.8\,\eta_\sigma \left(\frac{m_\sigma}{1\,\mathrm{TeV}}\right)^2 \qquad (16)$$

The reader may be concerned that by using $\ln(\Delta/\mu^2) = \pm 1$ we changed the definition of our finetuning requirement by evaluating μ away from the precise mass threshold. However, this was only done to maintain the intuitions of those who wish to determine the finetuning of the Higgs boson with respect to m_σ^2. There is no problem maintaining the strict requirement $\mu = m_\sigma$ and seeing that there is large finetuning. Recall, the finetuning is the maximum finetuning across a threshold with respect to any high-energy parameter. The mass m_σ^2 was one such parameter but so is the $H^\dagger H$ operator coefficient $m_{(+)}^2$ in the high-scale theory. The finetuning with respect to it is

$$\mathrm{FT}[m^2] = \mathrm{FT}[m_{(-)}^2 \,|\, m_{(+)}^2] = \left|\frac{m_{(+)}^2}{m_{(-)}^2}\frac{\partial m_{(-)}^2}{\partial m_{(+)}^2}\right| = \left|\frac{m_{(+)}^2}{m_{(-)}^2}\right|_{\mu^2=m_\sigma^2} = \left|1 - \frac{\eta_\sigma m_\sigma^2}{8\pi^2 m_h^2}\right|$$

$$\simeq \frac{\eta_\sigma m_\sigma^2}{8\pi^2 m_h^2} \quad \text{(for large } m_\sigma^2\text{)} \qquad (17)$$

which is the same result as Eq. 16. The reason for this high finetuning seeping into the $m_{(+)}^2$ dependence is that the large logarithm is none other than the β-function of the renormalization group flow for the $m_{(+)}^2$ mass parameter. Very large changes in $m_{(+)}^2$ over small changes in μ^2 is what ultimately led to the finetuning with respect to $m_{(+)}^2$ evaluated at the $\mu^2 = m_\sigma^2$ scale.

Therefore, we can with confidence use Eq. 16 to determine the finetuning across the m_σ^2 mass threshold. With this in hand, we find that to reach level-3 finetuning or above with $\eta_\sigma = 1$ the σ mass must be $m_\sigma > 36\,\mathrm{TeV}$. This Naturalness proscription against such high masses might upon first inspection seem remote and uninteresting, but it is a remarkably powerful constraint that in all the many mass scales of nature, from here to the Planck scale, the introduction of any scalar boson that interacts moderately well with the SM Higgs would create, according to our assessments, a theory that does not pass its Wilsonian Naturalness test across the mass threshold. This is another articulation of the serious "proliferation naturalness problem" that was discussed in Sect. 4, and it is why many wish nature to possess a deeper principle, such as supersymmetry, to automatically allow large numbers of additional states, beyond the SM states that we know about, to be present in nature without facing repeated worries about how the Higgs boson mass could be so light in the presence of them all.

6 Supersymmetry

The case of supersymmetry adds some complication to the algorithm of computing finetunings across thresholds. In the case of the minimal supersymmetric Standard Model (MSSM) the largest threshold finetuning is going from a full Higgs sector with two Higgs doublets to a low-energy theory with a single Higgs doublet that has the propagating $h125$ boson and the three Goldstone bosons that constitute the longitudinal components of the $W^{\pm}$ and Z^0. Our goal then is to describe the matching between the full MSSM lagrangian parameters with two Higgs doublet (H_u, H_d) and the effective theory below a heavy lepton multiplet $(\Phi = \{A, H^0, H^{\pm}\}$, where the low-energy theory contains the SM doublet H.

As is well known, as superpartner masses increase the heavy doublet decouples with it and can be thought increasingly as a vev-less heavy complex scalar. This can then be eliminated. This heavy state can then be integrated out and the low-scale lagrangian that results can match its parameters with those of the theory with the heavy state. The subtlety is what algorithm to use for the "lagrangian parameters" of the full MSSM theory. In this supersymmetric case the most straightforward choice, and the choice that has been considered for many applications before, are the gauge couplings $(g', g,$ and $g_s)$, superpotential parameters (Yukawa couplings and the μ term), and soft supersymmetry breaking terms $(m_{H_u}^2, m_{H_d}^2, m_{\tilde{Q}}^2$, etc.)

With these considerations we first write the theory above the heavy Higgs doublet threshold [16]:

$$V(H_u, H_d) = (|\mu|^2 + m_{H_u}^2)|H_u|^2 + (|\mu|^2 + m_{H_d}^2)|H_d|^2 - bH_u \cdot H_d + \text{c.c.}$$

$$+ \frac{1}{8}g'^2 \left(|H_u|^2 - |H_d|^2\right)^2 + \frac{1}{8}g^2 \left(H_u^{\dagger}\sigma^a H_u + H_d^{\dagger}\sigma^a H_d\right)^2 \quad (18)$$

This then needs to be matched to the theory below the heavy Higgs doublet threshold

$$V(H) = m^2|H|^2 + \lambda|H|^4 \quad (19)$$

After some manipulations one finds

$$m^2 = -\left(\frac{1 - \sin^2 2\beta}{2}\right)\left[\frac{|m_{H_d}^2 - m_{H_u}^2|}{\sqrt{1 - \sin^2 2\beta}} - m_{H_d}^2 - m_{H_u}^2 - 2|\mu|^2\right] \quad (20)$$

$$\lambda = \frac{1}{8}\left(g'^2 + g^2\right)(1 - \sin^2 2\beta) \quad (21)$$

where

$$\sin 2\beta = \frac{2b}{m_{H_d}^2 + m_{H_u}^2 + 2|\mu|^2}. \tag{22}$$

Given this definition of the angle β, in the limit of large superpartner mass scale we can identify the light SM Higgs boson H and the heavy decoupled doublet state Φ with

$$H = \cos\beta H_d + \sin\beta \overline{H}_u \tag{23}$$

$$\Phi = -\sin\beta H_d + \cos\beta \overline{H}_u. \tag{24}$$

where $\overline{H}_u = i\sigma^2 H_u^*$.

We are left with the below-threshold lagrangian of Eq. 19 and its matching equations (Eqs. 20 and 21) entirely in terms of the parameters of the above-threshold theory once the dependence of that angle β on the supersymmetry parameters are substituted (Eq. 22). If we identify the heavy supersymmetry masses as the collection of $\tilde{m}_k^2 = \{m_{H_u}^2, m_{H_d}^2, |\mu|^2, b\}$ that have typical scale values of Λ_{susy}, one can see readily from Eq. 20 that

$$\mathrm{FT}[m^2] = \max_k \left| \frac{\tilde{m}_k^2}{m^2} \frac{\partial m^2}{\partial \tilde{m}_k^2} \right| \sim \frac{\Lambda_{susy}^2}{m_Z^2}. \tag{25}$$

The precise values of the FT depends on the exact choices of parameters but it is generic that the result is as shown, FT $\sim \Lambda_{susy}^2/m_Z^2$.

This scaling of finetuning matches the intuitions that have been present in the supersymmetry community for quite some time now. Traditionally the calculation was to check on the finetuning of the small value of m_Z^2 given all the heavy superpartner masses in the scalar potential. The equation for m_Z^2 for electroweak symmetry breaking at leading order is

$$m_Z^2 = -2|\mu|^2 + \frac{2(m_{H_d}^2 - m_{H_u}^2 \tan^2\beta)}{\tan^2\beta - 1} \tag{26}$$

Finetunings are then computed and the result is generically FT $\sim \Lambda_{susy}^2/m_Z^2$. So, although the $O(1)$ factors will be different between our algorithm for computing the threshold finetuning and the finetuning computed from considering superpartner mass dependences on m_Z, the results are the same within $O(1)$ factors. The main reason for this is that at tree-level $m^2 = -\frac{1}{2}m_Z^2 \cos^2 2\beta$, and so computations of finetuning on $\tilde{m}^2$ should be very similar to that of m_Z^2.

Roughly speaking using the guide that FT $\sim \Lambda_{\text{susy}}^2/m_Z^2$ one can contemplate declaring some supersymmetric theories[6] with large Λ_{susy} to fail the Naturalness test and thus should be relegated to lower status. Less probable, and perhaps even improbable, level-4 finetuning, for example, would put $\Lambda_{\text{susy}}^2/m_Z^2 \sim 10^4$ and thus $\Lambda_{\text{susy}} \sim 10\,\text{TeV}$. LHC limits of 1–3 TeV for the mass of superpartners puts the current level of finetuning at about 10^2–10^3, which is borderline for when someone would wish to confidently relegate the remaining supersymmetric theories to the trash bin. Note, there still remains an infinite number of supersymmetric theories that perfectly satisfy all known data and are not in conflict with any expectations. Only a high level of finetuning across thresholds (i.e., failing its Wilsonian Naturalness test) can cast a shadow on any of those otherwise good theories. It is with trepidation that one throws out otherwise perfectly good theories that are consistent with all known data; nevertheless, we reiterate adherence to the notion that such high finetunings are correlated with low a priori probabilities for realization [3].

In concluding this section, let us briefly remark that our approach differs significantly from some others in assessing finetuning in supersymmetric theories in that here we only assess finetunings among parameters "locally" across a threshold. We look at parameters above and below a heavy mass threshold and ask if there is finetuning in that matching. Some have implemented "non-local" finetuning assessments, by considering a supersymmetric theory with input parameters, say $m_{1/2}$ (universal gaugino mass) and others [17], defined at the GUT scale and then tracking how $m_{H_u}^2$ and $m_{H_d}^2$ scale with it: $m_{H_{u,d}}^2 = m_{H_{u,d}}^2(m_{1/2})$. Finetunings are then computed with respect to the input parameter $m_{1/2}$:

$$\text{FT}[m_Z^2] = \left| \frac{m_{1/2}}{m_Z^2} \frac{\partial m_Z^2}{\partial m_{1/2}} \right|. \qquad (27)$$

In this case there is implicitly a calculation of a parameter at one scale (m_Z at $q^2 = m_Z^2$) in terms of a parameter $m_{1/2}$ defined far away at the scale $q^2 = M_{\text{GUT}}^2$. Related schemes for identifying independent vs. dependent variables with correlations at various scale choices among the supersymmetric parameter choices may be introduced, which in our language would be useful for turning a theory that fails to be endo-Natural into one that is exo-Natural, and thus respects Wilsonian Naturalness. We are neutral as to the value of this activity, although we recognize that conceivably it could be used to both artificially and inappropriately lower the finetuning of a theory, but also to identify underlying parameter correlations that yield lower finetuning, which a deeper theory might be able to justify. With that

---

[6] Remember, a theory in our definition here does not have variable parameters. Theories are indexed values of a theory class. A supersymmetric theory then has specific single values for each of its parameters $m_{H_u}^2$, μ, etc. defined at some convenient scale.

said, the intuition at play here that requires a speculative theory to be Wilsonian Natural is laudable and justified.

7   Grand Unified Theories

Another example of an improbability of parameter cancellations that has been discussed in the literature for years is the so-called doublet-triplet splitting problem in grand unified theories [18, 19]. For example, minimal $SU(5)$ theory breaks down to the SM gauge groups via the condensation of the 24 dimensional representation Σ. The vacuum expectation value of this field is

$$\langle \Sigma \rangle = v_\Sigma \cdot \text{diag}(2, 2, 2, -3, 3) \tag{28}$$

where the value of the vev v_Σ is determined by parameters $\mathbf{w}$ in GUT-scale Higgs potential: $v_\Sigma = v_\Sigma(\mathbf{w})$.

In the supersymmetric case the Σ also couples to the 5- and $\bar{5}$-dimensional Higgs representation H_5 and $H_{\bar{5}}$ respectively. Within the $H_{5,\bar{5}}$ are the Higgs doublets $H_{u,d}$ and the Higgs triplet $H_{3,\bar{3}}$ representations. The relevant GUT-scale superpotential for H_5 is

$$W_{(+)} = \mu_5 H_{\bar{5}} H_5 + \lambda H_{\bar{5}} \Sigma H_5 \tag{29}$$

After symmetry breaking the superpotential splits the $H_{5,\bar{5}}$ into $H + u, d, 3, \bar{3}$ terms:

$$W = \mu_3 H_{\bar{3}} H_3 + \mu H_u H_d \implies W_{(-)} = \mu H_u H_d + \cdots \tag{30}$$

where

$$\mu_3 = \mu_5 + 2\lambda v_\Sigma, \quad \text{and} \tag{31}$$

$$\mu = \mu_5 - 3\lambda v_\Sigma. \tag{32}$$

We know that $v_\Sigma \simeq 10^{16}$ GeV for the unification of couples, and we also know that μ needs to be 10^{2-3} GeV for weak scale supersymmetry. Thus, there is an extraordinary finetuning in the cancellation that must occur in Eq. 32 to realize these constraints. Upon symmetry breaking and assessing the finetuning of μ with respect to the high-scale theory parameter μ_5 one finds

$$\text{FT}[\mu] = \left| \frac{\mu_5}{\mu} \frac{\partial \mu}{\partial \mu_5} \right| = \left| \frac{\mu_5}{\mu} \right| \simeq \left| \frac{2\lambda v_\Sigma}{\mu} \right| \sim 10^{13} \tag{33}$$

Thus, minimal supersymmetric $SU(5)$ GUTs have level-13 finetuning and do not pass their Wilsonian Naturalness test. This is why it is often referred to in the literature as the doublet-triplet splitting problem. It really is simply a Wilsonian Naturalness problem of the theory across matching EFT thresholds.

Of course there are many interesting ideas on how to solve this problem in Grand Unified theories, but the point here is that it is identified as a problem immediately from the perspective of our algorithmic finetuning tests for Wilsonian Naturalness, and the magnitude of the problem (level-13 finetuning) matches intuitions of early researchers in GUT theories who appreciated the seriousness of this deficiency in minimal GUT theories.

8   Extra Dimensions and the Hierarchy Problem

Let us make a few brief remarks about theories of extra dimensions [20–22]. Researchers have added extra spatial dimensions, either flat or warped, in order to recast and perhaps solve the hierarchy problem. It is worthwhile making a rather precise definition of the hierarchy problem here in order to give our perspective on the worth of introducing extra spatial dimensions to solve it.

If the SM does not have a Wilsonian Naturalness problem, as we argued Sect. 4, then what can we mean that it has a hierarchy problem, especially since the terms are often used interchangeably in the literature? The definition proposed here is that a theory has a hierarchy problem if it fails to pass its Naturalness tests (low finetunings across thresholds) *or* the simple introduction of heavy states into the spectrum immediately creates a Naturalness problem for the theory ("the proliferation of states problem" of Ref. [14]). The SM passes its Wilsonian Naturalness test, but if we add an additional scalar with mass $100\,\text{TeV}$, our results from Sect. 5 tell us that this new theory will fail its Wilsonian Naturalness test. In that sense the SM has a hierarchy problem—its passing of the Wilsonian Naturalness test is precarious in the face of nature having additional heavy particles well above the weak scale, which we generically expect nature to possess.

The introduction of supersymmetry solves the hierarchy problem by introducing superpartners that cancel out the quadratic sensitivities of new heavy mass scales to the Higgs parameter. Composite Higgs theories solve the hierarchy problem by eliminating this most special quadratically sensitive parameter associated with a fundamental Higgs boson mass [23]. And now, here, extra dimensions also could solve the hierarchy problem by disallowing any state to have mass that is well above the weak scale. There is no singlet scalar mass of $45\,\text{PeV}$ that would destabilize the hierarchy because the highest, fundamental scale Λ_F accessible to field theory is the weak scale, by virtue of, for example, an exponential suppression $\Lambda_F = M_{\text{Pl}}e^{-y}$, where $y \simeq 34$ associated with a compactified warped extra dimension.

Therefore, large extra dimension or compactified extra dimensions do not cure the SM of its Naturalness problem, because the SM has no Naturalness problem, but it does replace our Standard Theory of the SM with a different theory of extra dimensions, which, unlike the SM, has no hierarchy problem if the scale of extra dimensions is small enough. Extra dimensional theories have little to speak for themselves otherwise, so their advocacy has implicitly assumed in it that nature should have other states besides the SM states, and when those other states are added to the SM there develops a significant finetuning problem across its EFT matching thresholds, and thus the SM will not pass its Wilsonian Naturalness test without the introduction of extra spatial dimensions to cut off the offending high scales.

9   Twin Higgs Theories

Another class of theories that is reported to stabilize the weak scale is Twin Higgs theories [24–26]. The idea is to introduce a global $U(4)$ in the Higgs sector that the Φ scalar transforms under. The original potential for Φ admits spontaneous symmetry breaking,

$$V(\Phi) = -m_\Phi^2 |\Phi|^2 + \lambda_\Phi |\Phi|^4 \text{ where } \langle \Phi \rangle = \frac{m_\Phi^2}{2\lambda_\Phi} = f^2, \tag{34}$$

which breaks $U(4) \to U(3)$, giving 7 Goldstone bosons in the process.

Now, assume that the $SU(2)_A \times SU(2)_B$ subgroup of $U(4)$ is gauged and the Φ field can be split into

$$\Phi = \begin{pmatrix} H_A \\ H_B \end{pmatrix} \tag{35}$$

where $SU(2)_A$ is ultimately identified with the SM $SU(2)_L$ gauge symmetries and $SU(2)_B$ is a "twin $SU(2)$". H_A then will become the SM Higgs boson and H_B the "twin partner boson." One also assumes a Z_2 "twin symmetry" that enforces invariance under exchange of $H_A \leftrightarrow H_B$. This requirement implies equivalence of the $SU(2)_A \times SU(2)_B$ gauge couplings, $g_A = g_B = g$.

Radiative corrections from gauge fields to the potential of this theory will yield terms proportional to

$$\Delta V_2(H_A, H_B) \sim \frac{g^2 \Lambda^2}{16\pi^2}(|H_A|^2 + |H_B|^2) = \frac{g^2 \Lambda^2}{16\pi^2}\Phi^\dagger \Phi \text{ and} \tag{36}$$

$$\Delta V_4(H_A, H_B) = \frac{g^4}{16\pi^2} \ln\left(\frac{\Lambda^2}{f^2}\right)(|H_A|^4 + |H_B|^4). \tag{37}$$

where $\Lambda \sim 4\pi f$. The first terms ΔV_2 do not violate $U(4)$ symmetry but the ΔV_4 terms do, turning the Goldstone bosons into pseudo-Nambu-Goldstones (pNGB). The estimate of the physical Higgs mass is then

$$m_h^2 \sim \frac{g^4}{16\pi^2} f^2 \sim \left(\frac{g^2}{16\pi^2}\right)^2 \Lambda^2. \tag{38}$$

Therefore, even for very large Λ of several TeV—high-mass "new physics"—the Higgs is two-loop suppressed compared to the scale Λ and stays light.

The problem with the scenario above is that the there is complete alignment of the vevs for H_A and H_B, where both have $\langle H_{A,B} \rangle \sim f$, which is too high. Also, the SM Higgs, which is a pNGB of the breaking, is an equal admixture of the $SU(2)_A = SU(2)_L$ charged state and the $SU(2)_B$ charged state, and so its couplings to the SM states will be greatly reduced compared to the SM expectations. This is unacceptable given current constraints on the discovered $h125$ state at CERN which shows that the couplings of $h125$ to SM states are at least within 10% of SM couplings.

The solution is to introduce a soft breaking potential

$$V_{\text{soft}}(\Phi) = \mu_\Phi^2 \Phi^\dagger \Phi \qquad (39)$$

which explicitly breaks the Z_2 twin symmetry of $H_A \leftrightarrow H_B$. It is called "soft" because this term is technically natural in that when $\mu_\Phi \to 0$ one recovers a higher symmetry (the Z_2 twin symmetry). The effect of this is to misalign the vevs such that $\langle H_A \rangle \ll \langle H_B \rangle$, and the light Higgs becomes much more SM-like.

Now, what about the naturalness qualities of this theory? One can readily see by the discussion above that f can be in the multi-TeV range and this Twin Higgs theory would still passes its Wilsonian Naturalness test of computing the finetuning of the low-scale theory parameter m_h (coefficient of $|H|^2$ operator) with respect to the high-scale theory parameters m_Φ^2, μ_Φ, etc. Thus, the twin Higgs theory does not have a Naturalness problem *as long as* the theories mass parameters are not above tens of TeV.

But what has the Twin Higgs theory done for us, vis-à-vis naturalness and finetuning? From the algorithmic Wilsonian Naturalness test introduced here in this work the Twin Higgs theory has no impact. It does not solve the SM's Naturalness problem, because the SM has no Naturalness problem to begin with. And it does not solve the Hierarchy problem because if one adds another singlet σ to the spectrum and attaches it to $|\Phi|^2\sigma^2$ then a heavy m_σ will cause the Twin Higgs+σ theory to fail its Wilsonian Naturalness test at the same disastrous level as SM+σ theory does.

What the Twin Higgs theory does do is soften the quadratic divergences from a cutoff regulation scale from top quark and gauge boson loops. Instead of the Higgs boson being sensitive to $m_h^2 \sim (y_t^2/16\pi^2)\Lambda^2$ it is softened to $m_h^2 \sim (g^2/16\pi^2)^2\Lambda^2$, enabling Λ to be much higher before the destabilizing finetuning effects of a large Λ manifest themselves. For those who believe that there is meaning in tracking cutoff scale dependences in an effective theory, and taking them as serious indications of a destabilizing impact of quantum corrections *within the theory itself*, the Twin Higgs theory delays the requirement of new physics (UV completion) that will take care of the issue once and for all, such as supersymmetry, conformal symmetries or extra dimensional theories could do. For those who do not put stock in gaining intuitions by naive applications of cutoff dependent regulated quantum corrections, the Twin Higgs theory is merely another interesting theory that has little to do with Naturalness.

10 Conclusions

In this article we have presented a straightforward methodology for testing if a theory is Natural based on finetuning assessments across effective theories above and below massive particle thresholds. The resulting Wilsonian Naturalness test satisfies the requirement of being an a priori and unambiguous algorithm, which is required for it to have a connection to probability, as argued in [3].

In the process we have articulated the difference between a theory that has a Naturalness problem and one that has a Hierarchy problem. A theory has a Naturalness problem if it fails to pass its tests of low finetunings across EFT thresholds. A theory has a Hierarchy problem if it immediately develops a Naturalness problem in the presence of very massive additional particles added to the spectrum, such as a massive real scalar or a vectorlike fermions.

This approach unifies the understanding of traditionally claimed Naturalness problems based on what before appeared to be different criteria, such as within supersymmetry (finetuning of m_Z), the SM theory augmented by a massive real scalar (finetuning of m_h), and grand unified theories (finetuning of μ-term in GUT superpotential). All three of these Naturalness concerns have straightforward interpretations and are unavoidably accounted for within Wilsonian Naturalness.

Another implication of the Wilsonian Naturalness test described here is the unambiguous conclusion that the Standard Model does not suffer from a Naturalness problem, in contrast to many statements in the literature. Furthermore, within the Wilsonian Naturalness approach the twin Higgs theories neither solves the SM's Naturalness problem, because it does not have one, nor does it solve the Hierarchy problem since twin Higgs theories are just as susceptible to failing its Wilsonian Naturalness test in the presence, for example, of a PeV real scalar as is the SM. Nevertheless, the claim here is not that the Wilsonian Naturalness test is necessarily the only possible Naturalness assessment, but the rigor of other assessments that claim a theory is not Natural must explain its a priori algorithms and its connection to probabilities, which no other assessments do.

In summary, Wilsonian Naturalness is firmly grounded in its algorithmic a priori formulation, and it is connected to probability tests (failing strict Naturalness test is low probability). Therefore, we believe the "next good theory" beyond the Standard Model whose peculiar features will be corroborated by experiment will very likely be Wilsonian Natural, which in turn can be a guide to whether a new theory competing for attention should have high status among all the theories compatible with known data.

Acknowledgments This work is supported in part by DOE grant DE-SC0007859. I wish to thank S. Martin and Z. Zhang for enlightening conversations on these topics.

References

1. G.F. Giudice, Naturally speaking: the naturalness criterion and physics at the LHC. https://doi.org/10.1142/9789812779762_0010 [arXiv:0801.2562 [hep-ph]]
2. A. de Gouvea, D. Hernandez, T.M.P. Tait, Criteria for natural hierarchies. Phys. Rev. D **89**(11), 115005 (2014). https://doi.org/10.1103/PhysRevD.89.115005 [arXiv:1402.2658 [hep-ph]]
3. J.D. Wells, Finetuned cancellations and improbable theories. Found. Phys. **49**(5), 428–443 (2019). https://doi.org/10.1007/s10701-019-00254-2 [arXiv:1809.03374 [physics.hist-ph]]
4. M. Farina, D. Pappadopulo, A. Strumia, A modified naturalness principle and its experimental tests. J. High Energy Phys. **08**, 022 (2013). https://doi.org/10.1007/JHEP08(2013)022 [arXiv:1303.7244 [hep-ph]]
5. See, for example, C.P. Burgess, *Introduction to Effective Field Theory* (Cambridge University Press, 2021)
6. R. Barbieri, G.F. Giudice, Upper bounds on supersymmetric particle masses. Nucl. Phys. B **306**, 63–76 (1988). https://doi.org/10.1016/0550-3213(88)90171-X
7. C.P. Burgess, The cosmological constant problem: why it's hard to get dark energy from microphysics. https://doi.org/10.1093/acprof:oso/9780198728856.003.0004 [arXiv:1309.4133 [hep-th]]
8. A. Hebecker, *Naturalness, String Landscape and Multiverse* (Springer, New York, 2021)
9. N. Arkani-Hamed, K. Harigaya, Naturalness and the muon magnetic moment. J. High Energy Phys. **09**, 025 (2021). https://doi.org/10.1007/JHEP09(2021)025 [arXiv:2106.0173 [hep-ph]]
10. G.W. Bennett et al. [Muon g-2], Final report of the muon E821 anomalous magnetic moment measurement at BNL. Phys. Rev. D **73**, 072003 (2006). https://doi.org/10.1103/PhysRevD.73.072003 [arXiv:hep-ex/0602035 [hep-ex]]
11. T. Aoyama et al., The anomalous magnetic moment of the muon in the Standard Model. Phys. Rep. **887**, 1–166 (2020). https://doi.org/10.1016/j.physrep.2020.07.006 [arXiv:2006.04822]
12. B. Abi et al. [Muon g-2], Measurement of the positive muon anomalous magnetic moment to 0.46 ppm. Phys. Rev. Lett. **126**(14), 141801 (2021). https://doi.org/10.1103/PhysRevLett.126.141801 [arXiv:2104.03281 [hep-ex]]
13. A. Borrelli, E. Castellani. The practice of naturalness. Found. Phys. **49**(9), 860–878 (2019)
14. J.D. Wells, Higgs naturalness and the scalar boson proliferation instability problem. Synthese **194**(2), 477–490 (2017). https://doi.org/10.1007/s11229-014-0618-8 [arXiv:1603.06131 [hep-ph]]
15. D.B. Kaplan, Effective field theories. [arXiv:nucl-th/9506035 [nucl-th]]
16. Our conventions follow those of S. P. Martin, A Supersymmetry primer. Adv. Ser. Direct. High Energy Phys. **21**, 1–153 (2010). https://doi.org/10.1142/9789812839657_0001 [arXiv:hep-ph/9709356 [hep-ph]]
17. M. Bastero-Gil, G.L. Kane, S.F. King, Fine tuning constraints on supergravity models. Phys. Lett. B **474**, 103–112 (2000). https://doi.org/10.1016/S0370-2693(00)00002-2 [arXiv:hep-ph/9910506 [hep-ph]]
18. R.N. Mohapatra, Supersymmetric grand unification. [arXiv:hep-ph/9801235 [hep-ph]]
19. S. Raby, Supersymmetric grand unified theories: from quarks to strings via SUSY GUTs. Lect. Notes Phys. **939**, 1–308 (2017). https://doi.org/10.1007/978-3-319-55255-2
20. N. Arkani-Hamed, S. Dimopoulos, G.R. Dvali, The Hierarchy problem and new dimensions at a millimeter. Phys. Lett. B **429**, 263–272 (1998). https://doi.org/10.1016/S0370-2693(98)00466-3 [arXiv:hep-ph/9803315 [hep-ph]]
21. I. Antoniadis, N. Arkani-Hamed, S. Dimopoulos, G.R. Dvali, New dimensions at a millimeter to a Fermi and superstrings at a TeV. Phys. Lett. B **436**, 257–263 (1998). https://doi.org/10.1016/S0370-2693(98)00860-0 [arXiv:hep-ph/9804398 [hep-ph]]
22. L. Randall, R. Sundrum, A Large mass hierarchy from a small extra dimension. Phys. Rev. Lett. **83**, 3370–3373 (1999). https://doi.org/10.1103/PhysRevLett.83.3370 [arXiv:hep-ph/9905221 [hep-ph]]

23. G. Panico, A. Wulzer, The composite Nambu-Goldstone Higgs. Lect. Notes Phys. **913**, 1–316 (2016). https://doi.org/10.1007/978-3-319-22617-0 [arXiv:1506.01961 [hep-ph]]
24. Z. Chacko, H.S. Goh, R. Harnik, The Twin Higgs: Natural electroweak breaking from mirror symmetry. Phys. Rev. Lett. **96**, 231802 (2006). https://doi.org/10.1103/PhysRevLett.96.231802 [arXiv:hep-ph/0506256 [hep-ph]]
25. M. Low, A. Tesi, L.T. Wang, Twin Higgs mechanism and a composite Higgs boson. Phys. Rev. D **91**, 095012 (2015). https://doi.org/10.1103/PhysRevD.91.095012 [arXiv:1501.07890 [hep-ph]]
26. R. Barbieri, L.J. Hall, K. Harigaya, Minimal mirror twin Higgs. J. High Energy Phys. **11**, 172 (2016). https://doi.org/10.1007/JHEP11(2016)172 [arXiv:1609.05589 [hep-ph]]

The Geometric Phase: Consequences in Classical and Quantum Physics

Roberta Citro and Ofelia Durante

Contents

Abstract

Whenever a classical or a quantum system undergoes a cyclic evolution governed by a slow change of parameters, it acquires a phase factor: the geometric phase. The most common denomination is the Aharonov-Bohm, Pancharatnam, and Berry phases. Although this phase is attributed to the foundations of quantum mechanics, geometric phase has become increasingly influential in many areas of physics, from condensed matter physics, optics, high energy physics, fluid mechanics to gravity and cosmology. Moreover, the geometric phase offers a unique opportunity for the realization of quantum information. In this chapter, we first review the geometric phase in the classical context, in the

R. Citro (✉) · O. Durante
Department of Physics, "E.R. Caianiello", University of Salerno, Fisciano, Italy
e-mail: rocitro@unisa.it; odurante@unisa.it

© The Author(s), under exclusive license to Springer Nature Switzerland AG 2023
R. Citro et al. (eds.), *Sketches of Physics*, Lecture Notes in Physics 1000,
https://doi.org/10.1007/978-3-031-32469-7_3

Foucault's pendulum, and then discuss one of the first quantum manifestations, the Aharonov-Bohm effect. In addition, we will briefly present some examples, in optics and in condensed matter, where geometric phase is manifested through the polarization of light through an optical fiber or birefringent sheet, in a three-level systems in interferometry, and through the effect of topological pumping, quantum spin pump, and phase battery.

1   Introduction

In classical and quantum mechanics, the geometric phase is the phase difference acquired by the system when it is subject to a cyclic evolution governed by a slow change of parameters [15]. The existence of this phase was discovered independently by S. Pancharatnam [33] and by H. C. Longuet-Higgins [25], and finally formalized by M. Berry [48]. In fact, it is also known as the Pancharatnam-Berry phase. The geometric phase can be found in classical systems, as the Foucault pendulum, or in the Aharonov-Bohm effect and the conic intersection of potential surfaces, to cite some quantum situations [21, 25]. In the case of Aharonov-Bohm effect, the changed parameter is the magnetic field whose flux is enclosed between two closed interfering paths. Apart from mechanics, the geometric phase emerges in a great variety of contexts, such as classical and quantum optics. It manifests itself whenever there are at least two parameters that characterise a wave near a singularity or a hole (in topology). That is, it occurs when these parameters are varied simultaneously, but so slowly that the system is always in an equilibrium state (adiabatic modulation) and brought back to the initial configuration in order to describe a closed path in the parameter space. At the end of the cycle, one might expect the waves to returns to their initial state, however the final and initial states differ by a phase. This phase difference is the geometric phase that indicates how the dependence on system parameters is singular. The measurement of geometric phase, in general, requires an interference experiment, as we will see later. The Foucault pendulum is a classical mechanics example to illustrate the geometric phase and the mechanical analogue of it is known as the Hannay angle, as we will see in the next paragraph.

2   The Geometric Phase in Foucault's Pendulum

Since the first public presentation in 1851, Léon Foucault's pendulum has played a prominent role in physics and in the history of science [1]. Foucault's pendulum is a long pendulum, suspended high above the ground and in planar motion. The phenomenon that we want to describe concerns the plane of oscillation of the pendulum. First, we define the orientation of the pendulum as the direction of the vector orthogonal to the plane of oscillation of the pendulum and passing through the suspension point. An observer, outside the Earth, can testify that the orientation

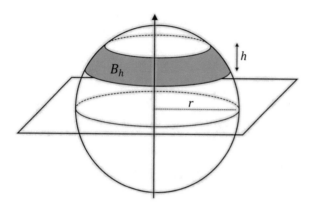

Fig. 1 The spherical corona of height h

of the pendulum varies during the course of the day and in general does not return to its original ones. This observation brings to two different aspects of the same phenomenon: the first is a dynamic aspect, and the second is the geometric one. The difference between initial and final orientation is called phase shift. The phase shift is positive if it points in the anti-clockwise direction. Since the phase shift is an angle, it is defined to be less than an angle of 2π. We will derive this phase shift and show how it applies to other phenomena. The derivation will be done using only geometric arguments to allow everyone to understand it, following the treatment of J. and H. von Bergmann [44]. Furthermore, the discussion will be done with the following approximations: the forces are considered isotropic and the motion of the pendulum is taken planar. The assumption of planar motion is sometimes called *adiabaticity hypothesis*, and it is justified as long as the period of the pendulum is small compared to the period of rotation of the earth. The last condition is not satisfied if the centrifugal force is taken into account. The centrifugal force comes into play because the suspension point of a pendulum follows a circular path on the Earth, however it is small compared to gravity and therefore has a small effect on the phase shift. We will therefore neglect it for a simplified treatment. Before introducing the topic, we recall a well-known geometric relation for the area of a spherical corona B_h of height h on a sphere of radius r:

$$S(B_h) = 2\pi r h, \tag{1}$$

as shown in Fig. 1.

2.1   The Mathematical Model

In order to understand Foucault's pendulum geometrically, we can consider an inertial reference system outside the Earth, in which the center of the Earth is fixed

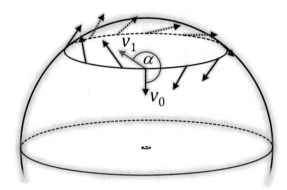

Fig. 2 The pendulum orientation rotates slowly during the course of the day. In general, it does not return to its original orientation v_0, but goes to the final one v_1 which differs from the initial one by a certain angle α

and the Earth rotates around an axis passing through the poles, and we record the motion of the pendulum with respect to the sphere. From this perspective, it appears that in one day the point of suspension of the pendulum, because of the rotation of the Earth, traces a closed path on the sphere. At different times of the day, the pendulum is located at different points on the sphere and after one day the pendulum returns to its original position on the sphere, and we can compare the initial and final orientations to obtain the phase shift (see Fig. 2). In accordance with the previous assumptions, the centrifugal force is not considered and no external forces that can change its orientation act on the pendulum (gravity acts only radially).

2.2   The Geometric Phase

Before deriving the phase shift relation of the orientation of the pendulum, it is useful to consider a pendulum whose suspension point follows an arbitrary path on the sphere (not a path with fixed latitude). For the sake of simplicity, we consider a pendulum that is slowly moved along a Euclidean plane (as shown in Fig. 3), it acts as a compass [47]. In fact, a symmetry argument tells us that the orientation of the pendulum remains fixed while the pendulum moves along the path because gravity acts only downwards, therefore it does not change the orientation of the pendulum that is forced to lie in the plane. When the path is a straight line, the angle between the path and the orientation of the pendulum remains constant. If the path is not a straight line, the orientation of the pendulum still does not change, but the path direction changes, so that the angle between the orientation and the path is also changing (see Fig. 3). This observation is crucial to understand the motion of a pendulum on a sphere. First, we define a *straight* path based on Newton's first law. A straight path is the one along which a particle moves in the absence of external forces. In our case, particles move along a sphere and the equator is an example

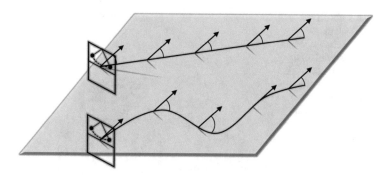

Fig. 3 The pendulum as a compass. When a pendulum in oscillation is held at its suspension point and moves in the plane, the angle that the orientation forms with a straight line remains constant, while the angle that it forms with a curved path changes

of a straight line: it divides the sphere into two mirror-symmetrical parts. In the absence of external forces, a particle moving along the equator cannot distinguish the two hemispheres and therefore remains on the equator. After a rigid rotation of the sphere, the image of the equator is still a straight line and it is called *great circle* or *geodesic*. Given a particle and its position, there exists at least one geodesic tangent to the direction of motion of the particle. Other paths, with latitude fixed beyond the equator, are not straight lines. If a pendulum is taken along a great circle then the angle between the orientation of the pendulum and the path not change, but if it is taken along a path with a fixed latitude, which is not the equator, then this angle changes. Now, we consider a triangular path, like the one in Fig. 4, in which each path belongs to a great circle and $\theta_1, \theta_2, \theta_3$ are the vertex angles. Suppose that the pendulum initially moves along the segment between θ_1 and θ_3 that belongs to a great circle, so the angle of orientation of the pendulum does not change. At the vertex of the θ_1 angle, the orientation has acquired an angle $\theta_1 - \pi$. Similarly, upon arriving to the second vertex of the triangle, the orientation acquires an additional angle $\theta_2 - \pi$ (as we start using the third segment as reference), and finally acquires an angle $\theta_3 - \pi$ when it returns to the starting point when we again use the original segment as reference. In total, the orientation has acquired a given phase shift $\alpha(C)$.

$$\alpha(C) = (\theta_1 - \pi) + (\theta_2 - \pi) + (\theta_3 - \pi) = \theta_1 + \theta_2 + \theta_3 - \pi, \tag{2}$$

remembering that the angles are defined in units of 2π. In a triangle on the Euclidean plane, the sum of the interior angles is π, so the phase shift is zero, but on a sphere the sum of the interior angles is greater than π since it has positive curvature.

A fundamental theorem of geometry links the phase shift for a triangular path C on the sphere of radius r, to the area S subtended by C. This is known as Gauss Bonnet's theorem:

$$\alpha(C) = \frac{S(C)}{r^2}. \tag{3}$$

Fig. 4 A triangle on the sphere: all its segments belong to large circles. When a pendulum is moved along the triangular path, the angle that the orientation of the pendulum forms with the segments remains constant along each segment, but acquires a phase shift

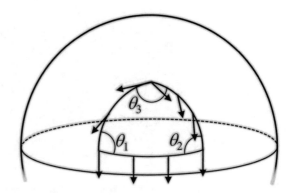

The demonstration can be performed according to Euler's argument, but can also be found in Ref. [29]. The phase shift $\alpha(C)$ is the geometric phase and it is related to the holonomy of the path C. In general, a vector that performs a parallel transport along a connection on the sphere never returns in the same starting vector and the difference between the two vectors is described by the holonomy. Obviously, this argument can be generalized to a generic path C that is a polygon on a sphere. Such a polygon can always be divided into triangles and the phase shift, acquired by moving along C, is the sum of the phase shifts associated to each triangle. Therefore, Gauss' theorem gives the geometric interpretation of the phase displacement. Thus concluding, the phase difference between the initial and final orientations of the pendulum after it has been moved along a closed path on the sphere, depends on the area enclosed by the path on the sphere, as we will see in the next paragraph.

2.3   The Foucault's Pendulum

Now, we consider a path corresponding to a latitude λ where the pendulum is located. If we take into account Eq. 1 with $h = r - r\sin(\lambda)$, the area enclosed by this path is given by:

$$S(C_\lambda) = 2\pi r^2[1 - \sin(\lambda)], \tag{4}$$

so that the phase shift, using Eq. (3) is given by:

$$\alpha(C_\lambda) = \frac{S(C_\lambda)}{r^2} = 2\pi[1 - \sin(\lambda)] = -2\pi\sin(\lambda), \tag{5}$$

considering that the phase shift is defined modulo 2π and that negative angles correspond to clockwise oriented angles. Equation (5) is the classical result of the Foucault pendulum. Now, we use the results previously discussed to understand how the orientation of the pendulum changes with time, i.e., we have to take into account the angle $\theta(t)$ that the pendulum makes with a path at time t, measured in days.

Symmetrically, the angle θ changes by an equal amount in equal times, so it follows the law: $\theta(t) = \theta_0 + ct$. After one day ($t = 1$), we know the phase shift which is equal to $\alpha(C_\lambda) = c = \theta(1) - \theta_0$, i.e., $c = -2\pi \sin(\lambda) + 2\pi k$, with k an integer. If the pendulum is on the equator, then $\lambda = 0$ and consequently $c = 0$ and the phase dynamics of the Foucault pendulum is described by $\theta(t) = \theta_0 - 2\pi \sin(\lambda)t$. This relation describes the geometric phase acquired by the Foucault pendulum.

3   The Aharonov-Bohm (AB) and the Berry Phases

As it is known, from the point of view of quantum mechanics, a wave function is a mathematical description of a quantum state of a particle as a function of momentum, time, position, and spin. The symbol used for a wave function is a Greek letter called ψ, and it is a complex function $\psi(\mathbf{r}, t) = |\psi(\mathbf{r}, t)|e^{i\varphi(\mathbf{r},t)}$, where $|\cdot|$, and ϕ are the modulus and the phase, respectively. By using a wave function, $|\psi(\mathbf{r}, t)|^2$ gives the probability of finding the particle at position $\mathbf{r}$ at time t. Now, we consider a charged particle moving in space: is it sufficient to know the local electromagnetic field to predict the evolution of the wave function of the particle? The answer is negative as explained by Yakir Aharonov and David Bohm about 60 years ago. In fact, we consider a charged particle travelling around a long and impenetrable cylinder inside which there is a magnetic field parallel to its axis. The path of the particle encloses a magnetic flow, as shown in Fig. 5. Even if the magnetic field along the path of the particle is zero, the vector field $\mathbf{A}$ is different from zero and, using Stokes' theorem, we can derive the vector potential to produce the field along the z direction parallel to the axis of the cylinder. The particle, therefore, besides having a kinetic energy, will have a contribution of potential energy given by $\frac{e}{c}\mathbf{v} \cdot \mathbf{A}$, where e is the electric charge, and $\mathbf{v}$ is the velocity of the particle, c the speed of light. The wave function acquires a phase along the above path and is given by:

$$\psi_A(\mathbf{r}, t) = e^{i\frac{e}{\hbar c}\int_{upper} \mathbf{A}\cdot d\mathbf{s}}\psi_0(\mathbf{r}, t), \tag{6}$$

where $\hbar$ is Planck's constant, ψ_0 is the wave function in the absence of the vector potential. The sign of the phase is opposite for the path below and the wave function along this path is given by:

$$\psi_B(\mathbf{r}, t) = e^{-i\frac{e}{\hbar c}\int_{lower} \mathbf{A}\cdot d\mathbf{s}}\psi_0(\mathbf{r}, t). \tag{7}$$

The probability to find the particle to the right of the cylinder shown in Fig. 5 depends on the squared modulus of the global transition amplitude $|\psi_A(\mathbf{r}, t) - \psi_B(\mathbf{r}, t)|^2$, hence on the phase difference between the contributions coming from the upper and lower paths. This phase difference is given by the Aharonov and Bohm phase $\phi_{AB} = \frac{e}{\hbar c} \oint \mathbf{A} \cdot d\mathbf{s} = \frac{e}{\hbar c}\Phi_B$, where Φ_B is the magnetic flow inside the cylinder. This means that as the field strength changes there is a harmonic component in the probability of finding the particle in the interference region. The effect

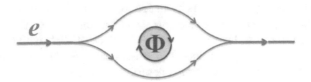

Fig. 5 An electron is circulating around a confined magnetic flow Φ. Although the magnetic field is zero along the path, the vector potential is non-zero. Therefore, the wave packet acquires a relative phase

AB highlights a *non-local* or *global* aspect of quantum mechanics that influences charge even in regions where electromagnetic fields are excluded. In the original paper, the ideal case of infinitely long flow lines was considered, but subsequently the effect was also demonstrated by relaxing these assumptions [24, 36, 46]. In fact, Michael Berry concluded that the effect Aharonov-Bohm is real, not ideal, physics. The AB effect has been demonstrated in systems such as superconducting films [43], metallic rings [45], quantum dots [50], topological insulators [34, 52], optical lattices [3] and trapped ions [32]. Furthermore, the AB phase is *topological*, i.e. it does not depend on the shape, or in general on the geometrical properties of the path, but only on the so-called *topologicalinvariants*. However, the AB phase is a special case of the more general geometric phase. The general form of the geometric phase was introduced but Michael Berry about 35 years ago [8]. It was noticed that when some parameters of a quantum system are varied slowly and cyclically along a closed path, the phase of its state does not return to the original value. Note that the state of the system is a vector in the vector space of the Hamiltonian of the system, called Hilbert space, the analogue of the orientation of the pendulum we saw in classical mechanics and, as such, it is characterised by modulus and phase. We consider a system described by a Hamiltonian H and a quantum state, a vector $\psi(\mathbf{r}, t)$. We suppose that $H(t)$ depends on a real parameter λ and that it varies in time, $\lambda = \lambda(t)$. Starting from an initial situation in which the system is in the m-th eigenstate of the Hamiltonian at time t_0, $H(\lambda(t_0))$, if the variation of the parameter is sufficiently slow, the adiabatic-theorem establishes that at a subsequent time t the system is still in the eigenstate of the new Hamiltonian $H(\lambda(t))$:

$$\psi(\mathbf{r}, t) = exp(i\phi(\mathbf{r}, t))exp(i\gamma(t))\psi(\mathbf{r}, t_0), \tag{8}$$

where the adiabatic theorem, the initial condition, and a convenient rewriting of the phase acquired by the state have been used to distinguish the dynamic contribution $\phi(\mathbf{r}, t)$, from the geometric one $\gamma(t)$. The latter defines the Berry phase [11]. The geometric phase is given by the theorem of Gauss and Bonnet as we have seen in Eq. 3, with the difference that it is applied to the parameter space and not to the fixed space. For example for a spin of an electron moving in the magnetic field B, we suppose that the time-dependent parameter for the Hamiltonian is the magnetic field and that it varies very slowly on the scale of all the other parameters of the system, travelling along a closed path C as t varies, then the Berry phase will be proportional to:

$$\gamma(C) = \int_\sigma \frac{dB}{B^2}, \tag{9}$$

where σ is the area enclosed by the curve C in the parameter space, and B is taken parallel to the outgoing direction.

The Berry phase is the manifestation of a curve holonomy, i.e., the impossibility to preserve geometric data during parallel transport along a closed curve [6] (as explained in section 2.2). Having a very wide applicability [48], the geometric phase has been anticipated in many areas of physics, we remember the work of Shivaramakrishnan Pancharatnam [33], who showed that the geometric phase can be obtained by a sequence of measurements, or better by measuring (and projecting) the polarization of a beam of light [9]. The geometric phase has been obtained also in the context of molecular electronics [25] and has been widely studied also in the context of conical refraction [20] to describe the parallel transport of light polarization along curved rays [12]. Aharonov and Anandan generalized the geometric phase (to the not necessarily adiabatic case) in terms of the closed curves of the quantum system, rather than the parameters of the Hamiltonian [2]. During the same years, Berry analysed the case in which the system returns to its starting state only in an approximate way, proposing an iterative technique to obtain corrections to the geometric phase [10]. Other generalizations arose later. Another important manifestation of the geometric phase in solids with crystalline symmetry is the Zak phase [51]. Thus, the geometric phase appears in a wide range of quantum systems and it continues to influence many areas of physics with an large number of applications. In the Table 1 some geometric phases and their field of application are reported.

Table 1 List of some relevant geometric phases

Phase	Year	Field	Parameter space
Pancharatnam	1956	Optics	Poincaré sphere
Aharonov-Bohm	1959	Quantum electrodynamics	Space-time
Berry	1983	Quantum mechanics	General
Aharonov-Casher	1984	Quantum electronics	Real space
Aharonov-Anandan	1987	Quantum mechanics	General
Zak	1989	Condensed Matter	Momentum space

4   Polarization of Light in an Optical Fiber or in a Birefringent Sheet

An example where the geometric phase plays a principal role is in the phenomenon of circularly polarized light entering an optical fiber with helix shape or crossing a birefringent medium. First, we remind that the polarization of light corresponds to the oscillation direction of the electric field, that is orthogonal to the wave propagation vector, as shown in Fig. 6a.

The representation of polarization states on the Poincarè's sphere was the fundamental observation that helped Pancharatnam to recognize the geometrical nature of the phase shift of polarization across a sheet. A polarization vector in two dimensions can be mapped onto the sphere surface, known as a Poincaré sphere. Conventionally, circular polarization points are positioned at the north and south poles to which correspond the helicity + and −. Linear polarization states are positioned on the equator, and elliptical polarization states are located all over the sphere. Let us assume that the fibre is rolled in such a way that the vector **k** draws a closed curve on the Poincarè's sphere. In this case, the Berry formula for the determination of the geometric phase acquired during the path through the fiber is:

$$\gamma_\sigma = -\sigma \Omega(C), \tag{10}$$

where $\Omega(C)$ is the solid angle determined by the curve C drawn in the momentum space (see Fig. 6b) and $\sigma = \pm$. If we consider a linear polarized light that is a

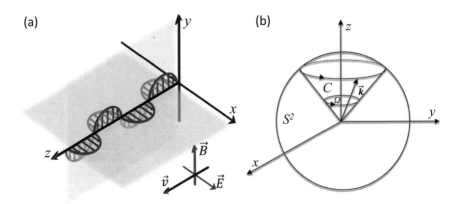

Fig. 6 (**a**) Electromagnetic field. The polarization corresponds to the oscillation plane of the electric field that is orthogonal to the direction of propagation along z. (**b**) Circularly polarised light passes through a half-wave sheet. The induced geometric phase varies in opposite directions for left-polarized and right-polarized light, as shown by the arrows in panels a and b. The incident light acquires a 2θ phase which is twice the angle of orientation of the optical axis of the material. The geometric phase is half the solid angle shown on the Poincaré sphere

superposition of helicity eigenstates, i.e.,

$$|\psi_i>=\frac{1}{\sqrt{2}}(|k,+>+|k,->),\tag{11}$$

the final state, after the propagation through the helix, is given by:

$$|\psi_f>=\frac{1}{\sqrt{2}}(e^{i\gamma_+}|k,+>+e^{i\gamma_-}|k,->).\tag{12}$$

Thus, the squared transition amplitude is given by:

$$<\psi_i|\psi_f>^2=cos^2(\gamma_+),\tag{13}$$

that can be explained as a rotation of the polarisation plane of an angle γ_+ given by Eq. 10. This expression is equivalent to the Malus' formula.

In conclusions, the optical fiber, wrapped in a helix shaped form, produces an effective optical activity. The amount of the rotation indeed does not depend on the light wavelength but on the solid angle and for this reason, it is a pure geometrical effect.

5 The Geometric Phase in Interferometry: Three-Level System

An interesting example of geometric phase manifestation is in the interferometry phenomenon of a three-level system, e.g., a system formed by three photons, described by a $|\Psi>$ state, which is $U(2)$-invariant. A three-level interferometer, formed by several beam splitters, realizes a general $SU(3)$ transformation, where each of the splitters correspond to an $SU(2)$ transformation and forms a group that is isomorphic to $U(2)$. Therefore, the space of states can be marked as $SU(3)/U(2)$ and the states are $U(2)$-invariant.

The geometric phase associated with an open curve C in $SU(3)/U(2)$ is given by Mukunda [30]:

$$\varphi_g[C]=\varphi_{tot}[\tilde{C}]-\varphi_{dyn}[\tilde{C}],\tag{14}$$

where φ_{tot} and φ_{dyn} are the total and the dynamical phases, respectively, and the $\tilde{C}\in SU(3)$ is an arbitrary increase of the curve C. In particular, φ_{tot} and φ_{dyn} are defined as follow:

$$\varphi_{tot}[\tilde{C}]:=arg<\varphi(s_1)|\varphi(s_2)>,\tag{15}$$

$$\varphi_{dyn}[\tilde{C}]:=Im\int_{s_1}^{s_2}ds<\varphi(s)|\dot{\varphi}(s)>,\tag{16}$$

where $s_1 \leq s \leq s_2$ is the $\tilde{C}$ parametrization that represents the evolution parameter. In the space of state $SU(3)/U(2)$, two points can be connected by a unique arc, called geodetic arc [30], where the geometric phase is zero. In the case a three-level system, the states are represented by $|\Psi_1 >$, $|\Psi_2 >$, and $|\Psi_3 >$ and connected by geodetic arcs. The geometric phase associated to the geodetic triangle is given by:

$$\varphi_g[\tilde{C}] = < \varphi_1|\varphi_2 > < \varphi_2|\varphi_3 > < \varphi_3|\varphi_1 > . \tag{17}$$

This is the well-know *Bargmann* invariant. Sanders et al. [37] introduced an experimental system formed by a three-channels optical interferometer and four experimentally adjustable parameters to observe the geometric phase (see Fig. 7). This optical scheme is adopted to produce and detect an abelian geometric phase shift, given by the appropriate Bargmann invariant, resulting from such transformation along a geodesic triangle. The experimental setting is outlined in [37]. In the interferometer, two orthogonally polarised beams travel in opposite directions

Fig. 7 Geodesic evolution in a three-level system in interferometry: By adjusting the parameters of the interferometer, the output state in the geometric space can be made to evolve along geodesic paths, from one vertex to the next, until the triangle is closed

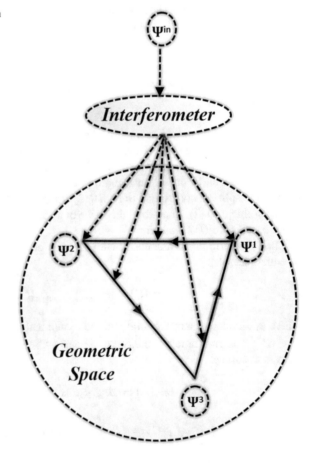

at the same time, and then they interfere at the output. The obtained interference pattern thus provides the relative phase difference $2\varphi_g$, which should corroborate the theoretical expectation.

6 The Quantum Pump

The quantum pump, originally proposed by David Thouless [42], is one of the most fascinating effects related to geometric phase. It is associated with the transport of charge in the absence of an external electric field, through the adiabatic and cyclic evolution of two parameters. Differently from the classical case [4], the transported charge is quantized and determined by the topology of the pump cycle, making it robust against disorder and effect of the interactions. Recently a topological pump has been reported using atoms immersed in a superlattice created with two optical lasers, see Fig. 8 [14]. A superlattice is formed by overlapping two lattices, i.e., two periodic potentials with different periodicity and phase shift between them. In a pumping cycle, the longer lattice is shifted with respect to the shorter one, varying the optical path difference over time.

At the end of the cycle, the potential returns to its original configuration, while the system always remains in its fundamental state during the cycle. At the end of

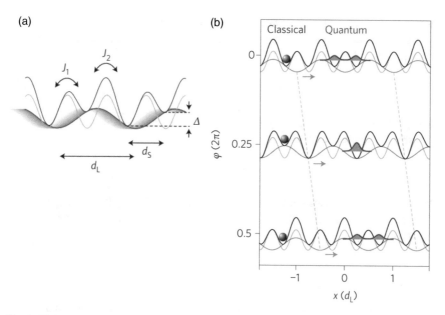

Fig. 8 The quantum pump: a superlattice $V(x, t) = V_0 \cos(x/d_0) + V_1 \cos(x/d_1 + \varphi(t))$ is moved to the right through the phase, which varies cyclically in time. After one cycle the lattice returns to the initial situation, while the particle has moved one lattice step by means of the tunnel effect. Reprinted by permissions from Nature: Nat. Phys. Citro, R. [14], copyright (2016)

the cycle, the state of the system has acquired a non-zero geometric phase (Berry). This results in a shift of center of mass of the atomic package during one cycle, i.e., the charge shifts by one lattice step. It is noteworthy that from the classical point of view, the transport of charges is intuitive, because particles are in the minima of the lattice and are dragged along by shifting the minima of the potential. Instead, in the quantum case, the minima are not shifted in space and therefore the pump is realized through a sequence of tunneling events between the minima of the potential. It should be noted that the single-valued wave function requires that the displacement of the center of mass be quantized in units of 2π. The pumped charge is like a magnetic flow enclosed in parameter space and can be expressed as an integral of the effective magnetic field over the area σ embedded by a closed curve C in parameter space, which is equal to a multiple of an integer, the so-called Chern number.

The origin of the center of mass shift and its relation to the geometric and Berry phase can be derived as follows. Consider first of all a period lattice with periodicity a and suppose the number of particles per site is equal to an integer $na = N$ in such a way that only the lowest band is occupied and the higher one empty (insulating phase). Now let us slide very slowly the potential in such a way that the potential returns to its original configuration after a time T. The Hamiltonian of the system is periodic and satisfies the condition:

$$H(q, T) = H(q, t + T),\qquad (18)$$

where q is the crystal momentum and T is the period.

Since the process is adiabatic, each eigenstate can at most acquire a phase after each pump cycle:

$$|\psi(q)\rangle = e^{\frac{i}{\hbar}\phi(q)}|\psi(q)\rangle.\qquad (19)$$

The phase ϕ can be different for each crystal momentum and this allows to define a single cycle evolution operator:

$$U_c = e^{i\phi(q)}.\qquad (20)$$

Independent on its origin this phase gives origin to a motion after a pump cycle. In fact, due to the fact that x and q do not commute and that $[x, F(q)] = i\hbar \partial F(q)/\partial q$, one has:

$$U_c^\dagger x U_c = x - \frac{\partial \phi(q)}{\partial q},\qquad (21)$$

i.e., one obtains the time-evolved position operator, whose average can be calculated to evaluate the shift of the center of mass. The phase can be derived from the time dependent Schroedinger equation, i.e. by substituting the Eq. (19) and has two contributions:

$$\phi(q) = \phi_B(q) - \frac{i}{\hbar} \int_0^T E(q, t) dt. \tag{22}$$

The latter is the dynamical phase coming from the integral over the Hamiltonian eigenvalue $E(q, t)$ and gives a zero position shift if the band is full, i.e. $\delta x = \frac{T}{\hbar} \frac{\partial \bar{E}}{\partial q}$, while the first one is the *geometric* phase:

$$\phi_B(q) = \frac{1}{\hbar} \int_0^T dt \langle \psi(q, t) | i\hbar \partial_t | \psi(q, t) \rangle, \tag{23}$$

where $|\psi(q, t)\rangle = e^{iqx} |u(q, t)\rangle$ where $u(q, t)$ is a Bloch wavefunction $u(q, t) = u(q + Q, t)$ periodic with the periodicity of the Bravais lattice. The geometric phase can be related to the time component of the so called Berry connection:

$$\mathbf{A}(q, t) = \langle \psi(q, t) | i\hbar \partial_q | \psi(q, t) \rangle \mathbf{e}_q + \langle \psi(q, t) | i\hbar \partial_t | \psi(q, t) \rangle \mathbf{e}_t; \tag{24}$$

giving the displacement:

$$\delta x = -\frac{1}{\hbar} \int_0^T dt \partial_q A_t(q, t). \tag{25}$$

By introducing the gauge field tensor, like in electrodynamics, the so-called *Berry cruvature*:

$$\Omega(q, t) = \partial_t A_q - \partial_q A_t = i\hbar \left(\langle \partial_t u_q | \partial_q u_q \rangle - \langle \partial_q u_q | \partial_t u_q \rangle \right), \tag{26}$$

according to Stokes theorem, the geometric phase can be written as a surface integral and the displacement is given by a loop integral, i.e.

$$\delta x = 2\pi a C, \tag{27}$$

$$C = \frac{1}{2\pi} \int_0^T \int_{-\pi/a}^{\pi/a} dq \Omega(q, t), \tag{28}$$

and a is the unit lattice vector. C is the topological invariant, called *Chern number* and is an integer. The integral is over a torus on the (q, t) space. Let is note that if one considers only the minimum of the band, as in the case of a Bose-Einstein condensate with momentum q_c at the minimum of the band, the center of mass shift is not an integer, as experimentally demonstrated in Ref. [26].

Thus a one-dimensional quantum charge pump transfers a quantized charge in each pumping cycle. This quantization is topologically robust, being analogous to the quantum Hall effect. The charge transferred in a fraction of the pumping period is instead generally unquantized. In particular, it has been shown that with specific symmetries in parameter space the charge transferred at well-defined fractions of the pumping period is quantized as integer fractions of the Chern number [27]. This can be shown exactly for a one-dimensional Harper-Hofstadter model where the fractional quantization of the topological charge pumping is independent of the specific boundary conditions taken into account. The realization of this pump has important applications in quantum computation [31], because the quantized charge works as a quantum bit (qu-bit), which is coherently transmitted. We remember that, while a classical bit can only take on two values 0 and 1, the qu-bit can simultaneously assume the value 0 and 1 and can therefore transmit more information at the same time [35].

7   The Quantum Spin Pump

A quantum spin-pump can be realized if one considers ultracold atoms in two hyperfine states in a spin-dependent controlled optical superlattice. First of all, to get a spin-dependent superlattice one needs to consider the fermionic atoms loaded in a superlattice in the limit of strong interaction, i.e. $U \gg t$ where U is the on-site atom interaction and t is the tunneling. In this limit the two spin components are coupled by an exchange interaction $\delta J_{exch} \propto U/t^2$ and the on-site energy Δ becomes spin-dependent.

A pump cycle can be realized by enclosing a circle around the degeneracy point ($\delta J_{exch} = 0$, $\Delta = 0$) as shown in Fig. 9a,b. After half a cycle the spins in a double well exchange their position and form a triplet state and after a full cycle the two spins have moved in opposite direction. Compared to the spinless case, since the system maintains the time-reversal symmetry, the topological invariant is the Z_2 Chern number:

$$C_s = C_\uparrow - C_\downarrow, \tag{29}$$

where C_σ with $\sigma = \pm$ is the spin Chern number. Evidence of the spin separation comes from the spin center of mass position which is measured from in situ absorption images [38] (see Fig. 9c). One observes jumps that correspond to quantized particle transport [49].

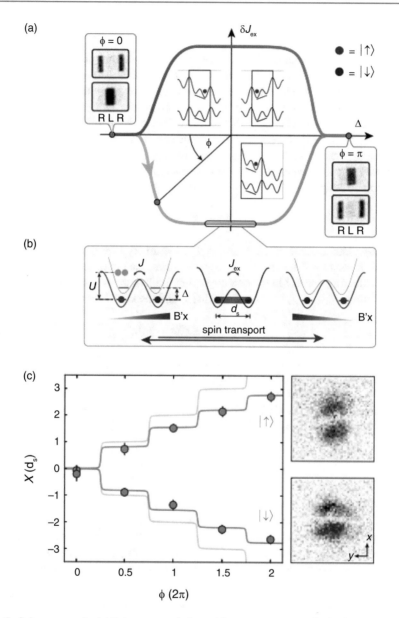

Fig. 9 Spin pump cycle. (**a**) Spin pump cycle (green) in parameter space of spin-dependent tilt Δ and exchange coupling dimerization δJ_{exch}. The path can be parametrized by the angle ϕ, the pump parameter. The insets in the quadrants show the local mapping of globally tilted double wells to the corresponding local superlattice tilts with the black rectangles indicating the decoupled double wells. Between $\phi = 0$ and π, the spins exchange their position, which can be observed by site-resolved band mapping images detecting the spin occupation on the left (L) and right (R) sites, respectively. (**b**) Evolution of the two-particle ground state in a double well around $\Delta = 0$ and the lattice constant d_s. (**c**) Center-of-mass position of up (red) and down (blue) spins as a function of the pump parameter ϕ. The points show the center-of-mass position averaged over ten data sets

8   The Phase Battery

Finally, we describe another important example in quantum physics involving the geometric phase, that is the so-called *Josephson phase battery* [40]. In order to explain the quantum phenomenon, we have to introduce the Josephson effect. In 1962, D. Josephson predicted a flow of dissipationless current through two superconductors sandwiching a thin layer (oxide, or semiconductor) [22]. In this system, that can work at very low temperatures ($\sim mK$) so as to make the thermal noise negligible, the so-called Josephson current (I_j) is related to the macroscopic phase difference between the two superconductors, φ, via the following current-phase relationship (CPR):

$$I_j(\varphi) = I_C \sin(\varphi), \tag{30}$$

where I_C is the junction critical current. The symmetry is preserved, for both time-reversal ($t \rightarrow t$) and inversion ($\mathbf{r} \rightarrow -\mathbf{r}$) symmetries impose the rigidity on the superconducting phase. On the contrary, when both these symmetries are broken, a finite phase shift $0 < \varphi_0 < \pi$ can be induced and the CPR becomes:

$$I_j(\varphi) = I_C \sin(\varphi + \varphi_0). \tag{31}$$

In this case, we obtain a φ_0-junction, which can generate a constant phase polarization in an open circuit configuration and an anomalous Josephson current when the junction is embedded into a closed superconducting loop [19]. In fact, in a superconducting quantum interference device (SQUID) formed by inserting two φ_0-Josephson junctions into a superconducting ring, as in Fig. 10a–c, the current is:

$$I_s(\Phi) = 2I_C \left| \cos\left(\pi \frac{\Phi}{\Phi_0} + \frac{\varphi_{tot}}{2} \right) \right|, \tag{32}$$

where Φ is the flux piercing the ring area of the out-of-plane magnetic field, B_z, and $\varphi_{tot} = 2\varphi_0$ resulting from the φ_0-shifts in each of the junctions.

Recently, anomalous Josephson effect has been demonstrated experimentally in a few hybrid Josephson devices fabricated with a topological insulator Bi_2Se_3 and Al/InAs heterostructures and nanowires [5, 18, 28, 40, 41]. In addition, attention has

---

Fig. 9 (continued) of a spin-selectively imaged atom cloud; the error bars show the error of the mean. Each data set consists of an average of ten pairs, which contain an image obtained by a sequence with pumping and one using a reference sequence with the same length but constant pump parameter $\phi = 0$. Difference in situ absorption images of both sequences for up and down spins are shown on the right side. The solid lines depict the calculated motion of a localized spin for the ideal case (light gray) and for a reduced ground state occupation and a pump efficiency per half pump cycle that was determined independently through a band mapping sequence (gray). Readapted from [38] by permissions from APS Publishing

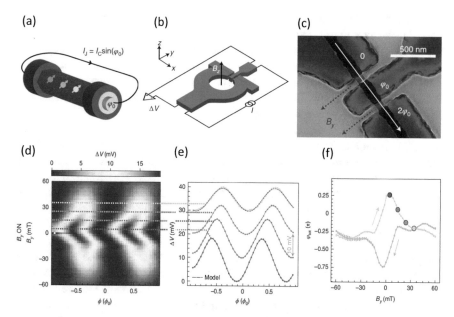

Fig. 10 Josephson phase battery device: (**a**) cartoon of phase battery. InAs nanowire in red, ferromagnetic impurities in yellow, and superconducting poles in blue. (**b**) Schematic illustration with the representation of out-of-plane magnetic field, B_z. (**c**) False-colored scanning electron microscopy image of the device with in-plane magnetic field, B_y, orthogonal to the nanowire. (**c**) Voltage drop $\Delta V(\Phi)$ at constant current bias $I = 1\,\mu A$ as a function of the in-plane magnetic field B_y applied orthogonally to the nanowire axis. (**d**) Selected traces $\Delta V(\Phi)$ at different B_y. Data are vertically staggered for clarity. (**e**) phase shift φ_{tot} derived from data obtained backward (green) and forward (yellow) sweeping B_y. Readapted by permissions from Nature: Nature Nanotech., A Josephson Phase Battery, Strambini E. et al. [40], copyright (2020)

also been turned to the theoretical study of the magnetic response of anomalous ferromagnetic φ_0-junctions, which are essentially superconductor-ferromagnet-superconductor JJs with an intrinsic spin-orbit coupling, with a ground state corresponding to a finite phase shift, $0 < \varphi_0 < \pi$, in the CPR [13, 23]. In these systems, the ferromagnetic magnetization can be electrically manipulated, for the charge current induces an in-plane magnetic moment that in turn acts as a torque on the out-of-plane magnetization, inducing eventually its switching [16,17,23,39]. From the experimental side, Strambini et al. [40] have conceived the first Josephson phase battery through a SQUID formed by two anomalous junctions made of an InAS nanowire and Al superconducting poles. These one-dimensional structures (nanowires) with strong Rashba spin-orbit coupling are characterized by surface oxides or defects (see yellow arrows in panel a), which behaving as ferromagnetic impurities can be polarized along the y-direction through an in-plane magnetic field, B_y, so as to induce a persistent exchange interaction leading to the anomalous phase φ_0 across the wire. This phase shift is governed by the Lifshitz-type invariant in the free energy, $F_L \approx f(\alpha, h)(n_h \times \hat{z}) \cdot v_s$, where $f(\alpha, h)$ is an odd function of both the strength of the Rashba α coupling and the exchange or Zeeman field h, n_h is a unit vector pointing in the h direction, and v_s is the superfluid velocity [7]. The triple scalar product in F_L defines the vector symmetries of φ_0 while its magnitude is a function of specific microscopic details and macroscopic quantities, such as temperature. The geometry of the Josephson phase battery is designed to maximize the symmetry of the two JJs in order to sum the two anomalous φ_0 shifts, when a uniform in-plane magnetic field is applied.

In the case of the magnetic field directed only out of the superconducting ring plane, the magnetic impurities are not polarized and there is no anomalous phase. However, when an in-plane magnetic field is applied orthogonally to the nanowire axis causes the SQUID interference patterns to shift, see Fig. 10d,e, due to the φ_0 influence. This is a clear signature of the magnetically-controllable anomalous Josephson effect. Moreover, it is possible to observe that the evolution of the anomalous phase shift extracted from the SQUID interference models reveals an important hysteretic behavior, see Fig. 10f, which cannot be merely ascribed to the trapping of fluxons in the superconductors [19]. In fact, this is the result of the ferromagnetic coupling between the intrinsic magnetic impurities in the nanowire due to two competitive mechanisms, namely the effective exchange field created by the uncoupled surface pins and the Zeeman field generated by the in-plane magnetic field, each of which makes a distinct "intrinsic" or "extrinsic" contribution to the anomalous phase.

In summary, the presence of in-plane magnetic field allows the continuous tuning of the anomalous phase, which leads to the charge and discharge of this "battery", just like a classical battery converts chemical energy into a persistent voltage bias that can supply electronic circuits. The Josephson phase battery that can represent a key element for quantum technologies was indeed finally demonstrated.

9   Conclusion

In this chapter, we have illustrated the implications of geometric phase in classical and quantum physics, connecting purely geometric and topological properties of parameter space to measurable properties, such as the Aharonov-Bohm effect. Other illustrated evidence for geometric phase is related to the polarization of light through an optical fiber or birefringent sheet, the three-level systems in interferometry, the quantum pump, the quantum spin pump, and the phase battery. The geometric phase, if appropriately engineered, can be useful in the various fields of physics involving quantum computation and quantum technologies.

Acknowledgments R.C. and O.D. thanks F. Giazotto and C. Guarcello for the great and fruitful scientific collaboration.

References

1. A.D. Aczel, *Pendulum*, New York (2003)
2. Y. Aharonov, J. Anandan, Phase change during a cyclic quantum evolution. Phys. Rev. Lett. **58**(16), 1593 (1987)
3. M. Aidelsburger et al., Experimental realization of strong effective magnetic fields in an optical lattice. Phys. Rev. Lett. **107**(25), 255301 (2011)
4. Archimedes, *The Works of Archimedes* (Cambridge University Press, Cambridge, 1987)
5. A. Assouline et al., Spin-orbit induced phase-shift in Bi_2Se_3 Josephson junctions. Nat. Commun. **10**, 126 (2019)
6. J.E. Avron et al., Topological invariants in Fermi systems with time-reversal invariance. Phys. Rev. Lett. **61**(12), 1329 (1988)
7. F.S. Bergeret, I.V. Tokatly, Theory of diffusive φ_0 Josephson junctions in the presence of spin-orbit coupling. Europhys. Lett. **110**, 57005 (2015)
8. M.V. Berry, Quantal phase factors accompanying adiabatic changes. Proc. R. Soc. A. **392**(1802), 45–57 (1984)
9. M.V. Berry, The adiabatic phase and Pancharatnam's phase for polarized light. J. Mod. Opt. **34**(11), 1401–1407 (1987)
10. M.V. Berry, Quantum phase corrections from adiabatic iteration. Proc. R. Soc. Lond. A, **414**(1846), 31–46 (1987)
11. M. Berry, Anticipations of the geometric phase. Phys. Today **43**(12), 34–40 (1990)
12. E. Bortolotti, Memories and notes presented by fellows. Rend. R. Acc. Naz. Linc. **4**, 552 (1926)
13. A. Buzdin, Direct coupling between magnetism and superconducting current in the Josephsonφ_0 junction. Phys. Rev. Lett. **101**, 107005 (2008)
14. R. Citro, A topological charge pump. Nat. Phys. **12**(4), 288–289 (2016)
15. M.L. Foucault, Démonstration physique du mouvement de rotation de la terre au moyen du pendule. C. R. Acad. Sci. Hebd Seances Acad. Sci. D **32**, 135 (1851)
16. C. Guarcello, F.S. Bergeret, Cryogenic memory element based on an anomalous Josephson junction. Phys. Rev. Appl. **13**(3), 034012 (2020)
17. C. Guarcello, F.S. Bergeret, Thermal noise effects on the magnetization switching of a ferromagnetic anomalous Josephson junction. Chaos Solitons Fract. **142**, 110384 (2021)
18. C. Guarcello, R. Citro, Progresses on topological phenomena, time-driven phase transitions, and unconventional superconductivity. Europhys. Lett. **132**, 60003 (2020)
19. C. Guarcello et al., rf-SQUID measurements of anomalous Josephson effect. Phys. Rev. Res. **2**(2), 023165 (2020)
20. W.R. Hamilton, Théories of systems of rays. Trans. R. Irish Acad. **17**, 1–144 (1837)
21. J.B. Hart et al., A simple geometric model for visualizing the motion of a Foucault pendulum. Am. J. Phys. **55**(1), 67–70 (1987)
22. B.D. Josephson, Possible new effects in superconductive tunnelling. Phys. Lett. **1**(7), 251–253 (1962)
23. F. Konschelle, A. Buzdin, Magnetic moment manipulation by a Josephson current. Phys. Rev. Lett. **102**, 017001 (2009)
24. M. Kretzschmar, Aharonov-Bohm scattering of a wave packet of finite extension. Z. Phys. **185**(1), 84–96 (1965)
25. H.C. Longuet Higgins et al., Studies of the Jahn-Teller effect. II. The dynamical problem. Proc. R. Soc. A **244**(1236), 1–16 (1958)
26. H.-I. Lu et al., Geometrical pumping with a Bose-Einstein condensate. Phys. Rev. Lett. **116**, 200402 (2016)
27. P. Marra et al., Fractional quantization of the topological charge pumping in a one-dimensional superlattice. Phys. Rev. B **91**, 125411 (2015)
28. W. Mayer et al., Gate controlled anomalous phase shift in Al/InAs Josephson junctions. Nat. Commun. **11**, 212 (2020)

29. J. McCleary, *Geometry from a Differentiable Viewpoint* (Cambridge University Press, Cambridge, 1994)
30. N. Mukunda, Quantum kinematic approach to the geometric phase I. General formalism. Ann. Phys. **228**, 205–268 (1993)
31. C. Nayak et al., Non-Abelian anyons and topological quantum computation. Rev. Mod. Phys. **80**(3), 1083–1159 (2008)
32. A. Noguchi et al., Aharonov–Bohm effect in the tunnelling of a quantum rotor in a linear Paul trap. Nat. Commun. **5**(1), 1–6 (2014)
33. S. Pancharatnam, Generalized theory of interference, and its applications. Proc. Indian. Acad. Sci. **44**, 247–262 (1956)
34. H. Peng et al., Aharonov–Bohm interference in topological insulator nanoribbons. Nat. Mater. **9**(3), 225–229 (2010)
35. M. Rasetti, Dal bit al qu-bit: per sfidare la complessità. Le Scienze **385**, 82–88 (2000)
36. S.M. Roy, Condition for nonexistence of Aharonov-Bohm effect. Phys. Rev. Lett. **44**(3), 111 (1980)
37. B.C. Sanders et al., Geometric phase of three-level systems in interferometry. Phys. Rev. Lett. **86**, 369–372 (2001)
38. C. Schweizer et al., Bloch, spin pumping and measurement of spin currents in optical superlattices. Phys. Rev. Lett. **117**, 170405 (2016)
39. Yu.M. Shukrinov et al., Magnetization reversal by superconducting current in $\varphi 0$ Josephson junctions. Appl. Phys. Lett. **110**, 182407 (2017)
40. E. Strambini et al., A Josephson phase battery. Nat. Nanotechnol. **15**(8), 656–660 (2020)
41. D.B. Szombati et al., Josephson φ_0-junction in nanowire quantum dots. Nat. Phys. **12**, 568 (2016)
42. D.J. Thouless, Quantization of particle transport. Phys. Rev. B **27**(10), 6083–6087 (1983)
43. A. Tonomura et al., Evidence for Aharonov-Bohm effect with magnetic field completely shielded from electron wave. Phys. Rev. Lett. **56**(8), 792 (1986)
44. J. Von Bergmann, H. Von Bergmann, Foucault pendulum through basic geometry. Am. J. Phys. **75**(10), 888–892 (2007)
45. R.A. Webb et al., Observation of h/e Aharonov-Bohm oscillations in normal-metal rings. Phys. Rev. Lett. **54**(25), 2696 (1985)
46. V.F. Weisskopf, *Lectures in Theoretical Physics*, vol. III (Interscience, New York, 1961), pp. 67–70
47. H. Weyl, Naturwissenschaften **12**, 197 (1924)
48. F. Wilczek, A. Shapere, *Geometric Phases in Physics* (World Scientific, Singapore, 1989)
49. D. Xiao et al., Berry phase effects on electronic properties. Rev. Mod. Phys. **82**, 1959 (2010)
50. A. Yacoby et al., Coherence and phase sensitive measurements in a quantum dot. Phys. Rev. Lett. **74**(20), 4047 (1995)
51. J. Zak, Berry's phase for energy bands in solids. Phys. Rev. Lett. **62**(23), 2747 (1989)
52. Y. Zhang, A. Vishwanath, Anomalous Aharonov-Bohm conductance oscillations from topological insulator surface states. Phys. Rev. Lett. **105**(20), 206601 (2010)

The Coming Decades of Quantum Simulation

Joana Fraxanet, Tymoteusz Salamon, and Maciej Lewenstein

Contents

J. Fraxanet · T. Salamon
ICFO - Institut de Ciencies Fotoniques, The Barcelona Institute of Science and Technology,
Barcelona, Spain
e-mail: joana.fraxanet@icfo.eu; tymoteusz.salamon@icfo.eu

M. Lewenstein (✉)
ICFO - Institut de Ciencies Fotoniques, The Barcelona Institute of Science and Technology,
Barcelona, Spain
ICREA, Barcelona, Spain
e-mail: maciej.lewenstein@icfo.eu

Abstract

Contemporary quantum technologies face major difficulties in fault tolerant quantum computing with error correction, and focus instead on various shades of quantum simulation (Noisy Intermediate Scale Quantum, NISQ) devices, analogue and digital quantum simulators and quantum annealers. There is a clear need and quest for such systems that, without necessarily simulating quantum dynamics of some physical systems, can generate massive, controllable, robust, entangled, and superposition states. This will, in particular, allow the control of decoherence, enabling the use of these states for quantum communications (e.g. to achieve efficient transfer of information in a safer and quicker way), quantum metrology, sensing and diagnostics (e.g. to precisely measure phase shifts of light fields, or to diagnose quantum materials). In this Chapter we present a vision of the golden future of quantum simulators in the decades to come.

1   Introduction and Outline

This Chapter has a form of an essay on quantum simulators (QS) and represents personal opinions of the authors. It begins with Sect. 2, which provides a brief review of the current status of quantum computing both in the scope of future fault tolerant quantum computers as well as Noisy Intermediate Scale Quantum (NISQ) devices, focusing on their general limitations. The content we present is based on the lectures and papers by many authors and which has been recently reviewed during ICFO Theory Lecture series on Quantum Computing by Alba Cervera-Lierta [19, 33, 34].

Next, in Sect. 3, we introduce the concept of quantum simulation, and discuss with some details and examples their: (1) Ideology; (2) Platforms and architectures; and finally (3) New challenges for the coming decades.

Sections 4 and 5 are devoted to a more detailed discussion of various achievements of quantum simulation. There, we discuss modern problems of physics that have been addressed in various systems. We divided them in two categories: (1) Fundamental problems of physics and (2) Novel systems that exhibit novel physics. The former aims at utilizing quantum simulators as tools for better understanding of vital problems in condensed matter physics, high energy physics and quantum field theory. The latter addresses the progress of quantum simulation in the left-field and exotic areas such as ultra-fast processes, twisted multi-layer materials, strongly correlated phases of matter and Rydberg atoms.

In Sect. 6, we focus on novel methods in diagnostics and design of quantum many body systems, specific for QS and crucial for studying the phenomena mentioned above. In particular, we discuss: (1) Detection with single site/single "particle" resolution; (2) Entanglement/topology characterization (with random unitaries); (3) Entanglement characterization with experiment-friendly approaches; (4) Topology characterization with experiment-friendly approaches; and last, but not least (5) Synthetic dimensions. Finally, we list some of the new and old (but renewed) methods of theoretical physics that are being intensively developed in the age of QS. We conclude in Sect. 7.

2 Quantum Computing

2.1 Classical Computers and Classical Information Processing

In classical computers, information processing is device independent—the mechanical machines and supercomputers using electronics use the same principles; only the speed of calculation and memories are different. The basic information processing unit in classical information theory [40, 80] is a BIT, which can take two values:

$$|0\rangle \text{ or } |1\rangle.$$

Classical computers have limitations: there exist difficult, if not impossible, tasks for them. Computer scientists classify problems according to complexity classes [9]. The simplest ones are L and NL, defined as the class of problems solvable with logarithmic (time or space) resources on a deterministic (non-deterministic) Turing machine, respectively. Similarly P and NP require polynomial time resources on deterministic (non-deterministic) Turing machines. On the other extreme, there are problems that are proven to require exponential resources in time and/or space, $EXPTIME$ and $EXPSPACE$. While it is known that

$$L \subseteq NL \subseteq P \subseteq NP \subseteq PSPACE \subseteq EXPTIME \subseteq EXPSPACE,$$

it is not known if the inclusions are strict: one of the most challenging problems of computer science is whether $P = NP$.

Another thing worth mentioning is that "device independence" of classical computation is also an illusion. As first discussed by Rolf Landauer, computation is a physical process that costs energy, work, and heat. In fact, a lot of these resources: contemporary supercomputers pay astronomic costs for cooling the installations. Landauer's ideas are summarized in his famous Landauer's Principle, according to which erasing a bit in a system of temperature T costs $k_B T/2$, where k_B is the Boltzmann constant [71] (Fig. 1).

Fig. 1 Rolf Landauer
(1927–1999). This photo is
taken from https://ethw.org/
File:landauer.jpg, reprinted
with permissions of IEEE
History Center

Fig. 2 Charles H. Bennet
(born 1943). This photo was
taken on February 28, 2018.
© SmugMug+Flickr/IBM
Research/51002548905 under
CC BY-SA 2.0

2.2   Quantum Information Processing

According to the father of Quantum Information (QI) science, Charles H. Bennett,
information processing in QI is apparently strongly device dependent (quantum
mechanics decides about the laws of information processing) (Fig. 2).

The basic information processing unit in QI is a QUBIT, which may attain
superposition states

$$\alpha|0\rangle + \beta|1\rangle.$$

Quantum computers are intrinsically parallel and may be much faster than
classical ones; but they also have have limitations: quantum super-positions live
short, due to interactions with the external world.

2.3   Universal Quantum Computers

A universal quantum computer realizes arbitrary unitary operators U ("isometry")
acting on the vector space of N qubits (qutrits, qudits, ...). The device acts in a space

Fig. 3 Example of a
quantum circuit

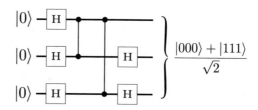

which is exponentially large: the so called Hilbert space has (for qubits) dimension 2^N. A quantum computer can be regarded as a quantum circuit. According to Wikipedia: "In quantum information theory, a quantum circuit is a model for quantum computation, similar to classical circuits, in which a computation is a sequence of quantum gates, measurements, initializations of qubits to known values, and possibly other actions. The minimum set of actions that a circuit needs to be able to perform on the qubits to enable quantum computation is known as DiVincenzo's criteria [44]."

A universal quantum computer requires a universal set of quantum gates. A possible universal set contains all single qubit ones and the so called CNOT gate, which entangles two qubits (Fig. 3). Calculations correspond to applying quantum gates—single qubit gates or two qubit gates—but also to applying quantum measurements.

Unfortunately, calculations in quantum computers are very fragile and require fault tolerant error correction. This is similar to classical machines, but it is much more difficult to realize with quantum devices. For instance, to have one logical qubit we need an overhead of 10^4 additional qubits to ensure error correction (10^3 errors for the typical, so called, surface codes). Thus, a quantum computer with 100 logical qubits requires 10^6 qubits to function well.

Moreover, error scaling should be such that the error per gate is independent of the size of the device and of the number of times the gates are used, but this is not the case in the presently available devices [19]. Obviously, these requirements are much beyond current technological feasibility, and constitute a true scientific and technological challenge.

2.4 Noisy Intermediate Scale Quantum Devices

All of this stimulated John Preskill [98] to propose seeking for quantum advantage with Noisy Intermediate Scale Quantum (NISQ) devices, which: (1) operate with 50–100 "dirty" qubits without fault tolerant error correction; (2) still, are capable of show some advantage over classical computers. So far, NISQ devices have been used to:

• Demonstrate **Quantum Supremacy** in systems that have been rigorously proven to potentially provide advantage by quantum computer scientists (cf. [3]). So far, two experiments claim such advantage: an experiment by Google with superconducting qubits [10] and an experiment by the group of Jianwei Pan with photons [123]. Both essentially perform sampling (the outcome measurements) from random unitary circuits. In the case of Josephson junction superconducting qubits, the random unitary circuits are nonlinear, while in the case of boson sampling in a photonic circuit they are linear. Three things should be stated clearly: (1) the experiments are beautiful and mark amazing progress in experimental capabilities; (2) the demonstrated quantum advantage concerns purely academic problems without having technological applications so far; (3) it is not yet clear whether the advantage is true, i.e., we do not know how far we can optimize classical algorithms that describe sampling problems, although we know that they are NP-hard.

- Realize alternative forms of quantum computing, such as **Adiabatic Quantum Algorithms** [51] and/or **Quantum Annealing** [63]. In both cases: (1) the NISQ devices are used as special purpose quantum computers, *ergo* quantum simulators; (2) so far, the quantum advantage in the rigorous sense is, delicately speaking, disputable (cf. [65, 96]).

- Realize **Quantum Variational Eigensolvers**. Quoting the recent review [114]: "The variational quantum eigensolver (or VQE) uses the variational principle to compute the ground state energy of a Hamiltonian, a problem that is central to quantum chemistry and condensed matter physics. Conventional computing methods are constrained in their accuracy due to the computational limits. The VQE may be used to model complex wavefunctions in polynomial time, making it one of the most promising near-term applications for quantum computing. Finding a path to navigate the relevant literature has rapidly become an overwhelming task, with many methods promising to improve different parts of the algorithm. Despite strong theoretical underpinnings suggesting excellent scaling of individual VQE components, studies have pointed out that their various pre-factors could be too large to reach a quantum computing advantage over conventional methods."

- Realize **Quantum Approximate Optimization Algorithms (QAOA)**, which, according to Wikipedia, are quantum algorithms that are used to solve optimization problems. "Mathematical optimization deals with finding the best solution to a problem (according to some criteria) from a set of possible solutions. Mostly, the optimization problem is formulated as a minimization problem, where one tries to minimize an error which depends on the solution: the optimal solution has the minimal error. The power of quantum computing may allow problems which are not practically feasible on classical computers to be solved, or suggest a considerable speed up with respect to the best known classical algorithm." Typically, QAOA are formulated as hybrid algorithms, where the quantum processor calculates the cost functions (cf. energy, energy gradients, etc.) which depend on variational parameters, and the classical computer performs a classical optimization of these parameters. Again, according to the recent Nature Physics [112]: "The bounds we obtain indicates that substantial quantum advantages are unlikely for classical optimization unless noise rates are decreased by orders of magnitude or the topology of the problem matches that of the device. This is the case even if the number of available qubits increases substantially."

- Realizing **Quantum Machine Learning**. Also in this application the prospects and even reason for searching quantum advantage is questionable, as discussed in the recent paper of Maria Schuld and Nathan Killoran [106]: "In this perspective we explain why it is so difficult to say something about the practical power of quantum computers for machine learning with the tools we are currently using. We argue that these challenges call for a critical debate on whether quantum advantage and the narrative of "beating" classical machine learning should continue to dominate the literature the way it does, and provide a few examples for alternative research questions."

As Sankar Das Sarma writes in [102]: "Quantum computing has a hype problem. Is quantum computing thus just a hype? No, it is a fair and serious effort, but realistically practical quantum supremacy of importance for science and technology remains in the domain of special purpose quantum computers, i.e. **quantum simulators**". In particular, there already exists IBM quantum computing infrastructure that people can access and run their quantum algorithms there. In fact, there have been quite a few of applications in chemistry and materials run in that platform with interesting results, not yet at the point of overruling what can be done with classical computers. It is worth noticing that recently many of the cases of the claimed quantum supremacy [10] have been beaten by contemporary classical simulations using tensor networks [92, 93]. Also, the prospects for exponential quantum supremacy in quantum chemistry are questionable [74]. The very near future will bring surely answers to some these questions.

3   Quantum Simulators

3.1   Idea and Main Concepts

A simulator is a device that imitates desired properties of a given system such as an airplane, train or a physical system of quantum particles. Airplane simulator cannot fly, nor take passengers on board but provides realistic reproduction of all navigational, mechanical and electronic apparatus necessary for the efficient and safe training of the crew. Similarly, simulators of the quantum systems do not reproduce all properties of the original, simulated systems, but imitate only their specific features that are required to understand the phenomena of interest occurring in the original system. The general idea and the concept of such quantum simulators (QSs) can be shortly sketched as follows:

- There exist many interesting **quantum phenomena** with highly important applications(such as, for instance, superconductivity).
- These phenomena are often complex to describe and understand with the help of standard or even super computers.
- Maybe we can design another, simpler and more controllable quantum system to simulate, understand and control these phenomena, as proposed originally by

Manin [81] and Feynman [52]. Such a system would thus work as a quantum computer of one specific purpose, i.e. a **quantum simulator**.

Designing such a simulator is however not an easy task at all. The research line on QS goes back to the beginning of this century and there are numerous valuable reviews covering the various platforms and types of QS's (cf. [116]). In fact, the beginning of the practical concept to QS goes back to the proposal for simulating strongly correlated systems in optical lattices [60] and the first experiments [55]. Nowadays, QS are commonly used for the following tasks and goals:

- **Fundamental problems of physics.** This is the most developed application, in which many achieved results are believed to reach quantum advantage; this is particularly true for the studies of quantum dynamics, or quantum disordered systems, such as the ones that exhibit many body localization (MBL).
- **Quantum chemistry.** Applications of quantum NISQ devices and QS to quantum chemistry has only started [19, 29, 59, 84] and, although promising, it is still far from achieving the precision and accuracy of contemporary theoretical quantum chemistry. There is a growing evidence that the expect of exponential quantum advantage in quantum simulations for quantum chemistry is not such (see recent works by Chan [113], and his talks at the APS meeting in 2022).
- **Classical/quantum optimization problems for technology.** Applications of quantum NISQ devices and QS to optimization problems are also in an initial phase [85] and cannot yet compete with the classical supercomputer methods (cf. [112]).

3.2   Platforms and Architectures

Now is the time to mention how the quantum simulators can be actually built. There are several very well developed platforms where QSs are being realized and "practical" quantum advantage has been achieved. We should stress that QSs can be analog or digital.

In the latter case, any platform that offers tools for universal quantum computing can also be used for quantum simulation. Here comes a definitely incomplete list of already developed QS architectures:

- **Superconducting qubits**: These are the same systems as the ones used by Google [10] or D-Wave [63]. Even though in principle they allow for "noisy" but universal quantum computing, they can be and very often are used as digital QSs (cf. [94]). They can be coupled to microwave cavities, resulting into circuit QED systems [20].
- **Ultracold atoms**: They mostly offer the possibility for analog quantum simulation. This can be realized in the continuum or in optical lattices [77]. They are very flexible and they allow to simulate complex Hubbard models, as well as spin systems.

- **Trapped ions**: Similarly to superconducting qubits, trapped ions allow for universal quantum computing, but they caalso be used as perfect analog or digital QSs [86, 122]. Similarly to superconducting qubits, they can be used to simulate spin 1/2 systems, rather than Hubbard models. Very recently a qudit (or spin 1 and 2) quantum computer/simulator was realized with ions [100].
- **Rydberg atoms**: These are atoms where the electron has been excited to a high principal quantum number, and which are trapped in optical tweezers. They mimic spin systems with long range interactions [18, 21, 105].
- **Photonic systems**: These are typically linear optics systems which, combined with photon counting, may mimic a universal quantum computer, according to the famous paper from Knill et al. [64]. Achieving strong non-linearity with photons is very challenging, but there are ongoing attempts and proposals [11].
- **Light and Cavity materials**: Quantum Simulators based on Cavity Quantum Electrodynamics take advantage of the coupling between quantum system and the coherent light field of the cavity, in which such system has been placed. This branch of quantum simulation is commonly named as "cavity quantum materials", since one in principle could place a many-body quantum system into a cavity and control its properties via light-matter interactions. Currently, however, the experimental studies are mainly conducted in the scope of Jaynes—Cummings and Dicke models [104]. Another research path was taken by engineering materials entirely from light with resulting photon-photon interactions[31, 39, 79, 103]. Such systems, however, require a mediator (for example Rydberg-dressed atoms) facilitating the light-light interactions.
- **Twistronics systems**: Twistronics deals with twisted bilayer graphene or other two-dimensional materials [28, 111]. For small "magic" angle, such systems lead to periodic moiré patterns at a length scale much larger than the typical scale of condensed matter systems: in this sense, they can themselves be considered as condensed matter quantum simulators of condensed matter [62]. Such approach has been explained in detail in Chapter 1 of this book. Twisted bilayer materials can, however, also be mimicked by ultracold atoms in a two-dimensional lattice with synthetic dimensions [101].
- **Polaritons**: Especially useful for non-equilibrium systems and quantum hydrodynamics simulation, as well as relativistic effects thanks to dual (half light half particle) nature of the polaritonic quasi-particles [16, 25, 58].

3.3 The Coming Decades of Quantum Simulation

In the coming decades, as clearly reflected in the current and future quantum programs such as National Quantum Initiative Act, Quantum Flagship, MIC 2025 and many others, the platforms that we use are not expected to change, but the challenges and focus are going to be very different. QSs of the future will have to be devices that are:

- **Robust**
- **Scalable**
- **Programmable**
- **Externally accessible**
- **Standardized**
- **Verifiable and certifiable**

Moreover, the priorities concerning the main goals will be revised, increasing the effort in optimization problems with application to technology, which we predict to be the main axis of interest. Quantum chemistry will most likely remain on the second place of the podium. Fundamental problems of physics will not loose importance, however they will most probably no longer be the unique problems where quantum advantage can be achieved. Below we highlight the main (in our opinion) efforts in each of above mentioned areas that will be undertaken in the upcoming decades.

- **Classical/quantum optimization problems for technology.** Quantum simulation is already starting to address practically relevant and computationally hard problems, such as the Maximum Independent Set problem, which was very recently addressed with Rydberg atom arrays [46]. We expect this type of applications to very strongly developed in the coming years.
- **Quantum chemistry.** This will for sure include novel methods of simulating quantum chemistry going beyond NISQ devices to analog simulators (cf. [8]).
- **Fundamental problems of physics.** Besides the simulation of Hamiltonian systems for condensed matter and high-energy physics, the generation, manipulation and applications of massively entangled states, which useful for quantum communication, quantum metrology, sensing and detection, will gain momentum. In the coming years, we also expect QSs to focus on particularly complex and classically hard to simulate problems, including the Fermi-Hubbard model and the puzzle of high-temperature superconductivity, the simulation of lattice gauge theories and related problems in dynamical lattices (for reviews see [4, 14]) or frustrated and disordered systems, among others.

One should stress that currently, there is a huge effort led I. Bloch in Munich to make this a user service infrastructure in the short term ("sort of providing quantum computing access using atoms in optical lattices as quantum platform). In an even more general sense, the Quantum Flagship project PASQuanS—Programmable Atomic Large-Scale Quantum Simulation combines all the efforts for atomic systems: Rydberg atoms, trapped ions, ultracold atoms on optical lattices and traps, and more (https://pasquans.eu/).

4 Quantum Simulation of Fundamental Problems of Physics

In this Section, we present a selected list of some of the great achievements of QS over the last few years. Typically, we quote and describe some experimental papers from the leading groups over the years, and also present also some examples from our own groups. This Section is, by no means, a review. It is instead based on subjective choices and selections (for a recent review see [6]).

4.1 Relevant Paradigmatic Systems

4.1.1 Fermi-Hubbard Model

A "paradigmatic" and notoriously difficult to classically simulate systems is, of course, the Fermi-Hubbard model, believed by many to lie at heart of high-temperature superconductivity [73]. In recent years, major achievements in the quantum simulation of the Fermi-Hubbard model have been reported by the groups of Markus Greiner at Harvard and Immanuel Bloch in Munich, which are using quantum gas microscopes to explore antiferromagnetically ordered systems and the physics that emerges when perturbing them through doping. A good example of what QS can do in this context is the paper from the Harvard group [38], where the authors write: "Understanding strongly correlated quantum many-body states is one of the most difficult challenges in modern physics. For example, there remain fundamental open questions on the phase diagram of the Hubbard model, which describes strongly correlated electrons in solids. In this work we realize the Hubbard Hamiltonian and search for specific patterns within the individual images of many realizations of strongly correlated ultracold fermions in an optical lattice. Upon doping a cold-atom antiferromagnet we find consistency with geometric strings, entities that may explain the relationship between hole motion and spin order, in both pattern-based and conventional observables. Our results demonstrate the potential for pattern recognition to provide key insights into cold-atom quantum many-body systems." Interestingly, machine learning approaches can be exploited to analyze some of these experiments, cf. [23].

4.1.2 Systems with Fine-Tuned Interactions: Supersolids and Quantum Liquid Droplets

Another recent research direction in quantum simulation is the study of systems where interactions of different origins compete. When they have opposite signs, such interactions can be fine-tuned to a situation where they practically compensate each other. The resulting systems can then become completely dominated by quantum fluctuations and correlations despite remaining weakly interacting, leading to the emergence of novel phases of matter such as supersolids or quantum liquid droplets [24].

Supersolids are systems that combine the frictionless flow of a superfluid and the crystalline structure of a solid. Despite being predicted in the '50s, they remained elusive until quantum gases made their QS possible, using either atoms dressed

Fig. 4 Artistic view of a
quantum liquid droplet
formed by mixing two
Bose-Einstein condensates
with competing repulsive
intrastate and attractive
interstate interactions [27].
Reprinted with permission of
ICFO/PovarchikStudiosBarcelona

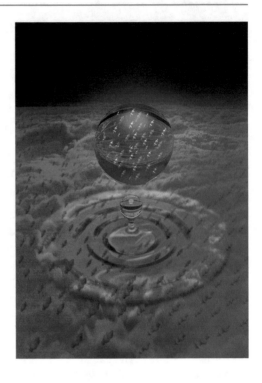

with light or dipolar quantum gases with competing interactions. Quantum liquid droplets, on the other hand, are ensembles of quantum particles that are hold together by attractive interparticle interactions, but stabilized against collapse by the repulsive effect of quantum fluctuations, and which have been observed in dipolar Bose gases and in mixtures of Bose-Einstein condensates (see Fig. 4).

4.2   Fundamental Systems of Condensed Matter, High Energy Physics and Quantum Field Theory

In the recent years, there has been a strong focus on QS of high energy physics (HEP) models, including lattice gauge theories (LGTs) and related models [4, 14]. The challenge here is to control many body interactions: "magnetic" interactions on the plaquettes of the lattice, and "electric" interactions, that are linked to matter through the local conservation laws of the gauge theory— the Gauss law—which ensures its local gauge invariance.

4.2.1   Schwinger Model

Due their versatility, trapped ions are particularly suitable to simulate LGTs in a digital way. These systems may serve as universal quantum computers and realize, in principle, any few body interactions and constraints. Here we quote two achievements of Rainer Blatt's group, where one-dimensional LGTs are implemented. In

Ref. [82], they "report the first experimental demonstration of a digital quantum simulation of a lattice gauge theory, by realising $1+1$-dimensional quantum electrodynamics (Schwinger model) on a few-qubit trapped-ion quantum computer. They are interested in the real-time evolution of the Schwinger mechanism, describing the instability of the bare vacuum due to quantum fluctuations, which manifests itself in the spontaneous creation of electron-positron pairs. To make efficient use of the quantum resources, they map the original problem to a spin model by eliminating the gauge fields in favour of exotic long-range interactions, which have a direct and efficient implementation on an ion trap architecture. They explore the Schwinger mechanism of particle-antiparticle generation by monitoring the mass production and the vacuum persistence amplitude. Moreover, they track the real-time evolution of entanglement in the system, which illustrates how particle creation and entanglement generation are directly related. Their work represents the first step towards quantum simulating high-energy theories with atomic physics experiments, the long-term vision being the extension to real-time quantum simulations of non-Abelian lattice gauge theories."

In the following paper [66], the authors consider hybrid classical-quantum algorithms that aim at solving optimisation problems variationally, using a feedback loop between a classical computer and a quantum co-processor, while benefiting from quantum resources. They "present experiments demonstrating self-verifying, hybrid, variational quantum simulation of lattice models in condensed matter and high-energy physics. Contrary to analog quantum simulation, this approach forgoes the requirement of realising the targeted Hamiltonian directly in the laboratory, thus allowing the study of a wide variety of previously intractable target models. Here, they focus on the lattice Schwinger model, a gauge theory of 1D quantum electrodynamics. Their quantum co-processor is a programmable, trapped-ion analog quantum simulator with up to 20 qubits, capable of generating families of entangled trial states respecting symmetries of the target Hamiltonian. They determine ground states, energy gaps and, by measuring variances of the Schwinger Hamiltonian, they provide algorithmic error bars for energies, thus addressing the long-standing challenge of verifying quantum simulation."

4.2.2 Bosonic Schwinger Model

Relativistic quantum gauge theories are fundamental theories of matter describing nature. Paradigmatic examples are quantum electrodynamics (describing electromagnetic interactions of charged particles and photons), chromodynamics (describing strong interactions of quarks and gluons), and the Standard Model, unifying the latter two with the weak interactions.

Despite enormous progress in our understanding of quantum gauge theories, the questions concerning the behaviors of systems described by such theories in the presence of strong correlations remain widely open: from the very nature of the quark confinement to the behavior of quark-gluon-plasma at high densities and temperatures. Moreover, quantum out-of-equilibrium dynamics of quantum gauge theories is out of reach of the present classical computers. For these reasons, there is a lot of effort to design and investigate quantum simulators of such systems.

The paradigmatic model of quantum gauge theories in one spatial dimension and time is the Schwinger model, in which "charged" electrons (fermions) interact with photons (bosons) in one dimension. Since quantum simulations with fermions are notoriously difficult, simulating a bosonic version of the Schwinger model is an interesting and more accessible goal [36].

Using state-of-the-art methods of theoretical physics, Titas Chanda and co-workers investigated in their work how the bosonic matter behaves when it is driven out-of-equilibrium by creating a pair of particle and antiparticle on top of the vacuum of the system. They obtain three important results for the understanding of quantum gauge theories in general: (1) the bosons undergo an evolution which is dominated by strong confinement of charges, responsible for only a partial screening of electric field, even in the massless limit; (2) the extended "meson" formed by the two charges and the electric-flux tube connecting them is very robust, leaving a strong footprint in the entanglement of the system; (3) the system fails to thermalize and generates exotic states at long times, characterized by two distinct space-time regions—one external region made by thermal mesons, and a central region between the two initial charges, where quantum correlations obey the, so called, area-law of entanglement (see Fig. 5). This work opens a path towards quantum simulations of quantum gauge theories in novel, unexplored regimes.

4.2.3 Abelian-Higgs Mechanism

The current focus of quantum simulation of fundamental models of high energy physics is lattice gauge theories and quantum field theories. The Higgs mechanism is an essential ingredient of the Standard Model of particle physics that explains the 'mass generation' of gauge bosons. Its seemingly simple one-dimensional lattice version may serve as an interesting novel quantum simulator.

In a recent study [35], a continuation of [36], we have taken up the challenge to fill this gap. Unlike the system in the continuum, two distinct regions in the lattice version are identified, namely the confined and Higgs regions. These two regions are separated by a line of first order phase transitions that ends in a second order critical point. Above this critical point the regions are smoothly connected by a crossover. The presence of a second order critical point allows one to construct an unorthodox continuum limit of the theory that is described by a conformal field theory (CFT) (see Fig. 6). This work is strongly motivated by the current prospects of quantum simulation of quantum gauge theories, and it opens a path towards observing the Higgs mechanism in experiments with ultracold atomic setups.

5   Novel Quantum Simulators for Novel Physics (NOQIA)

In this Section, we mention several novel directions/platforms for QS: (1) QS for strong laser-matter interacting (2) twistronics, which can be regarded as condensed matter QS of condensed matter, and photonic moiré 'patterns; (3) QS for strongly correlated phases of matter, which include indirect excitons and interaction induced

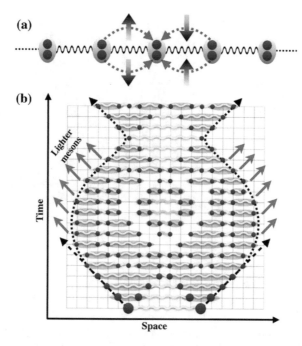

Fig. 5 (**a**) In the lattice version of bosonic Schwinger Model (BSM), sites are occupied by two kinds of bosons, corresponding to particles (red dots) and antiparticles (blue dots), which are coupled to U(1) gauge fields residing on the bonds of the lattice. Tunneling of bosons across a given bond change the internal state of the corresponding gauge field as dictated by the arrows. (**b**) The time-dependent dynamics of the BSM is greatly affected by strong confinement, where the light-cone of the particles bends inwards, producing exotic asymptotic states made of a central core of strongly correlated bosons and external regions populated by free mesons. Figure's author: Titas Chanda

topological order and (4) Rydberg atoms, which can simulate very interesting spin models with long-range interactions.

5.1 Quantum Simulation Based on Atto-Science and Ultra-Fast Processes

Recently, we have proposed a completely new platform for QS based on atto-science and ultra-fast processes. Over the past four decades, astounding advances have been made in the field of laser technologies and the understanding of light-matter interactions in the non-linear regime. Thanks to this, scientists have been able to carry out extremely complex experiments related to, for example, ultra-fast light-pulses in the visible and infrared range, and accomplish crucial milestones such as using a molecule's own electrons to image its structure, to see how it rearranges and vibrates or breaks apart during a chemical reaction.

Fig. 6 Phase diagram of the lattice Abelian-Higgs model as seen through the lens of entanglement entropy. At small couplings, the system occupies two qualitatively different regions, a confined and a Higgs region, separated by a line of first order quantum phase transitions. Figure's author: Titas Chanda

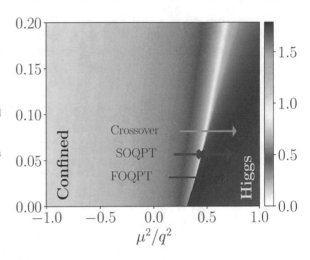

5.1.1 Optical Schrödinger Cat States

The development of high-power lasers allowed scientists to study the physics of ultra-intense laser-matter interactions which, in its standard version, treats ultra-strong ultra-short driving laser pulses only from a classical point of view. The famous theory coined as the "simple man's model" or the "three-step model" [75]—which had its 25th anniversary in 2019 [7]—dealt with the interaction of an electron with its parent nucleus sitting in a strong laser field environment, and elegantly described it according to classical and quantum processes. However, due to the fact that these laser pulses are highly coherent and contain huge numbers of photons, the description of the interaction in the strong field has so far been incomplete, because it treated the atomic system in a quantum way but the electromagnetic field in a classical way.

Now, in the description of the most relevant processes of ultra-intense laser-matter physics (such as high-harmonic generation, above threshold ionization, laser-induced electron diffraction, sequential and non-sequential multi-electron ionization, etc.) the quantum-fluctuation effects of the laser electric field, not even to mention the magnetic fields, are negligible. However, the quantum nature of the entire electromagnetic fields is always present in these processes, so a natural question arises: does this quantum nature exhibit itself? In which situations does it appear?

In the recent study [76], we have reported on the theoretical and experimental demonstration that intense laser-atom interactions may lead to the massive generation of highly non-classical states of light, one of the Holy Grails of the contemporary QS (see Fig. 7).

Such results have been obtained using the process of high-harmonic generation in atoms, in which large numbers of photons from a driving laser pulse of infrared frequency are up-converted into photons of higher frequencies in the extreme ultraviolet spectral range. The quantum electrodynamics theory formulated in this

Fig. 7 Optical cat states in intense laser-matter interaction. Reprinted with permission of ICFO/Scixel—E. Sahagun

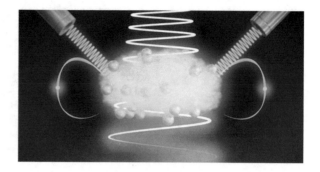

study, predicts that, if the initial state of the driving laser is coherent, it remains coherent but shifted in amplitude after interacting with the atomic medium.

Similarly, the quantum states of the harmonic modes become coherent with small coherent amplitudes. However, the quantum state of the laser pulse that drives the high-harmonic generation can be conditioned to account for this interaction, which transforms it into a, so-called, optical Schrödinger cat state. This state corresponds to a quantum superposition of two distinct coherent states of light: the initial state of the laser, and the coherent state reduced in amplitude that results from the interaction with the atoms.

We accessed the full quantum state of this laser pulse experimentally using quantum state tomography. To achieve this, the coherent amplitude of the light first needs to be reduced in a coherent way to only a few photons, on average, and then all of the quantum properties of the state can be measured.

The results of this study open the path for investigations towards the control of the non-classical states of ultra-intense light, and exploiting conditioning approaches on physical processes relevant to high-harmonic generation. This hopefully will link ultra-intense laser-matter physics and atto-science to quantum information science and quantum technologies in a novel and completely unexpected manner.

5.2   Twistronics and Moiré Patterns

5.2.1   Plethora of States in Magic-Angle Graphene

In the recent years, graphene made another major splash in the headlines when scientists discovered that by simply rotating two layers of this material one on top of the other, this material could behave like a superconductor where electrical currents can flow without resistance. This new phase of matter turned out to appear only when the two graphene layers were twisted between each other at an angle of $\approx 1.1°$ (no more and no less)—the so-called magic angle—and it is always accompanied by enigmatic correlated insulator phases, similar to what is observed in mysterious cuprate high-temperature superconductors.

The group of D. Efetov at ICFO has succeeded in vastly improving the device quality of this setup, and in doing so, has stumbled upon something totally

Fig. 8 Artistic illustration of the twisted bi-layer graphene and the different states of matter that have been discovered. Reprinted with permission of ICFO/ F. Vialla

unexpected. The researchers were able to observe a zoo of previously unobserved superconducting and correlated states, in addition to an entirely new set of magnetic and topological states, opening a completely new realm of rich physics (see Fig. 8).

Room temperature superconductivity is the key to many technological goals such as efficient power transmission, frictionless trains, or even quantum computers, among others. When discovered more than 100 years ago, superconductivity was only plausible in materials cooled down to temperatures close to absolute zero. Then, in the late '80s, scientists discovered high temperature superconductors by using ceramic materials called cuprates. In spite of the difficulty of building superconductors and the need to apply extreme conditions (very strong magnetic fields) to study the material, the field took off as something of a holy grail among scientists based on this advance. Since last year, the excitement around this field has increased. The double mono-layers of carbon have captivated researchers because, in contrast to cuprates, their structural simplicity has become an excellent platform to explore the complex physics of superconductivity.

In the recent experiment led by D. Efetov [78], using a "tear and stack" van der Waals assembly technique, the scientists at ICFO were able to engineer two stacked monolayers of graphene, rotated by only 1.1°— the magic angle. They then used a mechanical cleaning process to squeeze out impurities and to release local strain between the layers. In doing this, they were able to obtain extremely clean twisted graphene bilayers with reduced disorder, resolving a multitude of fragile interaction effects.

By changing the electrical charge carrier density within the device with a nearby capacitor, they discovered that the material could be tuned from behaving as an insulator to behaving as a superconductor or even an exotic orbital magnet with non-trivial topological texture—a phase never observed before. What is even more astounding is the fact that the device entered a superconducting state at the lowest carrier densities ever reported for any superconductor, a completely new breakthrough in the field. To their surprise, they observed that the system seemed to be competing between many novel states. By tuning the carrier density within the lowest two flat moiré bands, the system showed correlated states and superconductivity alternately, together with exotic magnetism and band topology. They also noted that these states were very sensitive to the quality of the device, i.e.

accuracy and homogeneity of the twist angle between two sheets of graphene layers. Last but not least, in this experiment, the researchers were also able to increase the superconducting transition temperature to above 3 Kelvin, reaching record values which are twice as high as previously reported studies for magic-angle-graphene devices.

What is exceptional about this approach is that graphene, a material that is typically poor on strongly interacting electron phenomena, now has been the enabling tool to provide access to this complex and exceptionally rich physics. So far, there is no unique theory that can explain the superconductivity in magic angle graphene at the microscopic level, however with this new discovery, it is clear that a new chance to unveil its origin has emerged.

5.2.2 Photonic Moiré Patterns

If you take two identical layers of semi-transparent material with the same structure, you put one on top of the other, rotate them and then look at them from above, hexagonal patterns start to emerge. These patterns are known as moiré patterns or moiré lattices (Fig. 9).

moiré patterns appear often in every-day life applications such as art, textile industry, architecture, as well as image processing, metrology and interferometry. They are a matter of major current interest in science, since they can be produced using coupled graphene -hexagonal boron nitride monolayers, graphene -graphene layers and graphene quasicrystals on a silicon carbide surface and have proven to generate different states of matter upon rotating or twisting the layers to a certain angle, opening to a new realm of physics. Scientists at MIT found a new type of unconventional superconductivity in twisted bilayer graphene that forms a moiré lattice, and a team of ICFO researchers recently unveiled a new zoo of unobserved states in the same structure (see above).

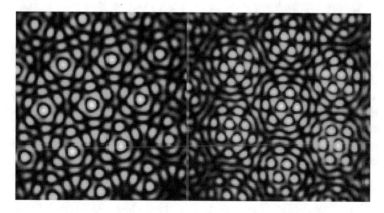

Fig. 9 moiré lattices created by superposition of two rotated hexagonal lattices. Reprinted with permission of ICFO/L. Torner

Now in a study published in Nature, Lluis Torner and collaborators have reported on the propagation of light in photonic moiré lattices, which, unlike their material counterparts, have readily controllable parameters and symmetry, allowing researchers to explore transitions between structures with fundamentally different geometries (periodic, general aperiodic and quasicrystalline—for the earlier theory paper see [57]). Note that both the theory and the experiments were done at the classical regime, with classical light, but the idea works for any kinds of waves, like matter waves, Bose-Einstein condensates, etc.

The paper shows the creation of moiré lattices by superimposing two periodic patterns with either square or hexagonal primitive cells and tunable amplitudes and twist angle. Depending on the twist angle, a photonic moiré lattice may have different periodic (commensurable) structure or aperiodic (incommensurable) structure without translational periodicity. The angles at which a commensurable phase (periodicity) of a moiré lattice is achieved are determined by Pythagorean triples or by another Diophantine equation depending on the shape of the primitive cell. Changing the relative amplitudes of the sublattices allowed researchers to smoothly tune the shape of the lattice without affecting its rotational symmetry.

Then, by using commensurable and incommensurable moiré patterns, researchers observed for the first time the two-dimensional localization-delocalization transition of light. The used photonic moiré lattices can be readily constructed in practically any arbitrary configuration consistent with symmetry groups, thus allowing the creation of potentials that may not be easily produced in tunable form using material structures. Therefore, in addition to their direct application to the control of light patterns, the availability of photonic moiré patterns allows the study of phenomena relevant to other areas of physics, particularly to condensed matter, which are harder to explore directly.

5.3   Quantum Simulation for Strongly Correlated Phases of Matter

5.3.1   Checkerboard Insulator with Semiconductor Dipolar Excitons

The Hubbard model constitutes one of the most celebrated theoretical frameworks of condensed-matter physics. It describes strongly correlated phases of interacting quantum particles confined in a lattice potential.In the last two decades the Hubbard Hamiltonian for bosons has been deeply scrutinised in the regime of short-range on-site interactions. On the other hand, extensions to longer-range interactions between neighbouring lattice sites have remained mostly elusive experimentally. Entering this regime constitutes a well identified research frontier where quantum matter phases can spontaneously break the lattice symmetry. In Ref. [70] we unveil one of such phases, precisely the long-sought-after checkerboard solid. It is accessed by confining semiconductors dipolar excitons in a two-dimensional square lattice. The exciton checkerboard is signalled by a strongly minimised compressibility of the lattice sites at half-filling, in quantitative agreement with theoretical expectations. Our observations thus highlight that dipolar excitons enable controlled implementations of extended Bose-Hubbard Hamiltonians.

5.3.2 Quantum Simulation of Interaction Induced Topological Phases

In the last decades, topological insulators have attracted great interest and also have promising applications in topics such as metrology or quantum computation. These exotic materials go beyond the standard classification of phases of matter: they are insulating in their bulk, conducting on their edges, and characterized by a global topological invariant, in contrast to a local order parameter as in the conventional Ginzburg-Landau theory of phases of matter. The further discovery of a great variety of unexpected properties has led to an intensive investigation of topological phases of matter in many areas of quantum physics ranging from solid state and quantum chemistry to high-energy physics.

The research on topological insulators reached its first peak in 1997, when Altland and Zirnbauern classified all the possible topological phases of non-interacting systems by means of symmetry arguments. Even though this classification shed light on many novel features of quantum matter, the picture is still far from being complete. The list of exotic topological phases that escape the current classification is only growing, with a special focus on the discovery of interaction-induced topology in strongly correlated systems.

The study of interaction-induced topological phases is particularly challenging due to the high level of complexity encoded in interacting systems, which nowadays needs to be tackled through quantum algorithms working on classical processors. Nevertheless, the advent of a new generation of quantum simulators made of particles at ultracold temperatures represents reliable a rout towards a complete understanding and control of topological phases in interacting systems.

In a recent article [53], the authors revealed that interacting processes can lead to topological properties persisting at the specific points separating two distinct phases, thus leading to the existence of topological quantum critical points (see Fig. 10). In

Fig. 10 The Haldane insulator is a symmetry protected topological phase which is found in the extended Bose Hubbard model at filling one. The topology extends to the critical point separating the Haldane insulator from a charge density wave. Reprinted with permission of ICFO/ J. Fraxanet

addition, they also propose a realistic scheme based on magnetic dysprosium atoms whose long-range dipolar interaction enables the implementation of topological quantum critical points on a quantum simulator.

More specifically, they reported the presence of two distinct topological quantum critical points with localized edge states and gapless bulk excitations. The results show that the topological critical points separate two phases, one topologically protected and the other topologically trivial, both characterized by a string correlation function which denotes a similar type of long-range ordering. In both cases, the long-range order persists also at the topological critical points and explains the presence of localized edge states protected by a finite charge gap. Finally, they introduce a super-resolution quantum gas microscopy scheme for dipolar dysprosium atoms for the experimental study of topological quantum critical points using cold atoms in optical lattices.

The quantum simulation of these exotic materials typically relies on the generation of artificial gauge fields. However, recent studies have shown that topological phases can also emerge from particle interactions. The latter mechanism leads to the concept of interaction-induced topological phases, in which topology is acquired through a spontaneous symmetry breaking process. The interplay of the spontaneous symmetry breaking with the global topological properties can lead to very interesting effects.

In another recent article [61], they report how such interplay can lead to new strongly-correlated topological effects in a 2D material. They shown how interactions can localize particles in the insulating bulk, leading to self-trapped polarons. Moreover, they have also shown how the interacting nature of the topological insulator gives rise to domains in the bulk. Interestingly, the nontrivial topology associated to each domain leads to the appearance of protected conducting states in the bulk, localized at the domain boundaries (see Fig. 11). Finally, they discuss the possibility of quantum simulating such phases with cold laser-excited Rydberg atoms in an optical lattice.

5.4 Quantum Simulation with Rydberg Atoms

The progress of this platform, led by the groups of Misha Lukin at Harvard and Antoine Browaeys at Institute d'Optique is truly amazing. Below, we include a lit of the most important applications and discoveries from Harvard-MIT:

5.4.1 Rydberg Atoms for Optimization Algorithms

In the recent preprint [46], the authors explain: "Realizing quantum speedup for practically relevant, computationally hard problems is a central challenge in quantum information science. Using Rydberg atom arrays with up to 289 qubits in two spatial dimensions, we experimentally investigate quantum algorithms for solving the Maximum Independent Set problem. We use a hardware-efficient encoding associated with Rydberg blockade, realize closed-loop optimization to test several variational algorithms, and subsequently apply them to systematically

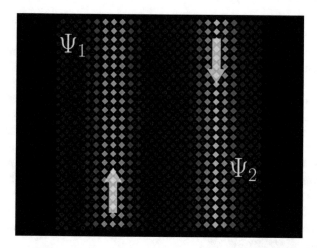

Fig. 11 Linear domain walls: a metastable self-consistent solution, in which the system develops two disconnected domain walls with ground state currents flowing in opposite directions. Reprinted with permission of ICFO/S. Julià

explore a class of graphs with programmable connectivity. We find the problem hardness is controlled by the solution degeneracy and number of local minima, and experimentally benchmark the quantum algorithm's performance against classical simulated annealing. On the hardest graphs, we observe a superlinear quantum speedup in finding exact solutions in the deep circuit regime and analyze its origins."

5.4.2 Quantum Spin Liquids

In an earlier paper [107], the group reported on quantum spin liquids, exotic phases of matter with topological order, that have been a major focus of explorations in physical science for the past several decades. As they explain: "Such phases feature long-range quantum entanglement that can potentially be exploited to realize robust quantum computation. We use a 219-atom programmable quantum simulator to probe quantum spin liquid states. In our approach, arrays of atoms are placed on the links of a kagome lattice and evolution under Rydberg blockade creates frustrated quantum states with no local order. The onset of a quantum spin liquid phase of the paradigmatic toric code type is detected by evaluating topological string operators that provide direct signatures of topological order and quantum correlations. Its properties are further revealed by using an atom array with nontrivial topology, representing a first step towards topological encoding. Our observations enable the controlled experimental exploration of topological quantum matter and protected quantum information processing."

5.4.3 Quantum Scars

Perhaps the most impressive result is the control of quantum dynamics in Rydberg QS. In the already mentioned paper [21], the authors state that "controlling nonequilibrium quantum dynamics in many-body systems is an outstanding challenge as interactions typically lead to thermalization and a chaotic spreading throughout Hilbert space. We experimentally investigate non-equilibrium dynamics following rapid quenches in a many-body system composed of 3–200 strongly interacting qubits in one and two spatial dimensions. Using a programmable quantum simulator based on Rydberg atom arrays, we probe coherent revivals corresponding to quantum many-body scars. Remarkably, we discover that scar revivals can be stabilized by periodic driving, which generates a robust subharmonic response akin to discrete time-crystalline order. We map Hilbert space dynamics, geometry dependence, phase diagrams, and system-size dependence of this emergent phenomenon, demonstrating novel ways to steer entanglement dynamics in many-body systems and enabling potential applications in quantum information science. "

5.4.4 Quantum Phases of Matter on a 256-Atom Programmable Quantum Simulator

The title of this article [47] is self explanatory. Still, it is worth to realize what the authors do: "Motivated by far-reaching applications ranging from quantum simulations of complex processes in physics and chemistry to quantum information processing, a broad effort is currently underway to build large-scale programmable quantum systems. Such systems provide unique insights into strongly correlated quantum matter, while at the same time enabling new methods for computation and metrology. Here, we demonstrate a programmable quantum simulator based on deterministically prepared two-dimensional arrays of neutral atoms, featuring strong interactions controlled via coherent atomic excitation into Rydberg states. Using this approach, we realize a quantum spin model with tunable interactions for system sizes ranging from 64 to 256 qubits. We benchmark the system by creating and characterizing high-fidelity antiferromagnetically ordered states, and demonstrate the universal properties of an Ising quantum phase transition in (2+1) dimensions. We then create and study several new quantum phases that arise from the interplay between interactions and coherent laser excitation, experimentally map the phase diagram, and investigate the role of quantum fluctuations. Offering a new lens into the study of complex quantum matter, these observations pave the way for investigations of exotic quantum phases, non-equilibrium entanglement dynamics, and hardware-efficient realization of quantum algorithms."

6 Design, Techniques and Diagnostics of Quantum Simulators

In this Section we shortly mention the most important diagnostics and design tools that have been developed in the recent years for QS, mostly with the field of atomic physics, providing interesting complementary methods for condensed matter physics.

6.1   Single Site and Single Particle Resolution

A quantum gas microscope is capable to detect single atoms in a single site of the (optical) lattice. In the first experiment of the Harvard group [13] (see also [108]), the device was capable only to distinguish between zero/two or one bosonic atom at the site. Still, it offered an unprecedented possibility of measuring density-density correlations far beyond the possibilities of the standard condensed matter physics. These devices are nowadays capable to measure single particles in a single site of the lattice with spin resolution [56] for both bosons and fermions [37], and even more (for recent reviews see [69]).

6.2   Entanglement and Topology Characterization Using Random Unitaries

Another aspect that is extremely important to characterize the quality of QSs, is clearly the verification of entanglement in the considered systems. This goes back to the theory and applications of entanglement witnesses (cf. [77]), but more recently focuses on measurement of entanglement entropies of the reduced density matrices and entanglement Hamiltonians (logarithm of of the reduced density matrix) of partial blocks of a given system. In this respect, there is a spectacular progress led by Peter Zoller's group collaborating with various experimental teams. In a series of works they propose to apply random unitary operations to a part of the system (a block) and then perform local measurements there [41, 49]. Experiments were realized with trapped ions [26], while the method generalizes to detect topological invariants [50].

This is how the authors describe their ideas in a recent review [48]: "Increasingly sophisticated programmable quantum simulators and quantum computers are opening unprecedented opportunities for exploring and exploiting the properties of highly entangled complex quantum systems. The complexity of large quantum systems is the source of their power, but also makes them difficult to control precisely or characterize accurately using measured classical data. We review recently developed protocols for probing the properties of complex many-qubit systems using measurement schemes that are practical using today's quantum platforms. In all these protocols, a quantum state is repeatedly prepared and measured in a randomly chosen basis; then a classical computer processes the measurement outcomes to estimate the desired property. The randomization of the measurement procedure has distinct advantages; for example, a single data set can be employed multiple times to pursue a variety of applications, and imperfections in the measurements are mapped to a simplified noise model that can more easily be mitigated. We discuss a range of use cases that have already been realized in quantum devices, including Hamiltonian simulation tasks, probes of quantum chaos, measurements of nonlocal order parameters, and comparison of quantum states produced in distantly separated laboratories. By providing a workable method for translating a complex quantum state into a succinct classical representation that preserves a rich variety of relevant physical properties, the randomized measurement toolbox strengthens our ability to grasp and control the quantum world."

6.3   Experiment-Friendly Approaches for Entanglement Characterization

Our approach for the characterization of entanglement and quantum correlations was different from that of Peter Zoller's group. Instead of looking for methods of measuring Renyi entropies and more, we focused on the characterization of quantum correlations via measurement of experimentally friendly low moments of local observables. In the first paper [117], we characterize all permutationally invariant Bell inequalities in many body (party) systems with two observables per party with two outcomes. Our results were immediately tested in experiments with spinor Bose Einstein condensates and more.

Entanglement is one of the most characteristic phenomena of quantum physics. The simplest and most studied form of entanglement is the bipartite case, in which two subsystems form a quantum composite (e.g., two entangled particles). However, systems with more than two particles can exhibit entanglement in a whole plethora of ways, presenting a much richer and challenging case. Contrary to the bipartite case, multipartite entanglement admits a hierarchy of definitions depending on the strength of the correlations between the subsystems forming the quantum system. Therefore, as the quantum system becomes larger, it becomes more challenging to characterize it.

Interestingly, the degree of violation of our inequalities gives direct information about the nature of entanglement inherent in the considered many body state [5,117], and provides a method to classify the different degrees of multipartite entanglement in a computationally and experimentally tractable way (see Fig. 12).

In this work, the researchers present a method to derive Device-Independent Witnesses of Entanglement Depth (DIWEDs) from Bell inequalities. Such DIWEDs quantify the strength of entanglement on a quantum many-body system by observing Bell non-local correlations. The difficultly of this approach lies in how one can

Fig. 12 Schematic representation of an Entanglement Depth Witness. Figure's author: J. Tura

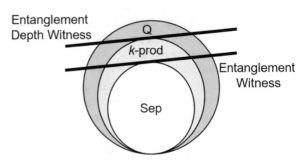

derive these DIWEDs from Bell inequalities. The researchers have been able to come up with an elegant solution that reduces the issue to an efficient optimization problem. In particular, their methodology finds the maximal amount of Bell non-local correlations that can be achieved given any quantum system that has at most k particles entangled. This provides a hierarchy of certification bounds, since such values can only be surpassed by quantum systems that have more than k particles entangled. In addition, they have also been able to show how the DIWEDs can be rewritten in terms of collective measurements, and then apply these DIWEDs on existing experimental data in order to certify an entanglement depth of 15 particles in a Bose-Einstein condensate of 480 particles.

Recently, we have continued to work on this approach to quantum correlations, and we were able to formulate Bell correlation and entanglement tests entirely based on accessible experimental or simulation data [88, 89].

6.4   Experiment-Friendly Approaches for Characterization of Topology

We quote here only two examples of recent experimental work. More experiment-friendly approaches will be mentioned in the context of synthetic dimensions below. The first example comes from yet another platform for QS: polaritons, the second from photonics.

Topological insulators attracted much interest in the last decades. These exotic materials are characterized by a global topological invariant, an integer, that cannot be deformed by local perturbations such as disorder or interactions. This robustness makes them ideal candidates for applications in metrology or quantum computation. For example, materials displaying the integer quantum Hall effect yield extremely robust plateaux in their transverse conductivity. Such robustness comes from the celebrated bulk-edge correspondence, which dictates that the topological invariant of the bulk is equal to the number of protected conducting edge states.

A distinct situation arises in 2D crystals like graphene, possessing both time reversal and chiral symmetry. Such systems do not have a gap. Nevertheless, they can present edge states which are robust against perturbations respecting the symmetry of the system. These states can be linked to a topological invariant defined over reduced one-dimensional subspaces of their Brillouin zone.

In a recent article published in Physical Review Letters [110] and highlighted as an Editor's Suggestion, the authors reported the measurement of such topological invariants in artificial graphene. They demonstrated a novel scheme based on a hybrid position- and momentum-space measurement to directly access these 1D topological invariants in lattices of semiconductor microcavities confining exciton-polaritons. They showed that such technique can be applied both to normal and strained graphene. This work opens, in our opinion, the door to a systematic study of such systems in the presence of disorder or interactions (see Fig. 13).

Yet another novel way of detecting topological properties relates to photonic systems. Research on topological insulators is moving at fast speed, promising a

Fig. 13 Scanning electron microscopy image of a honeycomb lattice of coupled micropillars. Source: ICFO webpage news

broad spectrum of applications ranging from metrology to quantum computing. Topological insulators are a new phase of matter in which the bulk of the material is an insulator, but its edges conduct electricity through what is called a "topological protected" edge states, that are a direct manifestation of the nontrivial topology hidden in their band structure.

To understand the effects of topology, physicists are working simultaneously on a plethora of experimental architectures. Among these, quantum walks are powerful models where topological phases of matter can be simulated in static and out-of-equilibrium scenarios. Most of the quantum walk architectures built so far generated one dimensional processes. However, there are currently efforts on increasing the dimensionality of these platforms to investigate the broad range of topological phenomena that exist in 2D and 3D.

In a study recently published in Optica [43], researchers reported on a novel photonic platform, capable of producing quantum walks of structured photons in two spatial dimensions, and exhibiting quantum Hall behavior. In their study, the researchers report on the realization of a photonic platform generating a quantum walk on a two-dimensional square lattice, that emulates a periodically-driven quantum Hall insulator. The apparatus consists of cascaded liquid crystal slabs, patterned to give polarization-dependent kicks to the impinging photons. Suitable combinations of these plates allow to manipulate dynamically the evolution of a light beam, realizing a quantum walk between light spatial modes carrying a variable amount of transverse momentum. The authors demonstrate the non-trivial topological character of their photonic system by directly reading out the anomalous displacement of an optical wavepacket when a constant force is introduced in the system (see Fig. 14).

The simulation of other condensed matter systems, the investigation of the evolution of quantum light and the study of dynamical phase transitions are among the possible paths that the researchers of the study intend to explore in the next future.

Fig. 14 A collimated beam crosses a sequence of liquid-crystal (LC) devices. Different LC patterns implement coin rotations and spin-dependent walker discrete translations. Reprinted with permission of SLAM research group (Naples)

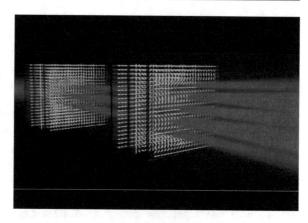

Fig. 15 Twistronics without a twist. Upper panel: synthetic bilayer of (pi-flux) square lattices with periodicity equal to 4 sites both in x and y axis. Such modulation of Ωx, y supports Dirac cones in single particle dispersion. Lower panel: Spatially modulated Raman coupling between the two layers mimics moiré patterns. Figure's author: T.Salamon/L.Tarruell

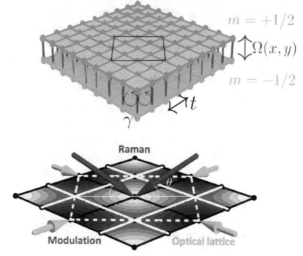

6.5   Synthetic Dimensions

The idea of synthetic dimensions was formulated by us [22], and was independently discovered in condensed matter physics. In AMO physics it gained particular interest when it was formulated directly in the contexts of using internal atomic states for synthetic dimension, and employing designed Raman couplings to create synthetic gauge fields [32]—indeed the first top level experiments came out few months after our paper. Amazingly, the synthetic dimensions approach allows, for instance, to measure Chern invariants in extremely narrow quasi-2D strips (see [87] and references therein) (Fig. 15).

Perhaps the most creative use of the idea of the synthetic dimension concerns, however, twistronics. The discovery of flat-bands in magic angle graphene has established a whole new concept—twistronics—which allows to induce strong correlation effects, anomalous superconductivity, magnetism and topology. The creation of this entirely novel field opens entirely new possibilities as well as previously unreachable degrees of manipulation and generation of the quantum many body states. Not only has it far reaching implications in the fundamental understanding of these exotic phases, but it is also likely to make far-reaching changes in technology, for instance in the fields of quantum metrology and sensing. The online symposium "Emergent phenomena in moiré materials", organized by ICFO and MIT illustrated perfectly the enormous interest and progress in this area.

The interesting phenomenology of twisted materials is apparently related to the formation of moiré patterns around small twist angles. Some of these twisting angles, the so-called magic angles, lead to vast band flattening already at the single-particle level. The geometrical moiré patterns induce spatially varying interlayer couplings that are behind the strong modification of the band structure. Emulating this physics beyond materials research can help in the identification of key minimal ingredients that give rise to the phenomenology of Twisted Bi-layer Graphene (TBLG), while also providing additional microscopic control. Photonic systems, for example, are well suited to explore this physics at the single-particle level as shown in a recent paper by Lluis Torner and his collaborators [119] (see above).

Ultracold atoms in optical lattices are, in principle, a very promising platform to experimentally explore also the corresponding emerging many body phenomena. In a recent letter [101], we have proposed an atomic quantum simulator of twisted bilayers without actual physical twisting between layers. The idea is to elegantly employ synthetic dimensions. Atoms are located in a single 2D lattice, but can occupy two internal states, mimicking two layers. Coupling between the layers is realized by Raman transitions that are spatially modulated so as to mimic the action of twists, or in general arbitrary desired moiré patterns. The advantages of this scheme are clear:

- Control of interlayer tunneling strength over wide ranges.
- Control of atom-atom interaction, so that the ratio of kinetic-to-interaction energy can be tuned over wide ranges.
- Magic angle physics appears at larger angles with smaller moiré supercells, implying less fluctuations of the twisting angle.

6.6   Methods for Verification and Certification of Quantum Simulators

To verify, validate and certify QSs, we need **theoretical methods** to design **experimental strategies** that are at least capable to describe qualitatively and even more importantly quantitatively the phenomena/systems in question. In this section we list the ones that have been developing particularly rapidly in the recent years:

6.6.1   Tensor Networks

Tensor networks (TN) are novel numerical methods suitable for studies of strongly correlated systems. For lattice models, they involve tensors defined at each lattice site that depend on the physical variables at this site (spin state, bosonic state, ...). These tensors live also in auxiliary (bond) spaces, through which they are connected (via index contraction) to other sites. In one dimension this approach goes back to the Density Matrix Renormalization Group (DMRG) method of Steve R. White [120]. More contemporary versions are termed Matrix Product States (MPS), or in time dependent version Time-Evolving Block Decimation (TEBD). While standard TN are perfectly describing states that fulfill the, so called, entanglement area law, stronger forms of entanglement are captured by the Multiscale Entanglement Renormalization Ansatz (MERA). The analog of MPS in 2D and higher dimensions is termed Projected Entangled Pair States (PEPS). There are many review and handbook about TN methods, published recently; we refer here to our own book, published with open access in Lecture Notes in Physics [99].

6.6.2   Exact Diagonalization

Obviously, exact diagonalization (ED) is limited to not-too-large systems, but is indispensable in some cases, such as for instance the studies of Many Body Localization (MBL). A state-of-art of ED is well represented by our recent study of disordered spin systems that exhibit MBL [109].

6.6.3   Polynomial Relaxations Based on Semi-Definite Programming

Many problems of interest are encoded in the ground state of an interacting many-bod system, classical or quantum. In this context, a quantum simulator provides a physical approximation to this unknown ground state. The measured energy therefore constitutes an upper bound to the ground-state energy. However, there is no certificate that the obtained solution is close to the unknown ground-state energy, which may significantly differ from the obtained value, for instance because of the presence of local minima. A complementary approach to verify the quality of the obtained upper bound is to construct methods that provide lower bounds to the ground-state energy. If the lower and upper bounds are close, one certifies that the approximation is close to the unknown solution. In the ideal case that the bounds coincide, one even concludes that the ground state has been reached.

A method to derive lower bounds to ground-state energy problem is given by hierarchies of relaxations to polynomial optimization problems based on semi-definite programming (SDP). The method was introduced by Lasserre [72] and Parrillo [95] for the classical case of commuting variables, and generalised to the quantum case of non-commuting operators by Navascués et al. [91] and Pironio et al. [97], see also [45]. Without entering into the details of the formalism, in the classical case, these methods can be used when dealing with polynomial optimization problem as follows:

$$\min_{\mathbf{x}} p(\mathbf{x}) \tag{1}$$

such that

$$g_j(\mathbf{x}) \geq 0, \quad j = 1, \ldots, m.$$

Here p and g_j are polynomials over some variables $\mathbf{x} = (x_1, \ldots, x_n)$. The form of the quantum problem is basically the same, reading

$$\min_{|\psi\rangle, \mathbf{X}} \langle \psi | p(\mathbf{X}) | \psi \rangle \tag{2}$$

such that

$$g_j(\mathbf{X}) \geq 0, \quad j = 1, \ldots, m.$$

Now, polynomials are defined over some operators $\mathbf{X} = (X_1, \ldots, X_n)$ and inequalities should be understood as operator inequalities. We denote the solution to these problems by p^*. Many classical and quantum ground-state problems can be written in this polynomial form, for instance, as polynomials of (1) two-value variables for classical spins, (2) pauli matrices for quantum spins or (3) creation and anihilation operators for fermions.

Without entering into the details, the previous methods provide a hierarchy of lower bounds to the searched solution $p_1 \leq p_2 \leq p_3 \ldots \leq p^*$. The interesting property is that each lower bound can be computed by semi-definite programming (SDP), a standard convex optimisation method [118] that is not affected by local minima. The complexity of the SDP problem increases with the so-called level of the hierarchy. Under mild conditions, the hierarchy is proven to asymptotically converge, $p_\infty \rightarrow p^*$. In our context, the method provides a sequence of improving lower bounds to the ground-state energy that can be used to benchmark the outputs of variational methods or quantum simulators.

SDP relaxations have been used in many different systems, for instance in quantum chemistry [83, 90], quantum spin systems [15, 17] or, more recently, to benchmark the outputs of a D-Wave quantum annealer [12].

6.6.4 Machine Learning

Machine learning (ML) and Artificial Intelligence (AI) can be found everywhere nowadays: from science through technology to politics and marketing. For the study of applications in physics and chemistry we recommend the recent overview lecture by the late Naftali Tishby on YouTube [115], the recent review [30], and our new book [42]. For the case of QSs, classical or quantum ML can be used to recognize quantum phases, quantum dynamics, and more, from data. The data can be generated in experiment or numerical simulation. A nice example of usefulness of ML to quantum many body physics is described in our recent Letter [68] (Fig. 16).

Fig. 16 Unsupervised
Machine Learning in Physics.
Figure's author: K. Kottmann

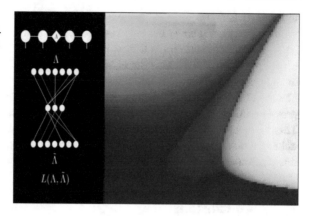

Machine Learning (ML) has the main goal of analyzing and interpreting data structures and patterns in order to learn from them, reason and carry out a decision-making task that is completely independent from human reasoning and engagement. Even though this field of study started in the mid 1900s, recent developments in the area have revolutionized the way how we process and find correlations in complex data.

Contrary to supervised learning, unsupervised learning seeks to discover patterns or classify information in large data sets into categories without prior knowledge. That is, it does not have labeled outputs, which means that it basically infers the natural structure that a dataset may have and extracts categorized information from it. This learning has proved to be very efficient for identifying phases and phase transitions of many-body systems. In our Letter, we report on a method that uses an unsupervised machine learning technique based on anomaly detection to automatically map out the phase diagram of a quantum many body system given unlabeled data.

The following example is very illustrative of what has been achieved. In machine learning the most common and known classification task example is to discriminate, for instance, images of cats and dogs. In particular, anomaly detection handles the classification task of discriminating dogs and everything that is not a dog, approaching the system is an entirely different perspective. The idea is to train a special neural network called an autoencoder to efficiently compress and reproduce images of dogs. If the network is later fed with images of cats, the network does not know how to efficiently compress the features of the cat image and it is possible to tell from the higher reconstruction loss that it is not a dog. In [68], the authors use this method in the context of quantum many-body systems. The images become observables, wave-functions or entanglement properties and the classes dogs and cats become different quantum phases. The model that they look at is the extended Bose-Hubbard model, which offers four different phases in the parameter space of interest. Since the researchers do not know the phases in their task a priori, they start by defining a region around the origin of the phase diagram as their starting point to train. Already from there, they are clearly capable of mapping the system in one training iteration, where all four phases of the system are easily distinguishable.

Finally, using this approach they have been able to reveal a phase-separated state between a supersolid and a superfluid, which appears in the system in addition to the standard superfluid, Mott insulator, Haldane-insulating, and density wave phases. The discovery of this new feature in this phase diagram is rather surprising provided that this model has been the investigated thoroughly for the past 20 years. This work is one of the first examples in which a machine detects a previously unknown phenomenon in a quantum many-body context. For a detailed analysis of this phenomenon and more, see the following Phys. Rev. B paper [67].

7   Conclusions

The area of quantum simulators is in any sense one of the most beautiful areas of contemporary physics, melting together in a common pot all possible branches and genres of physics and not only physics. The conclusion that we always propose is: Enjoy physics and beyond!

In particular, we at ICFO enjoy going beyond to try to interpret quantum mechanical processes, and more specifically quantum randomness in contemporary avantgarde music. We try indeed to incorporate quantum random processes, using the genuine quantum random number generators, provided to us by QUSIDE [2]. The highlight of our approach was the nearly 1 h long concert "Interpreting Quantum Randomness" at the famous SONAR Festival 2021 [121].

Finally, we would like to end this essay, by commenting on relation between quantum computing and quantum simulation. Clearly, we bet on the latter. But, we are sure that future of quantum computing is also bright; we expect it would be integrated with classical computing, combined in a symbiotic way with High Performance Computing (HPC) approaches. The distinction between quantum computing and simulation will probably become very diffuse, if not vanishing. The theoretical effort of computer scientists and theoretical physicists will play a pivotal role in this processes (cf. [54]).

Acknowledgments We acknowledge Antonio Acín and Leticia Taruell for their contributions to the manuscript. Over several years, we have been publishing mini-reviews of our "achieve-ments" on the ICFO Web Page in the News section [1]. These text were edited with the help of Alina Hirschmann, whose irreplaceable help is highly appreciated. We have adopted here some of these mini-reviews, mostly of the papers from our groups, but we also acknowledge courtesy of Adrian Bachtold, Alexandre Dauphin, Dima Efetov and Lluis Torner for being able to use their texts. JF, TS and ML acknowledge support from: ERC AdG NOQIA; Agencia Estatal de Investigación (R&D project CEX2019-000910-S, funded by MCIN / AEI / 10.13039 / 501100011033, Plan National FIDEUA PID2019-106901GB-I00, FPI, QUANTERA MAQS PCI2019-111828-2, Proyectos de I+D+I "Retos Colaboración" QUSPIN RTC2019-007196-7); Fundació Cellex; Fundació Mir-Puig; Generalitat de Catalunya through the European Social Fund FEDER and CERCA program (AGAUR Grant No. 2017 SGR 134, QuantumCAT U16-011424, co-funded by ERDF Operational Program of Catalonia 2014–2020); EU Horizon 2020 FET-OPEN OPTOlogic (Grant No 899794); National Science Centre, Poland (Symfonia Grant

No. 2016/20/W/ST4/00314); European Union's Horizon 2020 research and innovation programme under the Marie-Skołdowska-Curie grant agreement No 101029393 (STREDCH) and No 847648 ("La Caixa" Junior Leaders fellowships ID100010434: LCF/BQ/PI19/11690013, LCF/BQ/PI20/11760031, LCF/BQ/PR20/11770012, LCF/BQ/PR21/11840013).

References

1. Icfo qot website, https://www.icfo.eu/research-group/11/qot/news/437/
2. Quside website, https://www.linkedin.com/company/quside/?originalSubdomain=es
3. S. Aaronson, A. Arkhipov, The computational complexity of linear optics, in *Proceedings of the Forty-Third Annual ACM Symposium on Theory of Computing, STOC '11, New York, NY, USA, 2011* (Association for Computing Machinery), pp. 333–342
4. M. Aidelsburger, L. Barbiero, A. Bermudez, T. Chanda, A. Dauphin, D. Gonzá lez-Cuadra, P.R. Grzybowski, S. Hands, F. Jendrzejewski, J. Jünemann, G. Juzeliūnas, V. Kasper, A. Piga, S.-J. Ran, M. Rizzi, G. Sierra, L. Tagliacozzo, E. Tirrito, T.V. Zache, J. Zakrzewski, E. Zohar, M. Lewenstein, Cold atoms meet lattice gauge theory. Philos. Trans. R. Soc. A Math. Phys. Eng. Sci. **380**(2216) (2021)
5. A. Aloy, J. Tura, F. Baccari, A. Acín, M. Lewenstein, R. Augusiak, Device-independent witnesses of entanglement depth from two-body correlators. Phys. Rev. Lett. **123**, 100507 (2019)
6. E. Altman, K.R. Brown, G. Carleo, L.D. Carr, E. Demler, C. Chin, B. DeMarco, S.E. Economou, M.A. Eriksson, K.-M.C. Fu, M. Greiner, K.R. Hazzard, R.G. Hulet, A.J. Kollár, B.L. Lev, M.D. Lukin, R. Ma, X. Mi, S. Misra, C. Monroe, K. Murch, Z. Nazario, K.-K. Ni, A.C. Potter, P. Roushan, M. Saffman, M. Schleier-Smith, I. Siddiqi, R. Simmonds, M. Singh, I. Spielman, K. Temme, D.S. Weiss, J. Vučković, V. Vuletić, J. Ye, M. Zwierlein, Quantum simulators: architectures and opportunities. PRX Quantum **2**, 017003 (2021)
7. K. Amini, J. Biegert, F. Calegari, A. Chacón, M.F. Ciappina, A. Dauphin, D.K. Efimov, C.F. de Morisson Faria, K. Giergiel, P. Gniewek, A.S. Landsman, M. Lesiuk, M. Mandrysz, A.S. Maxwell, R. Moszyński, L. Ortmann, J.A. Pérez-Hernández, A. Picón, E. Pisanty, J. Prauzner-Bechcicki, K. Sacha, N. Suárez, A. Zair, J. Zakrzewski, M. Lewenstein, Symphony on strong field approximation. Rep. Progress Phys. **82**(11), 116001 (2019)
8. J. Argüello-Luengo, A. González-Tudela, T. Shi, P. Zoller, J.I. Cirac. Analogue quantum chemistry simulation. Nature **574**(7777), 215–218 (2019)
9. S. Arora, B. Barak, *Computational Complexity: A Modern Approach*, 1st edn. (Cambridge University Press, Cambridge, 2009)
10. F. Arute, K. Arya, R. Babbush, D. Bacon, J.C. Bardin, R. Barends, R. Biswas, S. Boixo, F.G.S.L. Brandao, D.A. Buell, B. Burkett, Y. Chen, Z. Chen, B. Chiaro, R. Collins, W. Courtney, A. Dunsworth, E. Farhi, B. Foxen, A. Fowler, C. Gidney, M. Giustina, R. Graff, K. Guerin, S. Habegger, M.P. Harrigan, M.J. Hartmann, A. Ho, M. Hoffmann, T. Huang, T.S. Humble, S.V. Isakov, E. Jeffrey, Z. Jiang, D. Kafri, K. Kechedzhi, J. Kelly, P.V. Klimov, S. Knysh, A. Korotkov, F. Kostritsa, D. Landhuis, M. Lindmark, E. Lucero, D. Lyakh, S. Mandrà, J.R. McClean, M. McEwen, A. Megrant, X. Mi, K. Michielsen, M. Mohseni, J. Mutus, O. Naaman, M. Neeley, C. Neill, M.Y. Niu, E. Ostby, A. Petukhov, J.C. Platt, C. Quintana, E.G. Rieffel, P. Roushan, N.C. Rubin, D. Sank, K.J. Satzinger, V. Smelyanskiy, K.J. Sung, M.D. Trevithick, A. Vainsencher, B. Villalonga, T. White, Z.J. Yao, P. Yeh, A. Zalcman, H. Neven, J. M. Martinis, Quantum supremacy using a programmable superconducting processor. Nature **574**(7779), 505–510 (2019)
11. S. Asban, S. Mukamel, Scattering-based geometric shaping of photon-photon interactions. Phys. Rev. Lett. **123**, 260502 (2019)
12. F. Baccari, C. Gogolin, P. Wittek, A. Ací n, Verifying the output of quantum optimizers with ground-state energy lower bounds. Phys. Rev. Res. **2**(4) (2020)

13. W. Bakr, J. Gillen, A. Peng, S. Fölling, M. Greiner, A quantum gas microscope for detecting single atoms in a hubbard-regime optical lattice. Nature **462**(7269), 74–77 (2009)
14. M.C. Bañuls, R. Blatt, J. Catani, A. Celi, J.I. Cirac, M. Dalmonte, L. Fallani, K. Jansen, M. Lewenstein, S. Montangero, C.A. Muschik, B. Reznik, E. Rico, L. Tagliacozzo, K. Van Acoleyen, F. Verstraete, U.-J. Wiese, M. Wingate, J. Zakrzewski, P. Zoller, Simulating lattice gauge theories within quantum technologies. Eur. Phys. J. D **74**(8), 165 (2020)
15. T. Barthel, R. Hübener, Solving condensed-matter ground-state problems by semidefinite relaxations. Phys. Rev. Lett. **108**, 200404 (2012)
16. D.N. Basov, A. Asenjo-Garcia, P.J. Schuck, X. Zhu, A. Rubio, Polariton panorama. Nanophotonics **10**(1), 549–577 (2021)
17. T. Baumgratz, M.B. Plenio, Lower bounds for ground states of condensed matter systems. New J. Phys. **14**(2), 023027 (2012)
18. H. Bernien, S. Schwartz, A. Keesling, H. Levine, A. Omran, H. Pichler, S. Choi, A.S. Zibrov, M. Endres, M. Greiner, V. Vuletic, M.D. Lukin, Probing many-body dynamics on a 51-atom quantum simulator. Nature **551**, 579 (2017)
19. K. Bharti, A. Cervera-Lierta, T.H. Kyaw, T. Haug, S. Alperin-Lea, A. Anand, M. Degroote, H. Heimonen, J.S. Kottmann, T. Menke, W.-K. Mok, S. Sim, L.-C. Kwek, A. Aspuru-Guzik, Noisy intermediate-scale quantum algorithms. Rev. Mod. Phys. **94**, 015004 (2022)
20. A. Blais, A.L. Grimsmo, S.M. Girvin, A. Wallraff, Circuit quantum electrodynamics. Rev. Mod. Phys. **93**, 025005 (2021)
21. D. Bluvstein, A. Omran, H. Levine, A. Keesling, G. Semeghini, S. Ebadi, T.T. Wang, A.A. Michailidis, N. Maskara, W.W. Ho, S. Choi, M. Serbyn, M. Greiner, V. Vuletic, M.D. Lukin, Controlling quantum many-body dynamics in driven rydberg atom arrays. Science **371**(6536), 1355–1359 (2021)
22. O. Boada, A. Celi, J.I. Latorre, M. Lewenstein, Quantum simulation of an extra dimension. Phys. Rev. Lett. **108**, 133001 (2012)
23. A. Bohrdt, C.S. Chiu, G. Ji, M. Xu, D. Greif, M. Greiner, E. Demler, F. Grusdt, M. Knap, Classifying snapshots of the doped hubbard model with machine learning. Nat. Phys. **15**(9), 921–924 (2019)
24. F. Böttcher, J.-N. Schmidt, J. Hertkorn, K.S.H. Ng, S.D. Graham, M. Guo, T. Langen, T. Pfau, New states of matter with fine-tuned interactions: quantum droplets and dipolar supersolids. Rep. Progress Phys. **84**(1), 012403 (2020)
25. T. Boulier, M.J. Jacquet, A. Maître, G. Lerario, F. Claude, S. Pigeon, Q. Glorieux, A. Amo, J. Bloch, A. Bramati, E. Giacobino, Microcavity polaritons for quantum simulation. Adv. Quantum Technol. **3**(11), 2000052 (2020)
26. T. Brydges, A. Elben, P. Jurcevic, B. Vermersch, C. Maier, B.P. Lanyon, P. Zoller, R. Blatt, C.F. Roos, Probing renyi entanglement entropy via randomized measurements. Science **364**(6437), 260–263 (2019)
27. C.R. Cabrera, L. Tanzi, J. Sanz, B. Naylor, P. Thomas, P. Cheiney, L. Tarruell, Quantum liquid droplets in a mixture of bose-einstein condensates. Science **359**(6373), 301–304 (2018)
28. Y. Cao, D. Rodan-Legrain, O. Rubies-Bigorda, J.M. Park, K. Watanabe, T. Taniguchi, P. Jarillo-Herrero, Tunable correlated states and spin-polarized phases in twisted bilayer–bilayer graphene. Nature **583**(7815), 215–220 (2020)
29. Y. Cao, J. Romero, J.P. Olson, M. Degroote, P.D. Johnson, M. Kieferová, I.D. Kivlichan, T. Menke, B. Peropadre, N.P.D. Sawaya, S. Sim, L. Veis, A. Aspuru-Guzik, Quantum chemistry in the age of quantum computing. Chem. Rev. **119**(19), 10856–10915 (2019)
30. G. Carleo, I. Cirac, K. Cranmer, L. Daudet, M. Schuld, N. Tishby, L. Vogt-Maranto, L. Zdeborová, Machine learning and the physical sciences. Rev. Mod. Phys. **91**, 045002 (2019)
31. I. Carusotto, A. Houck, A. Kollár, P. Roushan, D. Schuster, J. Simon, Photonic materials in circuit quantum electrodynamics. Nat. Phys. **16**, 03 (2020)
32. A. Celi, P. Massignan, J. Ruseckas, N. Goldman, I.B. Spielman, G. Juzeliūnas, M. Lewenstein, Synthetic gauge fields in synthetic dimensions. Phys. Rev. Lett. **112**, 043001 (2014)

33. A. Cervera-Lierta, Icfo theory lecture: noisy intermediate-scale quantum algorithms 1 (2022). https://youtu.be/pyB9d8fVTt0

34. A. Cervera-Lierta, Icfo theory lecture: Noisy intermediate-scale quantum algorithms 2 (2022). https://youtu.be/ooNEw64zfbg

35. T. Chanda, M. Lewenstein, J. Zakrzewski, L. Tagliacozzo, Phase diagram of $1 + 1D$ abelian-higgs model and its critical point. Phys. Rev. Lett. **128**, 090601 (2022)

36. T. Chanda, J. Zakrzewski, M. Lewenstein, L. Tagliacozzo, Confinement and lack of thermalization after quenches in the bosonic schwinger model. Phys. Rev. Lett. **124**, 180602 (2020)

37. L.W. Cheuk, M.A. Nichols, M. Okan, T. Gersdorf, V.V. Ramasesh, W.S. Bakr, T. Lompe, M.W. Zwierlein, Quantum-gas microscope for fermionic atoms. Phys. Rev. Lett. **114**, 193001 (2015)

38. C.S. Chiu, G. Ji, A. Bohrdt, M. Xu, M. Knap, E. Demler, F. Grusdt, M. Greiner, D. Greif, String patterns in the doped hubbard model. Science **365**(6450), 251–256 (2019)

39. L.W. Clark, N. Schine, C. Baum, N. Jia, J. Simon, Observation of laughlin states made of light. Nature **582**(7810), 41–45 (2020)

40. T.M. Cover, J.A. Thomas, *Elements of Information Theory*. Wiley Series in Telecommunications and Signal Processing, 2nd edn. (Wiley-Interscience, New York, 2006)

41. M. Dalmonte, B. Vermersch, P. Zoller, Quantum simulation and spectroscopy of entanglement Hamiltonians. Nat. Phys. **14**(8), 827–831 (2018)

42. A. Dawid, J. Arnold, B. Requena, A. Gresch, M. Płodzień, K. Donatella, K.A. Nicoli, P. Stornati, R. Koch, M. Büttner, R. Okuła, G. Muñoz Gil, R.A. Vargas-Hernández, A. Cervera-Lierta, J. Carrasquilla, V. Dunjko, M. Gabrié, P. Huembeli, E. van Nieuwenburg, F. Vicentini, L. Wang, S.J. Wetzel, G. Carleo, E. Greplova, R. Krems, F. Marquardt, M. Tomza, M. Lewenstein, A. Dauphin, Modern applications of machine learning in quantum sciences (2022), arXiv:2204.04198(quant-ph)

43. A. D'Errico, F. Cardano, M. Maffei, A. Dauphin, R. Barboza, C. Esposito, B. Piccirillo, M. Lewenstein, P. Massignan, L. Marrucci, Two-dimensional topological quantum walks in the momentum space of structured light. Optica **7**(2), 108–114 (2020)

44. D.P. DiVincenzo, The physical implementation of quantum computation. Fortschritte der Physik **48**(9–11), 771–783 (2000)

45. A. Doherty, Y.-C. Liang, B. Toner, S. Wehner, The quantum moment problem and bounds on entangled multi-prover games. Proceedings of the Annual IEEE Conference on Computational Complexity (2008), pp. 199–210

46. S. Ebadi, A. Keesling, M. Cain, T.T. Wang, H. Levine, D. Bluvstein, G. Semeghini, A. Omran, J. Liu, R. Samajdar, X.-Z. Luo, B. Nash, X. Gao, B. Barak, E. Farhi, S. Sachdev, N. Gemelke, L. Zhou, S. Choi, H. Pichler, S. Wang, M. Greiner, V. Vuletic, M.D. Lukin, Quantum optimization of maximum independent set using rydberg atom arrays. Science **376**(6598), 1209–1215 (2022). doi:10.1126/science.abo6587

47. S. Ebadi, T.T. Wang, H. Levine, A. Keesling, G. Semeghini, A. Omran, D. Bluvstein, R. Samajdar, H. Pichler, W.W. Ho, S. Choi, S. Sachdev, M. Greiner, V. Vuletić, M.D. Lukin, Quantum phases of matter on a 256-atom programmable quantum simulator. Nature **595**(7866), 227–232 (2021)

48. A. Elben, S.T. Flammia, H.-Y. Huang, R. Kueng, J. Preskill, B. Vermersch, P. Zoller, The randomized measurement toolbox. Nat Rev Phys **5**, 9–24 (2023). https://doi.org/10.1038/s42254-022-00535-2

49. A. Elben, B. Vermersch, M. Dalmonte, J.I. Cirac, P. Zoller, Rényi entropies from random quenches in atomic hubbard and spin models. Phys. Rev. Lett. **120**, 050406 (2018)

50. A. Elben, J. Yu, G. Zhu, M. Hafezi, F. Pollmann, P. Zoller, B. Vermersch, Many-body topological invariants from randomized measurements in synthetic quantum matter. Sci. Adv. **6**(15), eaaz3666 (2020)

51. E. Farhi, J. Goldstone, S. Gutmann, M. Sipser, Quantum computation by adiabatic evolution. Quantum Phys. (2000). arXiv

52. R.P. Feynman, Simulating physics with computers. Int. J. Theor. Phys. **21**(6), 467–488 (1982)

53. J. Fraxanet, D. Gonzá lez-Cuadra, T. Pfau, M. Lewenstein, T. Langen, L. Barbiero, Topological quantum critical points in the extended bose-hubbard model. Phys. Rev. Lett. **128**(4) (2022)
54. A. Glos, A. Krawiec, Z. Zimborás, Space-efficient binary optimization for variational quantum computing. npj Quantum Inf. **8**(1), 39 (2022)
55. M. Greiner, O. Mandel, T. Esslinger, T.W. Hänsch, I. Bloch, Quantum phase transition from a superfluid to a mott insulator in a gas of ultracold atoms. Nature **415**(6867), 39–44 (2002)
56. C. Gross, W.S. Bakr, Quantum gas microscopy for single atom and spin detection. Nat. Phys. **17**(12), 1316–1323 (2021)
57. C. Huang, F. Ye, X. Chen, Y.V. Kartashov, V.V. Konotop, L. Torner, Localization-delocalization wavepacket transition in pythagorean aperiodic potentials. Sci. Rep. **6**(1), 32546 (2016)
58. H. Hübener, U. De Giovannini, C. Schäfer, J. Andberger, M. Ruggenthaler, J. Faist, A. Rubio, Engineering quantum materials with chiral optical cavities. Nat. Mater. **20**(4), 438–442 (2021)
59. W.J. Huggins, J.R. McClean, N.C. Rubin, Z. Jiang, N. Wiebe, K.B. Whaley, R. Babbush, Efficient and noise resilient measurements for quantum chemistry on near-term quantum computers. npj Quantum Inf. **7**(1), 23 (2021)
60. D. Jaksch, C. Bruder, J.I. Cirac, C.W. Gardiner, and P. Zoller. Cold bosonic atoms in optical lattices. Phys. Rev. Lett. **81**, 3108–3111 (1998)
61. S. Julià-Farré, M. Müller, M. Lewenstein, A. Dauphin, Self-trapped polarons and topological defects in a topological mott insulator. Phys. Rev. Lett. **125**, 240601 (2020)
62. D. Kennes, M. Claassen, L. Xian, A. Georges, A. Millis, J. Hone, C. Dean, D. Basov, A. Pasupathy, A. Rubio, Moiré heterostructures as a condensed-matter quantum simulator. Nat. Phys. **17**, 1–9 (2021)
63. J. King, S. Yarkoni, J. Raymond, I. Ozfidan, A.D. King, M.M. Nevisi, J.P. Hilton, C.C. McGeoch, Quantum annealing amid local ruggedness and global frustration. J. Phys. Soc. Jpn. **88**(6), 061007 (2019)
64. E. Knill, R. Laflamme, G. Milburn, A scheme for efficient quantum computation with linear optics. Nature **409**, 46–52 (2001)
65. M.S. Könz, W. Lechner, H.G. Katzgraber, M. Troyer, Scaling overhead of embedding optimization problems in quantum annealing. Phys. Rev. X Quantum **2**, 040322 (2021)
66. C. Kokail, C. Maier, R. van Bijnen, T. Brydges, M.K. Joshi, P. Jurcevic, C.A. Muschik, P. Silvi, R. Blatt, C.F. Roos, P. Zoller, Self-verifying variational quantum simulation of lattice models. Nature **569**(7756), 355–360 (2019)
67. K. Kottmann, A. Haller, A. Acín, G.E. Astrakharchik, M. Lewenstein, Supersolid-superfluid phase separation in the extended bose-hubbard model. Phys. Rev. B **104**, 174514 (2021)
68. K. Kottmann, P. Huembeli, M. Lewenstein, A. Acín, Unsupervised phase discovery with deep anomaly detection. Phys. Rev. Lett. **125**, 170603 (2020)
69. S. Kuhr, Quantum-gas microscopes: a new tool for cold-atom quantum simulators. Natl. Sci. Rev. **3**(2), 170–172 (2016)
70. C. Lagoin, U. Bhattacharya, T. Grass, R. Chhajlany, T. Salamon, K. Baldwin, L. Pfeiffer, M. Lewenstein, M. Holzmann, F. Dubin, Extended Bose-Hubbard model with dipolar excitons. Nature **609**, 485–489 (2022). https://doi.org/10.1038/s41586-022-05123-z
71. R. Landauer, Irreversibility and heat generation in the computing process. IBM J. Res. Develop. **5**(3), 183–191 (1961)
72. J.B. Lasserre, Global optimization with polynomials and the problem of moments. SIAM J. Optim. **11**(3), 796–817 (2001)
73. P.A. Lee, From high temperature superconductivity to quantum spin liquid: progress in strong correlation physics. Rep. Progress Phys. **71**(1), 012501 (2007)
74. S. Lee, J. Lee, H. Zhai, Y. Tong, A.M. Dalzell, A. Kumar, P. Helms, J. Gray, Z.-H. Cui, W. Liu, M. Kastoryano, R.B. Babbush, J. Preskill, D.R. Reichman, E.T. Campbell, E.F. Valeev, L. Lin, G.K.-L.C. Chan, Evaluating the evidence for exponential quantum advantage in ground-state quantum chemistry. Nat Commun **14**, 1952 (2023). https://doi.org/10.1038/s41467-023-37587-6

75. M. Lewenstein, P. Balcou, M.Y. Ivanov, A. L'Huillier, P.B. Corkum, Theory of high-harmonic generation by low-frequency laser fields. Phys. Rev. A **49**, 2117–2132 (1994)
76. M. Lewenstein, M. Ciappina, E. Pisanty, J. Rivera-Dean, P. Stammer, T. Lamprou, P. Tzallas, Generation of optical schrödinger cat states in intense laser - matter interactions. Nat. Phys. **17** (2021)
77. M. Lewenstein, A. Sanpera, V. Ahufinger, *Ultracold Atoms in Optical Lattices: Simulating Quantum Many-Body Systems* (Oxford University Press, Oxford, 2012)
78. X. Lu, P. Stepanov, W. Yang, M. Xie, M.A. Aamir, I. Das, C. Urgell, K. Watanabe, T. Taniguchi, G. Zhang, A. Bachtold, A.H. MacDonald, D.K. Efetov, Superconductors, orbital magnets and correlated states in magic-angle bilayer graphene. Nature **574**(7780), 653–657 (2019)
79. R. Ma, B. Saxberg, C. Owens, N. Leung, Y. Lu, J. Simon, D.I. Schuster, A dissipatively stabilized mott insulator of photons. Nature **566**(7742), 51–57 (2019)
80. D.J.C. MacKay, *Information Theory, Inference, and Learning Algorithms* (Cambridge University Press, Cambridge, 2003)
81. I.I. Manin, *Vychislimoe i nevychislimoe* (Sov. radio, Moskva, 1980)
82. E.A. Martinez, C.A. Muschik, P. Schindler, D. Nigg, A. Erhard, M. Heyl, P. Hauke, M. Dalmonte, T. Monz, P. Zoller, R. Blatt, Real-time dynamics of lattice gauge theories with a few-qubit quantum computer. Nature **534**(7608), 516–519 (2016)
83. D.A. Mazziotti, Variational minimization of atomic and molecular ground-state energies via the two-particle reduced density matrix. Phys. Rev. A **65**, 062511 (2002)
84. A.J. McCaskey, Z.P. Parks, J. Jakowski, S.V. Moore, T.D. Morris, T.S. Humble, R.C. Pooser, Quantum chemistry as a benchmark for near-term quantum computers. npj Quantum Inf. **5**(1), 99 (2019)
85. N. Moll, P. Barkoutsos, L.S. Bishop, J.M. Chow, A. Cross, D.J. Egger, S. Filipp, A. Fuhrer, J.M. Gambetta, M. Ganzhorn, A. Kandala, A. Mezzacapo, P. Müller, W. Riess, G. Salis, J. Smolin, I. Tavernelli, K. Temme, Quantum optimization using variational algorithms on near-term quantum devices. Quantum Sci. Technol. **3**(3), 030503 (2018)
86. C. Monroe, W.C. Campbell, L.-M. Duan, Z.-X. Gong, A.V. Gorshkov, P.W. Hess, R. Islam, K. Kim, N.M. Linke, G. Pagano, P. Richerme, C. Senko, N.Y. Yao, Programmable quantum simulations of spin systems with trapped ions. Rev. Mod. Phys. **93**, 025001 (2021)
87. S. Mugel, A. Dauphin, P. Massignan, L. Tarruell, M. Lewenstein, C. Lobo, A. Celi, Measuring Chern numbers in Hofstadter strips. SciPost Phys. **3**, 012 (2017)
88. G. Müller-Rigat, A. Aloy, M. Lewenstein, I. Frérot, Inferring nonlinear many-body bell inequalities from average two-body correlations: Systematic approach for arbitrary spin-j ensembles. PRX Quantum **2**(3), (2021)
89. G. Müller-Rigat, M. Lewenstein, I. Frérot, Probing quantum entanglement from magnetic-sublevels populations: beyond spin squeezing inequalities, Quantum **6**, 887 (2022). https://doi.org/10.22331/q-2022-12-29-887
90. M. Nakata, H. Nakatsuji, M. Ehara, M. Fukuda, K. Nakata, K. Fujisawa, Variational calculations of fermion second-order reduced density matrices by semidefinite programming algorithm. J. Chem. Phys. **114**(19), 8282–8292 (2001)
91. M. Navascués, S. Pironio, A. Acín, A convergent hierarchy of semidefinite programs characterizing the set of quantum correlations. New J. Phys. **10**(7), 073013 (2008)
92. F. Pan, K. Chen, P. Zhang, Solving the sampling problem of the sycamore quantum. Phys. Rev. Lett. **129**, 090502 (2022)
93. F. Pan, P. Zhang, Simulation of quantum circuits using the big-batch tensor network method. Phys. Rev. Lett. **128**, 030501 (2022)
94. G.S. Paraoanu, Recent progress in quantum simulation using superconducting circuits. J. Low Temp. Phys. **175**(5–6), 633–654 (2014)
95. P.A. Parrilo, Semidefinite programming relaxations for semialgebraic problems. Math. Program. **96**(2), 293–320 (2003)
96. A. Perdomo-Ortiz, A. Feldman, A. Ozaeta, S.V. Isakov, Z. Zhu, B. O'Gorman, H.G. Katzgraber, A. Diedrich, H. Neven, J. de Kleer, B. Lackey, R. Biswas, Readiness of quantum optimization machines for industrial applications. Phys. Rev. Appl. **12**, 014004 (2019)

97. S. Pironio, M. Navascué s, A. Acín, Convergent relaxations of polynomial optimization problems with noncommuting variables. SIAM J. Optim. **20**(5), 2157–2180 (2010)

98. J. Preskill, Quantum Computing in the NISQ era and beyond. Quantum **2**, 79 (2018)

99. S.-J. Ran, E. Tirrito, C. Peng, X. Chen, L. Tagliacozzo, G. Su, M. Lewenstein, *Tensor Network Contractions* (Springer International Publishing, New York, 2020)

100. M. Ringbauer, M. Meth, L. Postler, R. Stricker, R. Blatt, P. Schindler, T. Monz, A universal qudit quantum processor with trapped ions. Nat. Phys. **18**, 1053–1057 (2022). https://doi.org/10.1038/s41567-022-01658-0

101. T. Salamon, A. Celi, R.W. Chhajlany, I. Frérot, M. Lewenstein, L. Tarruell, D. Rakshit, Simulating twistronics without a twist. Phys. Rev. Lett. **125**, 030504 (2020)

102. S.D. Sarma, Quantum computing has a hype problem. MIT Technology Review (2022). https://www.technologyreview.com/2022/03/28/1048355/quantum-computing-has-a-hype-problem/

103. N. Schine, A. Ryou, A. Gromov, A. Sommer, J. Simon, Synthetic landau levels for photons. Nature **534**(7609), 671–675 (2016)

104. F. Schlawin, D.M. Kennes, M.A. Sentef, Cavity quantum materials. Appl. Phys. Rev. **9**(1), 011312 (2022)

105. P. Scholl, M. Schuler, H.J. Williams, A.A. Eberharter, D. Barredo, K.-N. Schymik, V. Lienhard, L.-P. Henry, T.C. Lang, T. Lahaye, A.M. Läuchli, A. Browaeys, Quantum simulation of 2d antiferromagnets with hundreds of rydberg atoms. Nature **595**(7866), 233–238 (2021)

106. M. Schuld, N. Killoran, Is quantum advantage the right goal for quantum machine learning? PRX Quantum **3**, 030101 (2022)

107. G. Semeghini, H. Levine, A. Keesling, S. Ebadi, T.T. Wang, D. Bluvstein, R. Verresen, H. Pichler, M. Kalinowski, R. Samajdar, A. Omran, S. Sachdev, A. Vishwanath, M. Greiner, V. Vuletic, M.D. Lukin, Probing topological spin liquids on a programmable quantum simulator. Science **374**(6572), 1242–1247 (2021)

108. J.F. Sherson, C. Weitenberg, M. Endres, M. Cheneau, I. Bloch, S. Kuhr, Single-atom-resolved fluorescence imaging of an atomic Mott insulator. Nature **467**(7311), 68–72 (2010)

109. P. Sierant, M. Lewenstein, J. Zakrzewski, Polynomially filtered exact diagonalization approach to many-body localization. Phys. Rev. Lett. **125**, 156601 (2020)

110. P. St-Jean, A. Dauphin, P. Massignan, B. Real, O. Jamadi, M. Milicevic, A. Lemaître, A. Harouri, L. Le Gratiet, I. Sagnes, S. Ravets, J. Bloch, A. Amo, Measuring topological invariants in a polaritonic analog of graphene. Phys. Rev. Lett. **126**, 127403 (2021)

111. P. Stepanov, I. Das, X. Lu, A. Fahimniya, K. Watanabe, T. Taniguchi, F.H.L. Koppens, J. Lischner, L. Levitov, D.K. Efetov, Untying the insulating and superconducting orders in magic-angle graphene. Nature **583**(7816), 375–378 (2020)

112. D. Stilck França, R. García-Patrón, Limitations of optimization algorithms on noisy quantum devices. Nat. Phys. **17**(11), 1221–1227 (2021)

113. R.N. Tazhigulov, S.-N. Sun, R. Haghshenas, H. Zhai, A.T.K. Tan, N.C. Rubin, R. Babbush, A.J. Minnich, G.K.-L. Chan, Simulating challenging correlated molecules and materials on the sycamore quantum processor, PRXQuantum **3**, 040318 (2022). https://link.aps.org/doi/10.1103/PRXQuantum.3.040318

114. J. Tilly, H. Chen, S. Cao, D. Picozzi, K. Setia, Y. Li, E. Grant, L. Wossnig, I. Rungger, G.H. Booth, J. Tennyson, The Variational Quantum Eigensolver: A review of methods and best practices, Physics Reports **986**, (2022). https://doi.org/10.1016/j.physrep.2022.08.003

115. N. Tishby, Statistical physics and machine learning: a 30 year perspective. https://www.youtube.com/watch?v=BUfnIT92ukM

116. A. Trabesinger, Quantum simulation. Nat. Phys. **8**(4), 263–263 (2012)

117. J. Tura, R. Augusiak, A.B. Sainz, T. Vé rtesi, M. Lewenstein, A. Acín, Detecting nonlocality in many-body quantum states. Science **344**(6189), 1256–1258 (2014)

118. L. Vandenberghe, S. Boyd, Semidefinite programming. SIAM Rev. **38**(1), 49–95 (1996)

119. P. Wang, Y. Zheng, X. Chen, C. Huang, Y.V. Kartashov, L. Torner, V.V. Konotop, F. Ye, Localization and delocalization of light in photonic moiré lattices. Nature **577**(7788), 42–46 (2019)

120. S.R. White, Density matrix formulation for quantum renormalization groups. Phys. Rev. Lett. **69**, 2863–2866 (1992)
121. R. Yamada, M. Lewenstein, Interpreting quantum randomness. AI & Music S+T+ARTS (2021). https://www.youtube.com/watch?v=Z-GLxgg0Z18
122. J. Zhang, G. Pagano, P.W. Hess, A. Kyprianidis, P. Becker, H. Kaplan, A.V. Gorshkov, Z.-X. Gong, C. Monroe, Observation of a many-body dynamical phase transition with a 53-qubit quantum simulator. Nature **551**(7682), 601–604 (2017)
123. H.-S. Zhong, H. Wang, Y.-H. Deng, M.-C. Chen, L.-C. Peng, Y.-H. Luo, J. Qin, D. Wu, X. Ding, Y. Hu, P. Hu, X.-Y. Yang, W.-J. Zhang, H. Li, Y. Li, X. Jiang, L. Gan, G. Yang, L. You, Z. Wang, L. Li, N.-L. Liu, C.-Y. Lu, J.-W. Pan, Quantum computational advantage using photons. Science **370**(6523), 1460–1463 (2020)

Insights Into Complex Functions

Wolfgang P. Schleich, Iva Tkáčová, and Lucas Happ

Contents

W. P. Schleich (✉)
Institut für Quantenphysik and Center for Integrated Quantum Science and Technology (IQST), Universität Ulm, Ulm, Germany

Hagler Institute for Advanced Study, Institute for Quantum Science and Engineering (IQSE), and Texas A&M AgriLife Research, Texas A&M University, College Station, TX, USA
e-mail: wolfgang.schleich@uni-ulm.de

I. Tkáčová
Department of Physics, Faculty of Electrical Engineering and Computer Science, VSB-Technical University of Ostrava, Ostrava, Poruba, Czech Republic

L. Happ
Institut für Quantenphysik and Center for Integrated Quantum Science and Technology (IQST), Universität Ulm, Ulm, Germany

Few-body Systems in Physics Laboratory, RIKEN Nishina Center for Accelerator-Based Science, Wako, Saitama, Japan

© The Author(s), under exclusive license to Springer Nature Switzerland AG 2023
R. Citro et al. (eds.), *Sketches of Physics*, Lecture Notes in Physics 1000,
https://doi.org/10.1007/978-3-031-32469-7_5

Abstract

We introduce and employ two tools to gain insight into a function $f = f(s)$ with complex argument s: (i) the Newton flows corresponding to the lines of constant phase and constant height, and (ii) the Cauchy-Riemann differential equations in amplitude and phase.

1   Introduction

Complex-valued functions are ubiquitous in physics. The wave function [1] or path integral formulations [3] of quantum mechanics, the approach of solving problems in electrostatics with the help of a conformal mapping [6], or the identification of flow lines in hydrodynamics [9] represent only a few of the numerous applications of complex analysis to illuminate physical phenomena.

Due to this prevalence of complex functions we have become accustomed to them. Nevertheless, we often have troubles to fully grasp their behavior since they map the *two*-dimensional complex plane onto itself. It is exactly this higher-dimensional mapping which is sometimes challenging for us to envision. In the present lecture we provide a remedy for this complication and demonstrate that the concept of Newton flows as well as the Cauchy-Riemann differential equations yield considerable insight into the inner workings of complex functions.

1.1   Our Tools

A common and illustrative way to represent a complex-valued function f employs lines in the complex plane where the phase θ of f and its height $|f|$ are constant. The resulting network of curves exemplified in Fig. 1 for the function ξ containing the Riemann zeta function ζ and ruled by the Cauchy-Riemann differential equations [17] is reminiscent of electromagnetic field lines [15] connecting sources with sinks.

In our lecture we analyze the lines of constant phase and height from the point of view of a flow dictated by the Newton velocity field [10], and defined by the negative inverse of the logarithmic derivative of f. This interpretation brings out most clearly that the sources of the Newton velocity field $V = V(s)$ creating the phase lines are the poles of f and the sinks are the zeros. In the absence of poles the phase lines start at the edges of the complex plane where f is infinite.

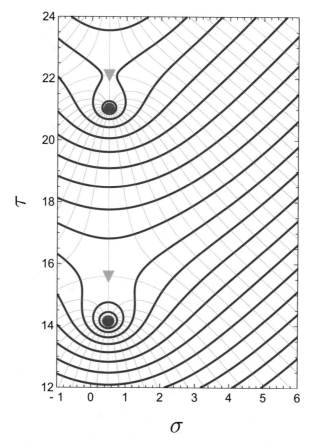

Fig. 1 Lines of constant phase (thin black curves) and lines of constant height (fat purple curves) of the function [2, 16] $\xi(s) \equiv \pi^{-s/2}(s - 1)\Gamma(s/2 + 1)\zeta(s)$ with complex argument $s \equiv \sigma + i\tau$ in the neighborhood of the two zeros (red dots) of ξ located on the critical line (orange curve) given by $\sigma = 1/2$. Here Γ and ζ denote the Gamma and the Riemann zeta function, respectively. Phase and contour lines are orthogonal on each other. Moreover, all phase lines terminate in zeros of ξ. Continental divides (green curves) separate the flows towards the individual zeros and merge with the critical line in zeros (green triangle) of the first derivative ξ' of ξ. Due to the functional equation $\xi(1 - s) = \xi(s)$ this network of phase and contour lines is symmetric with respect to the critical line

An interesting flow pattern emerges when there are more than one zero as shown in Fig. 1. In this case the flows approaching the individual zeros are separated by a special set of lines of constant phase which constitute continental divides. These so-called separatrices are characterized by the fact that, like any line of constant phase, they eventually terminate at a zero of f. However, they first traverse a zero of the first derivative f' of f, and only then end up in the zero.

Another powerful tool to gain insight into a complex-valued function f discussed in our lecture are the Cauchy-Riemann differential equations which are equivalent to the fact that f depends only on s and not on s and s^*. In this sense it is not a true two-dimensional mapping and there must exist an additional symmetry. The Cauchy-Riemann differential equations reflect this restriction on the class of all mappings.

The familiar Cauchy-Riemann differential equations relate the derivatives of the real and imaginary parts of f with respect to the two variables spanning the complex plane to each other. Since throughout our lecture we concentrate on lines of constant phase and constant amplitude we use the Cauchy-Riemann differential equations in a less familiar but for the present purpose more useful form. Indeed, we connect the derivatives of the *amplitude* and *phase* of f, rather than its real and imaginary parts, with respect to the two standard variables of the complex plane.

1.2   In a Nutshell

Our lecture is organized as follows. We dedicate Sect. 2 to a detailed discussion of Newton flows. In particular, we first obtain the lines of constant phase and constant height of a function f from Newton flows in the complex plane determined by a common velocity field $V = V(s)$. This identification clearly shows that poles of f as well as the edges of the complex plane serve as sources of the phase lines, and zeros of f are their sinks.

We also analyze the behavior of the phase lines in the neighborhood of a simple zero of f' and note a violation of the non-crossing rule for lines of constant phase. Indeed, for a simple zero of f' indicated in Fig. 1 for ξ by the filled green triangles on the critical line we find *two* ingoing and *two* outgoing lines which are orthogonal. Finally we address the problem of the inversion of the Newton flow condition which is intimately connected to the question if there is an inverse f^{-1} of f. Indeed, the separatrices are the borders of the individual domains of the complex plane in which f^{-1} exists.

We then turn in Sect. 3 to the illustration of the Cauchy-Riemann differential equations in amplitude and phase. Although our considerations are valid for any appropriately differentiable function, we throughout this section illuminate their consequences using an elementary function. In particular, we show that the slope of a line of constant phase is governed by the ratio of the derivatives of the amplitude of f with respect to the two variables of complex plane. Due to the orthogonality of phase lines and contour lines, their respective slopes are the inverse of each other. Thus the inverse of the ratio determining the slope of the phase lines governs the slope of the contour line.

We conclude in Sect. 4 by briefly summarizing our results and by providing an outlook addressing additional problems. Indeed, our notes do not intend to answer all questions about lines of constant phase and constant height. For example, we have only alluded to but not developed a criterion to identify a separatrix in the sea oh phase lines in a domain of the complex plane far away from the point where f' vanishes. Such a tool could open a door towards a proof of the long-standing

Riemann Hypothesis, as discussed in the Chapter "A Primer on the Riemann Hypothesis" of this volume.

In the spirit of lecture notes we keep our lecture self-contained and include in an appendix a detailed discussion of the Cauchy-Riemann differential equations in their various forms. Here we emphasize the ones connecting the amplitude and the phase of f.

2   Continuous Newton Method

The familiar Newton method to obtain the zero of a function f involves a discrete iteration of an expression which contains the ratio f/f'. The *continuous* Newton method [10] replaces this iteration by a *continuous trajectory*. The family of trajectories forming the Newton flow in the complex plane terminates like its discrete counterpart in the zeros of f.

In this section we rederive properties of the lines of constant phase as well as the ones of constant height of a complex function f by introducing and analyzing two Newton flows which differ by a phase $-\pi/2$. This feature reflects the well-known fact that these curves are orthogonal on each other as discussed in more detail in the appendix.

Throughout this section, we focus on a few essential points and refer for a more detailed discussion of the properties of Newton flows illustrated by several examples to Ref. [4, 5, 7, 8, 11]. Moreover, throughout this lecture we always remain in the familiar complex plane. Newton flows on the Riemann sphere are investigated in Ref. [12].

2.1   Newton Trajectories

We start by defining for the function $f = f(s)$ with the complex variable $s \equiv \sigma + i\tau$ trajectories $s = s(t)$ by the differential equation

$$\dot{s}(t) \equiv V(s(t)) \tag{1}$$

resulting from the Newton velocity field

$$V(s) \equiv -\frac{f(s)}{f'(s)} \tag{2}$$

given by the inverse of the logarithmic derivative of f. Here dot or prime denote differentiations with respect to the real-valued parameter t, or the complex variable s, respectively.

The trajectory $s = s(t)$ emerges from the Newton velocity field V in a rather intuitive way. Indeed, at every point s of the complex plane the ratio of $f(s)$ and

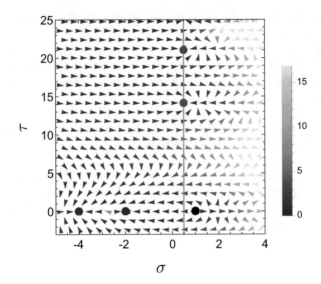

Fig. 2 Newton velocity field V, given by Eq. (2), for the Riemann zeta function ζ defined by Eq. (4). This field creates flows which originate from the left edge of the complex plane and approach either the trivial zeros (purple dots) located at the negative even numbers, or the non-trivial zeros (red dots) on the critical axis (orange line) with the real part 0.5. The flows to the latter are rather complicated approaching first the right edge of the complex plane and then turning around. The pole at $s = 1$ serves as a source with $s = -2$ being a sink

$f'(s)$ creates a complex number

$$\frac{f}{f'} = \left|\frac{f}{f'}\right| e^{i\beta} \tag{3}$$

which we represent by an arrow of length $|f/f'|$ under an angle β with respect to the real axis. However, due to the minus sign in the definition, Eq. (2), of V the Newton velocity field at s points in the direction opposite to f/f'.

In Fig. 2 we depict as an illustration, the Newton velocity field of the Riemann zeta function

$$\zeta(s) \equiv \sum_{n=1}^{\infty} \frac{1}{n^s} \tag{4}$$

represented here by its Dirichlet series. Obviously the arrangement of the arrows suggest trajectories which lead to the zeros of ζ indicated by the purple and red dots.

Indeed, these trajectories follow from the Newton velocity field V together with the definition, Eq. (1), of the flow. During the time interval dt we move in the direction of V by the amount $V dt$, and, according to Eq. (1), arrive at the point

$$s(t + dt) = s(t) + V(s(t))dt. \tag{5}$$

We then determine at $s(t + dt)$ the corresponding velocity field and move in this new direction. Repeating this process eventually creates in the limit of $dt \to 0$ a continuous trajectory.

More insight into the form of the Newton velocity field, Eq. (2), and in particular, into the question why the *inverse* of the logarithmic derivative of f rather than the logarithmic derivative itself appears, springs from the direct integration of the trajectory. Indeed, when we substitute Eq. (2) into Eq. (1), and multiply by f' we arrive at the equation

$$f'(s(t))\dot{s}(t) = -f(s(t)) \tag{6}$$

or

$$\frac{d}{dt} f(s(t)) = -f(s(t)), \tag{7}$$

leading us to the implicit expression

$$f(s(t; s_0)) = f(s_0)e^{-t} \tag{8}$$

for the solution of Eq. (7) where s_0 denotes the initial point at $t = 0$. Here we have included s_0 explicitly in the argument of the Newton trajectory $s = s(t; s_0)$ as to emphasize that the flow line depends crucially on its starting point.

This derivation brings out most clearly that it is the fact that f' is in the denominator of the expression for V that creates the complete derivative in time on the left-hand side of Eq. (7), which in turn allows us to immediately integrate it.

However, the deeper reason for the form of V is of course the analogy to the Newton method of finding the zero of a function which, in its most elementary version for a real-valued function $f = f(x)$, is based on the iteration

$$x_{n+1} \equiv x_n - \frac{f(x_n)}{f'(x_n)} \tag{9}$$

with an educated guess for the starting point x_0. Obviously replacing the difference $x_{n+1} - x_n$ by a differential leads us to the continuous Newton method of Eq. (1) with the Newton velocity field given by Eq. (2).

In complete accordance with the discrete Newton method the continuous one also yields a zero of the function f. Indeed, Eq. (8) shows that for $t \to \infty$ the trajectory approaches a zero of f, that is

$$\lim_{t \to \infty} f(s(t; s_0)) = 0. \tag{10}$$

Hence, the continuous Newton method is an effective numerical method to obtain zeros of a complex-valued function.

2.2 Lines of Constant Height and Constant Phase from Newton Flows

We are now in the position to define Newton flows which provide us with the lines of constant phase and constant height of a function f. The trajectories

$$\dot{s}_j(t) \equiv e^{-i(1+j)\pi/4} V(s_j(t)), \tag{11}$$

with the Newton velocity field given by Eq. (2), yield for the subscript $j = +1 \equiv h$, or $j = -1 \equiv p$ the lines of constant height, or constant phase of f, respectively.

Indeed, by direct differentiation we can verify that the solution of Eq. (11) reads

$$f(s_j(t; s_0)) = f(s_0) \exp\left[-e^{-i(1+j)\pi/4}t\right] = |f(s_0)| \exp\left[i\theta(s_0) - e^{-i(1+j)\pi/4}t\right] \tag{12}$$

where s_0 is the initial point of the trajectory $s_j = s_j(t; s_0)$ in the complex plane.

In the last step we have decomposed

$$f(s) \equiv |f(s)| e^{i\theta(s)} \tag{13}$$

into its absolute value $|f|$ and a phase θ.

2.2.1 Constant Phase
For $j = -1 = p$ we obtain from Eq. (12) the formula

$$f(s_p(t; s_0)) = |f(s_0)| e^{i\theta(s_0)} e^{-t}, \tag{14}$$

which indicates that along the path $s_p = s_p(t; s_0)$ the initial phase $\theta(s_0) \equiv \theta_0$ of f at the point s_0 of the complex plane is preserved. Hence, $s_p = s_p(t; s_0)$ is the line of constant phase corresponding to θ_0.

It is remarkable that the initial condition for the flow line determines with the help of the value $f(s_0)$ of f at the starting point s_0 only the phase value of the phase line. The absolute value $|f(s_0)|$ does not influence the line of constant phase at all.

For this reason we can simplify the notation and skip s_0 in the argument of s_p but replace the subscript p by the phase θ_0 of f at the starting point, that is

$$s_{\theta_0} = s_{\theta_0}(t). \tag{15}$$

Throughout our notes we shall follow this rule.

Moreover, we note from Eq. (14) that for increasing positive t the absolute value of $f(s_{\theta_0}(t))$ decreases exponentially, that is

$$|f(s_{\theta_0}(t))| = |f(s_0)|e^{-t}. \tag{16}$$

Hence, along a line of constant phase of f its absolute value $|f|$ decreases monotonically.

In particular, for $t \to \infty$ the right-hand side of Eq. (16) tends towards zero. Hence, the corresponding trajectory $s_{\theta_0} = s_{\theta_0}(t)$ has to approach a zero of f. Since we have not specified the initial value s_0 this property has to hold true for *every* line of constant phase. Indeed, all phase lines of f have to terminate in a zero of f.

In Fig. 3 we show two distinct representations of the same line of constant phase. In Fig. 3a we depict the curve $s_{\theta_0} = s_{\theta_0}(t)$ in the complex plane spanned by the real and imaginary parts of the argument s of f, and depicted by the Newton flow of Eq. (11) for $j = -1$. Obviously it starts at $t = t_0 \equiv 0$ at a point s_0 and terminates at a zero depicted here by the red dot. As the positive parameter t increases the absolute value $|f|$ decays exponentially. In general, the phase line $s_{\theta_0} = s_{\theta_0}(t)$ is not a straight curve.

In Fig. 3b we display the same behavior but now in the complex plane of the real and imaginary parts of f rather than those of the argument s. In this representation the line of constant phase θ_0 is obviously a straight line initiating at $f(s_0)$ and ending in the origin of the complex plane under the angle θ_0 with respect to the axis of Re f. The distance from the origin is $|f|$. Obviously, the origin is a zero of f. The ticks in the two pictures mark identical values of t.

Hence, as we traverse the curve $s_\theta = s_\theta(t)$ in the complex plane of s we follow the straight line in the complex plane of f given by Eq. (14). In particular, there exists a one-to-one relationship between $s_{\theta_0} = s_{\theta_0}(t)$ and $f = f(s_\theta(t))$.

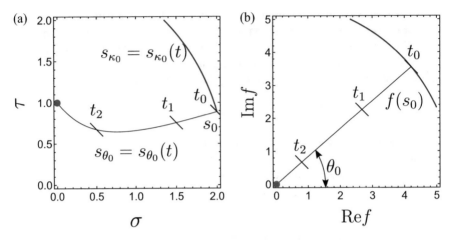

Fig. 3 Two distinct representations of a line of constant phase (thin black curves) or a line of constant height (fat purple curve): (**a**) as curved Newton flow lines $s_{\theta_0} = s_{\theta_0}(t)$ or $s_{\kappa_0} = s_{\kappa_0}(t)$ in the complex plane spanned by the argument $s \equiv \sigma + i\tau$ of the function f and parametrized by t, and (**b**) as a straight line or the segment of a circle in the complex plane of $\mathrm{Re}f$ and $\mathrm{Im}f$. The curve $s_{\theta_0} = s_{\theta_0}(t)$ determined by the phase θ_0 of f at the starting point s_0 and the condition $f(s_{\theta_0}(t)) = |f(s_0)| \exp(i\theta_0) \exp(-t)$ terminates at a zero of f indicated by the red dot at the end of the trajectory. Along the Newton trajectory $s_{\theta_0} = s_{\theta_0}(t)$ we mark three values $0 \equiv t_0 < t_1 < t_2$ of t by ticks where the initial value $|f(s_0)|$ corresponding to $t_0 = 0$ has decayed to its exponential parts $\exp(-t_1)$ and $\exp(-t_2)$ expressed in (**b**) as three different distances along the ray from the origin. The contour line $s_{\kappa_0} = s_{\kappa_0}(t)$ governed by the absolute value $|f(s_0)| \equiv \exp(\kappa_0)$ of f at the starting point s_0 and the condition $|f(s_{\kappa_0}(t))| = |f(s_0)| \equiv \exp(\kappa_0)$ initiates orthogonally to the phase line. This property is transferred to the complex plane of f where the contour line corresponding to (**a**) is a segment of a circle

2.2.2 Constant Height

Next, we turn to $j = +1 = h$ and obtain from Eq. (12) the expression

$$f(s_h(t; s_0)) = |f(s_0)| \exp\left[i(\theta(s_0) + t)\right],\tag{17}$$

that is

$$|f(s_h(t; s_0))| = |f(s_0)|.\tag{18}$$

Hence, the trajectory $s_h = s_h(t; s_0)$ in the complex plane is constructed in such a way as to keep the absolute value $|f|$ constant along the path parametrized by t. The constant is determined by $|f(s_0)| \equiv \exp(\kappa(s_0))$, that is by the absolute value of f at the initial point s_0. Thus, the curve $s_h = s_h(t; s_0)$ is a line of constant height, that is, a contour line of f.

Again, it is remarkable that the initial position s_0 enters into Eq. (18) only as the absolute value $|f(s_0)|$ of f at s_0, that is in the exponential representation of f, given by Eq. (13), by $\kappa_0 \equiv \kappa(s_0)$. For this reason it is reasonable to drop s_0 in the argument of the Newton trajectory s_h and replace the subscript h by κ_0, leading us to the concise notation

$$s_{\kappa_0} = s_{\kappa_0}(t) \qquad (19)$$

for the line of constant height.

In Fig. 3 we also show two distinct representations of the same line of constant height starting again at s_0. Whereas the contour lines in the complex plane of f, depicted in Fig. 3b are always circles, in the complex plane of the argument s shown in Fig. 3a they are orthogonal to the phase lines but can take a more complicated form. In Fig. 3a the Newton trajectory lives in the complex plane of the argument s and starts off being orthogonal on the phase line, the contour lines in the complex plane of f are circles.

Moreover, according to Eq. (17) in both cases the phase along $s_{\kappa_0} = s_{\kappa_0}(t)$ increases linearly with t. This feature has important consequences.

Indeed, due to the familiar periodicity relation

$$e^{i(\theta + 2\pi)} = e^{i\theta}, \qquad (20)$$

lines of constant phase corresponding to θ and $\theta + 2\pi$ are equivalent. Moreover, along a phase line, $|f|$ covers the complete positive real axis and zero. As a result, along two equivalent phase lines, $|f|$ assumes identical values, but at two different points in the complex plane of s. For this reason, f cannot be inverted. We shall return to the implications of this observation later in this section.

2.3   Sources and Sinks of Phase Lines

Every phase line terminates in a zero. This property immediately begs the question as to its origin, or source. Since $|f|$ decays in a monotonic way as we travel along a line of constant phase it must originate at a place where $|f|$ is infinite, for example at a pole, or at the edge of the complex plane.

Indeed, Eq. (16) supports this point of view since it leads us in the limit of $t \to -\infty$ to the relation

$$\lim_{t \to -\infty} |f(s_{\theta_0}(t))| = \lim_{t \to \infty} |f(s_0)| e^{|t|} = \infty. \qquad (21)$$

In order to illustrate the sources of phase lines in more detail, we now consider three elementary complex functions: (i) a simple pole, (ii) a simple zero, and (ii) the exponential function. We find the corresponding Newton velocity field and the associated phase lines.

2.3.1   Poles and Zeros

We start our discussion with the Newton velocity field $V^{(p)}$ of a simple pole located at the origin, and represented by the function

$$f^{(p)}(s) \equiv \frac{1}{s}. \qquad (22)$$

Here and throughout this section we include a superscript as to distinguish the different functions and the associated Newton velocity fields.

From the definition, Eq. (2), of V, we find

$$V^{(p)} = +s, \tag{23}$$

which up to the sign, is identical to the Newton velocity field

$$V^{(z)} = -s \tag{24}$$

of a simple zero

$$f^{(z)}(s) \equiv s \tag{25}$$

at the origin.

The difference in the signs in Eqs. (23) and (24) determines the direction in which the Newton flow lines are traversed. Indeed, for the simple zero, Eq. (25), we find from Eq. (11) the phase lines

$$s_{\theta_0}^{(z)}(t) = s_0 e^{-t} \equiv |s_0| e^{i\theta_0} e^{-t} \tag{26}$$

corresponding to rays in the complex plane aligned under the angle θ_0 with respect to the real axis, and terminating at the location of the zero, which in the present elementary example, is at the origin.

In contrast, for the pole, Eq. (22), the opposite sign in Eq. (23) leads us to the phase lines

$$s_{\theta_0}^{(p)}(t) = |s_0| e^{i\theta_0} e^{t}, \tag{27}$$

which for increasing positive values of t are traversed away from the origin, that is from the pole.

Hence, poles are sources and zeros are sinks of lines of constant phase. In this sense the phase lines are reminiscent of the field lines of electro- or magnetostatics [15] with one major distinction: They are not necessarily curl-free [12].

It is also interesting to analyze how many phase lines terminate in a zero, or emerge from a pole. For the example of a simple zero, θ_0 assumes the values $0 \leq \theta_0 < 2\pi$ which shows that phase lines of an interval of 2π must *end* there.

Likewise, at a simple pole phase lines from a 2π-interval *emerge*. Needless to say, for a double pole or a double zero a phase interval of 4π determines the lines that initiate or end there, respectively.

This restriction of the phase interval to an integer multiple of 2π has also an important consequence for the contour lines. Since, according to Eq. (17) the phase increases along a line of constant height, the contour lines are either closed surrounding a zero, or open traversing the complex plane.

2.3.2   No Poles and No Zeros

An intriguing problem arises for a complex-valued function which is free of poles and zeros, and therefore free of sources and sinks. Obviously such a function must also have lines of constant phase, but where do they start and where do they end?

Insight into this question arises from the exponential function

$$f^{(e)}(s) \equiv e^s \equiv e^{\sigma + i\tau} \tag{28}$$

which is free of poles and zeros. The lines of constant phase are obviously straight lines in the complex plane parallel to the real axis.

Indeed, at the right edge of the complex plane, that is for $\sigma \to +\infty$, the absolute value $|f^{(e)}|$ grows exponentially, whereas for $\sigma \to -\infty$, that is at the left edge, $|f^{(e)}|$ tends to zero. Hence, for $f^{(e)}$ phase lines run from the right to the left at a fixed value of τ.

This property which is evident from the definition, Eq. (28), of the exponential function also follows immediately from the corresponding Newton velocity field

$$V^{(e)} = -1, \tag{29}$$

and the associated phase line

$$s_{\theta_0}^{(e)}(t) = s_0 - t = i\tau_0 + (\sigma_0 - t), \tag{30}$$

or

$$s_{\theta_0}(t) = i\theta_0 + (\sigma_0 - t), \tag{31}$$

where we have made use of

$$f^{(e)}(s_0) = e^{s_0} = e^{\sigma_0} e^{i\tau_0} = \left| f^{(e)}(s_0) \right| e^{i\theta_0}. \tag{32}$$

Indeed, for increasing positive t the initial real part σ_0 of $s_{\theta_0}^{(e)}(t)$ decreases continuously but the imaginary part given by the value $\theta_0 = \tau_0$ of the initial phase of $f^{(e)}$ remains constant.

This elementary example brings out most clearly that the edges of the complex plane can serve as sources and sinks of lines of constant phase. However, which edge is a source and which one is a sink depends on the specific form of f.

2.4   Zero of the Derivative

According to the definition, Eq. (2), of the Newton velocity field V associated with the function f we find f to be in the numerator. Hence, at a simple zero of f, the velocity V vanishes. However, at a zero of the first derivative f' of f, the velocity field has a simple pole, since f' is in the denominator. Here V is infinite.

In the present section we concentrate on the behavior of the lines of constant phase in the neighborhood of a zero of the first derivative. We first show that due to the infinite velocity at this point the parameter T connecting $s_{\theta_0}(t)$ and $s_{\theta_0}(t+T)$ on a line of constant phase θ_0 is finite. However, when one of them is a zero of f it is logarithmically divergent. We then address the fact that zeros of f' prevent us from inverting f which leads us to a very special class of phase lines, that is separatrices. They mark the borders of the domains of the complex plane in which f^{-1} exists.

2.4.1 Parameter Connecting Two Points on a Phase Line

We start our discussion of the influence of a zero of f' by considering two points $s_{\theta_0}(t)$ and $s_{\theta_0}(t+T)$ along a line of constant phase θ_0 at the two different parameter values t and $t+T$. From Eq. (16) we find immediately the ratio

$$\frac{|f(s_{\theta_0}(t+T))|}{|f(s_{\theta_0}(t))|} = e^{-T}, \tag{33}$$

which yields the expression

$$T = \ln \frac{|f(s_{\theta_0}(t))|}{|f(s_{\theta_0}(t+T))|}, \tag{34}$$

for the difference T of the two parameters governed by the logarithm of the ratio of the absolute values of f at the two points of interest.

We emphasize that f' does not enter into the expression, Eq. (34), for T. Hence, even when at one of the two points f' vanishes, T remains finite and does not even feel the singularity of V.

In contrast, T diverges logarithmically as we approach a zero of f. Hence, the convergence of the continuous Newton method to identify a zero of a function is rather slow when we get close to it.

2.4.2 Violation of Non-crossing Rule

In general, lines of constant phase cannot cross. Indeed, a crossing of two lines corresponding to two distinct phases would indicate that at this point f assumes *two* distinct phases. However, two lines corresponding to equivalent phases defined by the periodicity relation, Eq. (20), can cross.

Indeed, such a situation occurs for example at a simple zero of the first derivative f' of f. Here we find *two* incoming and *two* outgoing phase lines which are orthogonal on each other, as discussed in Fig. 4. In this section we analyze the behavior of the lines of constant phase in the neighborhood of such a zero. Without any loss of generality, we consider the case when the zero is located at the origin of the complex plane, that is, at $s = 0$.

For this purpose, we represent f by the Taylor series

$$f(s) = f(0) + \frac{1}{2}f''(0)s^2 + \dots \tag{35}$$

(a)

(b)

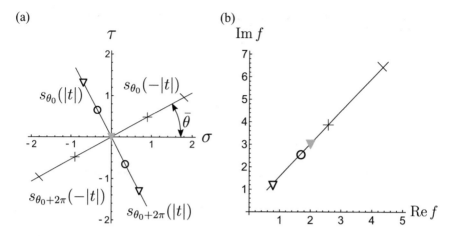

Fig. 4 Violation of the non-crossing rule and the associated problem of the existence of the inverse f^{-1} of f. In (a) we illustrate this phenomenon by the two lines $s_{\theta_0} = s_{\theta_0}(t)$ and $s_{\theta_0+2\pi} = s_{\theta_0+2\pi}(t)$ of equivalent phases θ_0 and $\theta_0 + 2\pi$ entering and leaving a simple zero of the first derivative f' of f which for the sake of simplicity is located at the origin and indicated by a filled green triangle in the complex plane of the argument s. For decreasing *negative* values of t represented by the ticks and crosses we enter the origin but for increasing *positive* values exemplified by the open circles and triangles we leave it again terminating eventually in zeros of f. The values of $|f|$ are identical at identical symbols. Since the phases are equivalent we find the identity $f(s_{\theta_0}(t)) = f(s_{\theta_0+2\pi}(t))$ valid for *all* t as indicated in (b) by a *single* straight line in the complex plane of f. No crossing of phase lines at the zero of f' marked again by the filled triangle occurs in this representation. Obviously, when f' has a zero there does not exist a one-by-one relationship between the phase lines in the s-plane and the f-plane. Indeed, we can go uniquely from s- to f-space, but the inverse direction is not unique since for every t we always find *two* values of s. Hence, f^{-1} cannot exist in the complete complex plane but only in the domains caught between the incoming lines $s_{\theta_0} = s_{\theta_0}(t)$ and $s_{\theta_0+2\pi} = s_{\theta_0+2\pi}(t)$

where we have used the fact that $f'(0) = 0$ and have assumed that $f(0)$ is non-vanishing.

We then follow this function along the line of constant phase $s_{\theta_0} = s_{\theta_0}(t)$, determined by the phase θ_0 of

$$f(0) \equiv |f(0)|e^{i\theta_0} \qquad (36)$$

at the origin.

Since we are interested in the behavior in the neighborhood of this point we choose it as the initial condition, that is $s_0 = 0$, and obtain from Eq. (14) the expression

$$f(s_{\theta_0}(t)) = |f(0)|e^{i\theta_0}e^{-t} \cong |f(0)|e^{i\theta_0}(1-t), \qquad (37)$$

where t is negative before we arrive at the origin, but positive when we leave.

When we substitute the explicit form of f given by Eq. (35), into the left-hand side of Eq. (37) we arrive at the condition

$$f(0) + \frac{1}{2}f''(0)s_{\theta_0}^2(t) = f(0) - |f(0)|e^{i\theta_0}t, \tag{38}$$

or

$$\frac{1}{2}f''(0)s_{\theta_0}^2(t) = -|f(0)|e^{i\theta_0}t, \tag{39}$$

which after solving the quadratic equation yields the two approximate lines

$$s_{\theta_0}(-|t|) \cong \sqrt{\epsilon}e^{i\bar{\theta}}\sqrt{|t|} \tag{40}$$

and

$$s_{\theta_0+2\pi}(-|t|) \cong \sqrt{\epsilon}e^{i(\bar{\theta}+\pi)}\sqrt{|t|} \tag{41}$$

of constant phase corresponding to the phases θ_0 and $\theta_0 + 2\pi$ of f.

Here we have introduced the abbreviations

$$\epsilon \equiv 2\left|\frac{f(0)}{f''(0)}\right| \tag{42}$$

and

$$\bar{\theta} \equiv \frac{1}{2}(\theta_0 - \theta_2) \tag{43}$$

following from the decomposition of $f(0)$ into its absolute value and its phase given by Eq. (36), and $f''(0) \equiv |f''(0)| \exp(i\theta_2)$.

Hence, for decreasing negative values of t we approach the origin of the complex plane of s, that is the zero of f' under the angles $\bar{\theta}$ and $\bar{\theta} + \pi$ as shown in Fig. 4. Here $\bar{\theta}$ is given by half the difference of the phases θ_0 and θ_2 determined by f and f'' at the zero of f'.

These are the lines of constant phase of f corresponding to θ_0 and $\theta_0 + 2\pi$. Indeed, when we substitute the expressions, Eqs. (40) and (41), for s_{θ_0} and $s_{\theta_0+2\pi}$ into the definition, Eq. (35), of f and use the abbreviations, Eqs. (42) and (43), we immediately arrive at the two identities

$$f(s_{\theta_0}(-|t|)) = |f(0)|e^{i\theta_0}(1 + |t|). \tag{44}$$

and

$$f(s_{\theta_0+2\pi}(-|t|)) = |f(0)|e^{(i\theta_0+2\pi)}(1 + |t|). \tag{45}$$

With the help of the periodicity relation, Eq. (20), we find the identity

$$f(s_{\theta_0}(-|t|)) = f(s_{\theta_0+2\pi}(-|t|)), \tag{46}$$

which implies that for every negative t the values of f along the two incoming phase lines are identical.

For positive values of t, that is after leaving the zero of f', the lines of constant phase read

$$s_{\theta_0}(|t|) \cong \sqrt{\epsilon}e^{i(\bar{\theta}+\pi/2)}\sqrt{|t|}, \tag{47}$$

and

$$s_{\theta_0+2\pi}(|t|) \cong \sqrt{\epsilon}e^{i(\bar{\theta}+3\pi/2)}\sqrt{|t|}, \tag{48}$$

and due to the well-known relation

$$\sqrt{-1} \equiv e^{i\pi/2}, \tag{49}$$

the outgoing lines are orthogonal on the incoming ones as depicted in Fig. 4.

Again we have a pair of lines which differ in their phases by 2π, that is

$$f(s_{\theta_0}(|t|)) = |f(0)|e^{i\theta_0}(1 - |t|) \tag{50}$$

and

$$f(s_{\theta_0+2\pi}(|t|)) = |f(0)|e^{i(\theta_0+2\pi)}(1 - |t|). \tag{51}$$

Again the periodicity property, Eq. (20), yields the identity

$$f(s_{\theta_0}(|t|)) = f(s_{\theta_0+2\pi}(|t|)). \tag{52}$$

When we combine this relation with Eq. (46) we arrive at

$$f(s_{\theta_0}(t)) = f(s_{\theta_0+2\pi}(t)), \tag{53}$$

which demonstrates that at the two points $s_{\theta_0}(t)$ and $s_{\theta_0+2\pi}(t)$ on the two lines of constant phase θ_0 and $\theta_0 + 2\pi$ the values of f are identical. Moreover, this identity holds true for every value of t.

As a consequence, the representation of the two phase lines in the complex plane of f shown in Fig. 4b is dramatically different. Instead of *two* lines entering the point where f' vanishes depicted by a filled green triangle, and *two* leaving, we find a *single* straight line inclined under the angle θ_0 with respect to the axis of Re f.

The ticks and crosses represent values of t identical to the ones in Fig. 4a before entering the filled greed triangle. Therefore, they correspond to values of $|f|$ larger than the ones associated with the open circles and triangles indicating values of t after having crossed the filled green triangle. Whereas in the complex plane of the argument s of f two phase lines annihilate and two new ones emerge, no such phenomenon takes place in the complex plane of f. Indeed, the trajectory passes through the green triangle without noticing the vanishing first derivative of f.

We conclude by briefly addressing the behavior of $|f|$ along the lines of constant phase traversing the zero of f'. According to Eq. (37) f decreases linearly with t. Moreover, due to the orthogonality of the two incoming and the two outgoing phase lines a non-differentiable saddle point emerges when we consider $|f|$ as a function of t.

2.4.3   Separatrices and the Inverse of the Function

The expression, Eq. (14) for the lines of constant phase is implicit, that is we need to invert the function f. Obviously this operation is only possible in domains of the complex plane where the inverse f^{-1} of f exists.

A function with more than one zero explains vividly the associated complication. Indeed, in this case we have more than one location in the complex plane where f assumes identical values. As a result, f^{-1} only exists in the domains associated with these zeros and special lines of constant phase separate the Newton flows approaching these zeros. These lines are called separatrices, but how to identify them in a sea of phase lines?

The answer to this question is contained in the discussion of the preceding section where we have analyzed the phase lines in the *neighborhood* of a simple zero of the first derivative f' of f. We have shown that in this case two phase lines enter this point and two leave again. The ones entering and leaving are always orthogonal on each other. Moreover, the phases corresponding to the entering lines, differ from each other by 2π with one being given by the phase θ_0 of f at the location of the zero of f'. The lines leaving this point display an identical behavior.

However, in this situation we do not only deal with equivalent phase lines but also for every t the corresponding points $s_{\theta_0}(t)$ and $s_{\theta_0+2\pi}(t)$ lead to identical values of $|f|$ as expressed by the identity, Eq. (53). As a result, for every point on these special phase lines we have a single value of f but for every value of f we have *two* different arguments. Hence, the two phase lines entering and leaving a zero of the first derivative play an important role in establishing the inverse f^{-1} of f.

On first sight these considerations, and especially the identity, Eq. (53), seem to be correct only in the neighborhood of the zero of f' since we have explicitly used the trajectories, Eqs. (40), (41), (47) and (48). However, Eq. (53) is valid along the complete trajectories.

This property follows directly from the definition, Eq. (14), of the phase line in the form

$$f(s_{\theta_0}(t)) = |f(s_0)|e^{i\theta_0}e^{-t} \tag{54}$$

when we take as the initial position s_0 the point of the complex plane where f' vanishes. In this case we need to consider increasingly negative values of t as to move away from it. Since here *two* trajectories enter whose behavior in the proximity of this point we have discussed before, these lines can now be represented in the complete complex plane by Eq. (54) and

$$f(s_{\theta_0+2\pi}(t)) = |f(s_0)|e^{i(\theta_0+2\pi)}e^{-t} \tag{55}$$

leading us to the identity

$$f(s_{\theta_0}(t)) = f(s_{\theta_0+2\pi}(t)) \tag{56}$$

valid for all t.

These two phase lines divide the complex plane into two domains each with a phase interval of 2π approaching the zeros located at the extensions of the lines emerging from the zero of f'. Hence, the separatrices are the phase lines which enter the zero of the first derivative of f.

3   Application of Cauchy-Riemann Differential Equations

So far our understanding of a complex function f has been guided by rather elementary properties of its lines of constant phase: (i) Their emergence from sources such as poles or the edges of the complex plane, (ii) their termination in zeros, and (iii) their orthogonality on contour lines. These features allow us to immediately obtain a *qualitative* picture of f.

However, to gain more insight into the distribution of phase and contour lines we now employ the Cauchy-Riemann differential equations

$$\frac{\partial \kappa}{\partial \sigma} = \frac{\partial \theta}{\partial \tau} \tag{57}$$

and

$$\frac{\partial \kappa}{\partial \tau} = -\frac{\partial \theta}{\partial \sigma} \tag{58}$$

in amplitude and phase of the function

$$f \equiv e^{\kappa+i\theta}, \tag{59}$$

discussed and derived in the appendix.

Indeed, we now illustrate their actions using the elementary function

$$h(s) \equiv s^2 + 1 = (s - i)(s + i) = \sigma^2 - \tau^2 + 1 + i2\sigma\tau \tag{60}$$

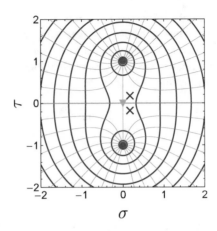

Fig. 5 Lines of constant phase (thin black curves) and lines of constant height (fat purple curves) of the elementary function $h(s) = s^2 + 1$, Eq. (60), with simple zeros at $\pm i$ (red dots) and a simple zero (green full triangle) of h' at $s = 0$. The real axis is a separatrix with phase $\theta = 0$ which is maintained along the imaginary axis up to the zeros. On the opposite side of the zeros the phase of h on the imaginary axis is π. Along the real as well as the imaginary axes the contour lines display a parabolic behavior which leads in the neighborhood of the σ-axis for $|\sigma| > 1$ to a focusing, and for $0 < |\sigma| < 1$ to a defocusing of the phase lines. Along the imaginary axis the contour lines of the neighborhood are always focusing. Around $\sigma = 1$ and close to the real axis the phase lines are parallel. In Fig. 6 we have a closer look at the behavior of $|h|$ close to the two points in the complex plane marked by the blue crosses

which provides us with yet another insight into the intimate interplay between the lines of constant phase and constant height shown for the example of h in Fig. 5.

We start by first briefly discussing the key features of h and the resulting distribution of lines of constant phase. In this analysis we mainly employ the properties derived in the previous section using the Newton flow. However, in the next step we show how the Cauchy-Riemann differential equations in amplitude and phase determine the slopes of the phase lines, and illustrate this mechanism using h. We conclude by turning to the lines of constant height and demonstrate that their slopes are also determined by the rate of changes of the amplitude.

3.1 Elementary Example

According to the definition, Eq. (60), of h, our function of interest is free of a pole. Hence, all phase lines have to start in infinity, that is at the edges of the complex plane, and terminate at the two zeros of h at $s = \pm i$ as shown in Fig. 5. In this example phase lines initiate from all sides.

At the crossing of the real and the imaginary axis, that is, at the origin of the complex plane, we have a simple zero of h', since

$$h'(s) = 2s. \tag{61}$$

Here *two* phase lines cross each other at right angles.

Along the real axis h is real and free of any zeros. Hence, the choice of the phase $\theta = 0$ or an integer of 2π is natural. We emphasize that they continue on the imaginary axis until we end up in the zeros.

Indeed, from the explicit form

$$h(\sigma = 0, \tau) = 1 - \tau^2 \tag{62}$$

following from Eq. (60) we note the property $h > 0$ for $0 < |\tau| < 1$. However, for $\tau > 1$ we find $h < 0$, corresponding to a phase change from $-\pi$ to $0 \cong 2\pi$ as we traverse the simple zeros at $\tau = \pm 1$ towards the origin of the complex plane.

The real axis is the separatrix of the flow of phase lines of h. Indeed, the flow lines of the upper half of the complex plane move to $s_0 = +i$, whereas the ones in the lower one terminate in $s_0 = -i$. In both cases, the right and the left halves of the respective half-planes contribute a total phase interval of π to supply a total interval of 2π, necessary for a simple zero.

3.2   Changes in Amplitude Determine Slopes of Phase Lines

After this first qualitative discussion of h we now employ the Cauchy-Riemann differential equations to develop a deeper understanding. Although our considerations are valid for any appropriately differentiable function f, the corresponding figures rely on h.

In Fig. 6 we show excerpts of the complex plane in which $|f|$, and by virtue of the exponential representation, Eq. (59), of f the function $\kappa = \kappa(\sigma, \tau)$ increases or decreases along the σ- or the τ-variable. Here we concentrate on regions small enough as to approximate $|f|$ by an inclined plane.

Obviously, the lines $\tau_\theta = \tau_\theta(\sigma)$ of constant phase have to run down the plane, but in which way? The Cauchy-Riemann differential equations, Eqs. (57) and (58) in amplitude and phase contain the answer to this question.

In order to bring out their crucial role most clearly, we consider in Fig. 6 two distinct cases for the behavior of f: (a) The amplitude κ of $|f|$ increases for increasing σ but decreases for increasing τ, and (b) κ increases for increasing σ and τ. Brighter colors indicate larger values of $|f|$ than darker colors. Here we use the function h given by Eq. (60) in the neighborhood of the two points in the complex plane marked in Fig. 5 by the blue crosses. The increase of κ with σ shown in Fig. 6a enforces by Eq. (57) an *increase* of the phase θ in the orthogonal direction, that is with τ, indicated by the three phase lines $\tau_\theta = \tau_\theta(\sigma)$ with $\theta_1 < \theta_2 < \theta_3$. In the considered domain of the complex plane, phase lines are approximately straight and parallel.

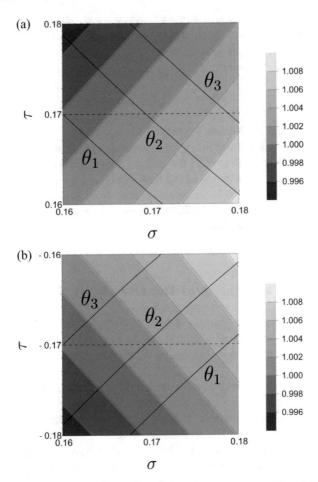

Fig. 6 Application of the Cauchy-Riemann differential equations (Eqs. (57) and (58)), in amplitude and phase in a region where $|h|$ is approximately an inclined plane. In (**b**) Eq. (57) predicts an increase in θ for increasing τ but fixed σ. However, Eq. (58) suggests that for increasing σ but fixed τ the phase *decreases*, which is apparent when we follow the dashed line from left to right. Hence, the phase lines have a positive slope, in order to be specific we have used here the function h given by Eq. (59) in the neighborhood of the two points marked in Fig. 5 by blue crosses. The partial derivatives of κ with respect to σ and τ are (**a**) positive and negative, respectively, or (**b**) both positive. In (**a**) we show the case where κ increases with σ but decreases with τ as depicted by the three phase lines corresponding to $\theta_1 < \theta_2 < \theta_3$. Here, Eq. (57) yields an increase for θ with τ but the decrease of κ with τ compensates the minus sign in Eq. (58) and θ *increases* with σ as we follow the dashed line. Consequently, the phase lines have a negative slope

Moreover, an increase of κ with the τ-variable leads, because of the minus sign in Eq. (58), to a *decrease* of the phase for increasing σ. Indeed, it is this condition which determines the slope of the phase lines to be positive, that is

$$\frac{\partial \tau_\theta}{\partial \sigma} > 0. \tag{63}$$

This behavior is apparent when we follow in Fig. 6b the dashed curve where we first cross the phase line corresponding to θ_3, then θ_2 and finally θ_1.

Next, we return to the case represented in Fig. 6a where κ decreases with τ but increases with σ. Again, Eq. (57) implies that θ increases with τ, but now Eq. (58) predicts that θ also increases with σ as indicated by the dashed line. As a result, in this domain of the complex plane the phase lines $\tau_\theta = \tau_\theta(\sigma)$ have a *negative* slope, that is

$$\frac{\partial \tau_\theta}{\partial \sigma} < 0. \tag{64}$$

The combination of the two Cauchy-Riemann differential equations, Eqs. (57) and (58), determines the slopes of the phase lines. This is a one-sentence summary of this discussion.

3.3   Expression for the Slope of a Phase Line

So far we have analyzed the slopes of phase lines by applying the *two* Cauchy-Riemann differential equations, Eq. (57) and (58). In the appendix we derive from them the relation

$$\frac{\partial \tau_\theta}{\partial \sigma} = \left(\frac{\partial \kappa}{\partial \sigma}\right)^{-1} \frac{\partial \kappa}{\partial \tau} \tag{65}$$

for the slope of the phase line $\tau_\theta \equiv \tau_\theta(\sigma)$ in terms of the rate of change of the amplitude κ as a function of the two directions in the complex plane. Obviously, when both derivatives of κ are positive, the slope of τ_θ is also positive, in complete agreement with the discussion of Fig. 6b. Moreover, when the derivatives of κ with respect to σ and τ are positive and negative, respectively, the slope of τ_θ is negative, in complete agreement with Fig. 6a. It does not matter which of the two derivatives of κ are positive or negative. Moreover, Eq. (65) shows that the slope of τ_θ is also positive when *both* derivatives of κ are negative.

An interesting situation occurs when the derivative of κ with respect to τ vanishes. In this case, Eq. (65) predicts that also the slope of τ_θ vanishes, that is the line of constant phase is parallel to the real axis. Likewise, when the derivative of κ with respect to σ vanishes the slope of τ_θ is infinite, that is the line of constant phase is vertical, that is parallel to the imaginary axis.

3.4   Parabolic Contour Lines

Another instructive case appears when the phase line at a point s_0 in the complex plane is parallel or even identical to the real axis, that is the right-hand side of Eq. (58) vanishes. As a result, also the derivative of κ with respect to τ vanishes. Indeed, this situation occurs for h along the real axis which is a line of constant phase with $\theta = 0$.

The representation

$$|h(\sigma, \tau)|^2 = \tau^4 + 2(\sigma^2 - 1)\tau^2 + (\sigma^2 + 1)^2 \tag{66}$$

of $|h|$ is symmetric in τ. For large values of τ the function $|h|^2$ grows with τ^4. However, for small values of τ the behavior depends crucially on the value of σ. Indeed, for $|\sigma| > 1$ the second term in Eq. (66) is always positive, and we have a quadratic increase with τ in addition to the quartic one. For $|\sigma| < 1$ this term is negative and dominates, at least for small values of τ, over τ^4, leading to a quadratic *decrease* of $|h|^2$ from the initial value $(\sigma^2 + 1)^2$. Needless to say, for large values of τ the positive quartic term takes over again. As a result, we find in the case of $|\sigma| < 1$ *two* local minima of $|h|^2$ as a function of τ.

Hence, for decreasing values of $|\sigma| > 1$ the function $|h|^2$ in its dependence on τ enjoys a local *minimum* on the real axis which turns into a local *maximum* for $|\sigma| < 1$. The point of transition is at $\sigma = 1$. Here we have chosen the word "transition" on purpose. Indeed, this behavior of $|h|^2$ along the real axis is analogous to the potential of a phase transition of second order.

This symmetric growth or decay of $|f|$ in the direction orthogonal to the phase line as exemplified by $|h|^2$ in Fig. 7 has immediate implications on the lines of constant height, that is contour lines. As mentioned already several times they are orthogonal on the phase lines. However, in this context only the *tangent* of the contour line is orthogonal on the phase line, and is of a parabolic form as shown in Fig. 5.

This feature becomes apparent when we recall that along the phase line $|f|$ decays. In the orthogonal direction $|f|$ grows symmetrically. Hence, for the case of growth the contour line curves to the right whereas for the decay it curves to the left as shown in Figs. 5 and 7.

We note that along the imaginary axis which for $\tau > 1$ is a line of constant phase π, the contour lines also display a parabolic character which is obvious when we cast the representation, Eq. (66), of $|h|^2$ into the form

$$|h(\sigma, \tau)|^2 = \sigma^4 + 2(\tau^2 + 1)\sigma^2 + (\tau^2 - 1)^2, \tag{67}$$

symmetric in σ. This expression also brings out the fact that in contrast to the real axis *no* bifurcation takes place along the imaginary axis since all terms are positive.

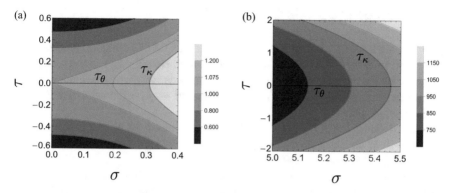

Fig. 7 Behavior of the contour lines crossing a phase line along the real axis where $|f|$ decays (**a**) or grows (**b**) symmetrically as $|\tau|$ increases. Since the real axis is a phase line, the absolute value $|f|$ decays for decreasing values of σ. For small values of σ, the absolute value $|f|$ *decreases* (**a**) symmetrically with increasing $|\tau|$, but *increases* (**b**) for large σ, as indicated by darker and brighter colors. Hence, the contour lines display a parabolic behavior in the neighborhood of the phase line. Whereas in (**a**) they curve to the right, in (**b**) they curve to the left. To be specific here we use the function $|h|^2$ given by Eq. (60) close to the real axis

3.5   Expression for the Slope of a Contour Line

We conclude this discussion of applications of the Cauchy-Riemann differential equations in amplitude and phase to gain insight into a complex-valued function f by presenting an expression for the slope of a contour line in terms of the steepness of the amplitude κ in the two directions given by τ and σ of the complex plane. This formula is intimately related to the one, Eq. (65), for the slope of a phase line.

Indeed, the orthogonality relation of phase and contour lines, expressed according to the appendix by the reciprocity relation

$$\frac{\partial \tau_\theta}{\partial \sigma} = -\left(\frac{\partial \tau_\kappa}{\partial \sigma}\right)^{-1} \tag{68}$$

between the slopes of the two curves allows us to cast Eq. (65) into the form

$$\frac{\partial \tau_\kappa}{\partial \sigma} = -\left(\frac{\partial \kappa}{\partial \tau}\right)^{-1} \frac{\partial \kappa}{\partial \sigma}. \tag{69}$$

Hence, also the slope of the contour lines is governed by the ratios of the derivatives of κ with respect to σ and τ. However, in comparison to Eq. (65), we now have a minus sign and the inverse of the derivatives.

4   Conclusions and Outlook

To view the lines of constant phase and constant height of a complex function f as the result of flows is the central idea of our lecture. Here the Newton velocity field V given by the negative inverse of the logarithmic derivative of f provides us with the phase lines, whereas an additional phase shift of $-\pi/2$ in the velocity field leads to the contour lines.

Due to the choice of the negative sign in the Newton field, poles are sources, and zeros sinks of the phase lines. Moreover, in case of several zeros the corresponding flows have to be separated by continental divides.

To distinguish separatrices from regular phase lines is a crucial ingredient of our geometric approach [11, 12] towards the Riemann hypothesis [13]. Indeed, our goal is to identify the separatrices at the edges of the complex plane, that is for $\sigma \to \pm\infty$ where the flow is rather quiet, and verify that the phase difference of two consecutive separatrices is π. This property enforces [14] the Riemann hypothesis. Unfortunately, this topic goes beyond the scope of the present lecture.

Acknowledgments We thank P. C. Abbott, W. Arendt, J. Arias de Reyna, M. V. Berry, K. A. Broughan, J. B. Conrey, C. Feiler, M. Freyberger, M.B Kim, M. Kleber, D. Lebiedz, H. Maier, H. Montgomery, H. Moya-Cessa, M. O. Scully, J. Steuding, A. Wünsche, and W. H. Zurek for many fruitful discussions. Moreover, we are most grateful to the late J.W. Neuberger for introducing us to the concept of the Newton flow. L.H. is supported by the RIKEN special postdoctoral researcher program. I.T. is grateful for the financial support from the Grant Agency of the Czech Technical University in Prague, Grant No. SGS16/241/OHK4/3T/14. W.P.S. thanks Texas A & M University for a Faculty Fellowship at the Hagler Institute for Advanced Study at Texas A& M University and Texas A& M AgriLife for the support of this work. The research of the IQST is financially supported by the Ministry of Science, Research and Arts Baden-Württemberg.

Appendix: Cauchy-Riemann Differential Equations

Central to our approach towards gaining an understanding of a complex-valued function are the Cauchy-Riemann differential equations for amplitude and phase. In this appendix we derive them in two different ways and make the connection to the more familiar version in terms of the real and imaginary parts of the function. We conclude by presenting an intriguing application of the Cauchy-Riemann differential equations in amplitude and phase and show that the slopes of the lines of constant phase and constant height are the inverse of each other.

Key Relations

The Cauchy-Riemann differential equations

$$\frac{\partial u}{\partial \sigma} = \frac{\partial v}{\partial \tau} \tag{70}$$

and

$$\frac{\partial v}{\partial \sigma} = -\frac{\partial u}{\partial \tau} \tag{71}$$

for an appropriately differentiable complex-valued function

$$f \equiv u + iv \tag{72}$$

expressed in the real and imaginary parts $u = u(\sigma, \tau)$ and $v = v(\sigma, \tau)$ of f of complex argument

$$s \equiv \sigma + i\tau \tag{73}$$

play a central role in complex analysis.

Their origin is the fact that f solely involves s but not s *and* s^*, that is,

$$f = f(s). \tag{74}$$

In order to bring this fact out most clearly we first rederive the Cauchy-Riemann differential equations, Eqs. (70) and (71), starting from the form, Eq. (74), of f, that is, f only involves s, and then show that in turn the Cauchy-Riemann differential equations enforce this dependence.

Throughout the lecture we focus on the analysis of f in terms of lines of constant phase and constant height. Deeper insight into their behavior springs from the Cauchy-Riemann differential equations

$$\frac{\partial \kappa}{\partial \sigma} = \frac{\partial \theta}{\partial \tau} \tag{75}$$

and

$$\frac{\partial \kappa}{\partial \tau} = -\frac{\partial \theta}{\partial \sigma}. \tag{76}$$

for the amplitude

$$|f| \equiv e^{\kappa} \tag{77}$$

of f determined by the real-valued function $\kappa = \kappa(\sigma, \tau)$, and the real-valued phase $\theta = \theta(\sigma, \tau)$ of

$$f \equiv |f|e^{i\theta}. \tag{78}$$

Here we pursue two different derivations: (i) We differentiate the explicit definitions of κ and θ in terms of u and v and use the corresponding Cauchy-

Riemann differential equations given by Eqs. (70) and (71), and (ii) we differentiate the exponential representation

$$f = e^{\kappa + i\theta} \tag{79}$$

of f.

The second approach does not explicitly use the Cauchy-Riemann differential equations of u and v but rather takes advantage of f being a function solely of the argument

$$\alpha \equiv \kappa + i\theta. \tag{80}$$

Indeed, the Cauchy-Riemann differential equations for amplitude and phase, Eqs. (75) and (76), are thus an elementary consequence of the ones for u and v, Eqs. (70) and (71), with $u \equiv \kappa$ and $v \equiv \theta$.

Representation in Real and Imaginary Parts

In this section we recall the Cauchy-Riemann differential equations for the real and imaginary parts of f. Crucial to their derivation is the fact that f depends solely on s, but not on s *and* s^*, as expressed by Eq. (74). This feature allows us to obtain a relation that is also at the very heart of the Cauchy-Riemann differential equations for amplitude and phase. We conclude by demonstrating that the Cauchy-Riemann differential equations for real and imaginary parts enforce that f only depends on s.

Derivation
We start by recalling with the help of the chain rule of differentiation the identities

$$\frac{\partial f}{\partial \sigma} = \frac{df}{ds} \frac{\partial s}{\partial \sigma} = \frac{df}{ds} \tag{81}$$

and

$$i\frac{\partial f}{\partial \tau} = i\frac{df}{ds} \frac{\partial s}{\partial \tau} = -\frac{df}{ds} \tag{82}$$

where we have used the decomposition, Eq. (73), of s in real and imaginary parts σ and τ.

When we equate Eqs. (81) and (82), we obtain the relation

$$\frac{\partial f}{\partial \sigma} = -i\frac{\partial f}{\partial \tau} \tag{83}$$

which with the representation, Eq. (72), of f in terms of real and imaginary parts reduces to

$$\frac{\partial u}{\partial \sigma} + i \frac{\partial v}{\partial \sigma} = \frac{\partial v}{\partial \tau} - i \frac{\partial u}{\partial \tau}. \tag{84}$$

The real and imaginary parts of this equation immediately yield the Cauchy-Riemann differential equations, Eqs. (70) and (71), in real and imaginary parts of f.

Dependence of f on s Rather Than s and s^*

Next we start from a complex-valued function

$$f(\sigma, \tau) \equiv u(\sigma, \tau) + iv(\sigma, \tau) \tag{85}$$

and show that f depends solely on s, provided u and v satisfy the Cauchy-Riemann differential equations.

For this purpose we evaluate the quantity

$$\left(\frac{\partial}{\partial \sigma} + i \frac{\partial}{\partial \tau} \right) (u + iv) = \frac{\partial u}{\partial \sigma} + i \left(\frac{\partial v}{\partial \sigma} + \frac{\partial u}{\partial \tau} \right) - \frac{\partial v}{\partial \tau} \tag{86}$$

which with the Cauchy-Riemann differential equations, Eqs. (70) and (71), reduces to

$$\left(\frac{\partial}{\partial \sigma} + i \frac{\partial}{\partial \tau} \right) (u + iv) = 0. \tag{87}$$

With the help of the identities

$$s^* = \sigma - i\tau \tag{88}$$

and

$$\frac{\partial}{\partial \sigma} = \frac{\partial}{\partial s} + \frac{\partial}{\partial s^*} \tag{89}$$

as well as

$$\frac{\partial}{\partial \tau} = i \left(\frac{\partial}{\partial s} - \frac{\partial}{\partial s^*} \right) \tag{90}$$

we find

$$\left(\frac{\partial}{\partial \sigma} + i \frac{\partial}{\partial \tau} \right) = 2 \frac{\partial}{\partial s^*}, \tag{91}$$

and thus Eq. (87) takes the form

$$\frac{\partial}{\partial s^*}(u + iv) = \frac{\partial}{\partial s^*}f = 0.$$

(92)

Hence, f is independent of s^*.

Representation in Amplitude and Phase

So far we have concentrated on the representation, Eq. (72), of f in terms of real and imaginary parts. In this section we turn to the exponential representation, Eq. (79), and obtain the Cauchy-Riemann differential equations, Eqs. (75) and (76), in amplitude and phase.

Derivation
From the definition

$$\theta \equiv k\pi + \arctan \frac{v}{u}$$

(93)

of the phase θ based on the representation Eq. (79) of f where k is an integer, we find by differentiation the relation

$$\frac{\partial \theta}{\partial \tau} = \frac{1}{u^2 + v^2}\left(u\frac{\partial v}{\partial \tau} - v\frac{\partial u}{\partial \tau}\right),$$

(94)

which with the help of the Cauchy-Riemann differential equations, Eqs. (70) and (71), for u and v leads us to the identity

$$\frac{\partial \theta}{\partial \tau} = \frac{1}{2}\frac{1}{|f|^2}\frac{\partial}{\partial \sigma}\left(|f|^2\right) = \frac{1}{2}\frac{\partial}{\partial \sigma}\left(\ln |f|^2\right) = \frac{\partial \kappa}{\partial \sigma},$$

(95)

in complete agreement with Eq. (75). Here we have made use of the identity

$$|f|^2 = u^2 + v^2 = e^{2\kappa},$$

(96)

following from Eqs. (72) and (77).

Similarly, we find by direct differentiation

$$\frac{\partial \theta}{\partial \sigma} = \frac{1}{u^2 + v^2}\left(u\frac{\partial v}{\partial \sigma} - v\frac{\partial u}{\partial \sigma}\right),$$

(97)

which with the help of the Cauchy-Riemann differential equations, Eqs. (70) and (71) takes the form

$$\frac{\partial \theta}{\partial \sigma} = -\frac{1}{u^2 + v^2}\left(u\frac{\partial u}{\partial \tau} + v\frac{\partial v}{\partial \tau}\right) = -\frac{\partial}{\partial \tau}(\ln|f|) = -\frac{\partial \kappa}{\partial \tau}, \tag{98}$$

that is Eq. (76). The last step follows from Eq. (77).

Direct Derivation

The Cauchy-Riemann differential equations for amplitude and phase follow also directly from the identity, Eq. (83), when we use the exponential representation, Eq. (79) of f.

Indeed, we find the relation

$$e^{\kappa+i\theta}\left(\frac{\partial \kappa}{\partial \sigma} + i\frac{\partial \theta}{\partial \sigma}\right) = -ie^{\kappa+i\theta}\left(\frac{\partial \kappa}{\partial \tau} + i\frac{\partial \theta}{\partial \tau}\right) \tag{99}$$

or

$$\frac{\partial \kappa}{\partial \sigma} + i\frac{\partial \theta}{\partial \sigma} = \frac{\partial \theta}{\partial \tau} - i\frac{\partial \kappa}{\partial \tau} \tag{100}$$

whose real and imaginary parts yield the desired equations, Eqs. (75) and (76).

This derivation brings out most clearly that the Cauchy-Riemann differential equations in amplitude and phase are a consequence of the application of the Cauchy-Riemann differential equations in real and imaginary parts to $u \equiv \kappa$ and $v \equiv \theta$.

Reciprocity Relation for Slopes of Phase and Contour Lines

Lines of constant phase and constant amplitude are always orthogonal on each other which manifests itself directly in the slopes of the two curves. Indeed, they must be the reciprocal of each other. In this section we rederive this familiar result using the Cauchy-Riemann differential equations, Eqs. (75) and (76), for amplitude and phase.

Lines $\tau_\theta = \tau_\theta(\sigma)$ in the complex plane along which the function f represented by the exponential form, Eq. (79), assumes the constant phase θ are often given implicitly, that is by a relation of the form

$$\theta = \theta(\sigma, \tau_\theta(\sigma)). \tag{101}$$

Nevertheless, we can find the slope of the line τ_θ at σ from this condition by differentiation taking into account that θ is constant. Indeed, we immediately arrive at the identity

$$0 = \frac{d\theta}{d\sigma} = \frac{\partial\theta}{\partial\sigma} + \frac{\partial\theta}{\partial\tau}\frac{d\tau_\theta}{d\sigma} \tag{102}$$

or

$$\frac{d\tau_\theta}{d\sigma} = -\left(\frac{\partial\theta}{\partial\tau}\right)^{-1}\frac{\partial\theta}{\partial\sigma}. \tag{103}$$

Likewise, we find from the implicit condition

$$\kappa = \kappa(\sigma, \tau_\kappa(\sigma)) \tag{104}$$

on the lines $\tau_\kappa = \tau_\kappa(\sigma)$ in the complex plane along which f assumes the constant value e^κ the relation

$$0 = \frac{d\kappa}{d\sigma} = \frac{\partial\kappa}{\partial\sigma} + \frac{\partial\kappa}{\partial\tau}\frac{d\tau_\kappa}{d\sigma}, \tag{105}$$

or

$$\frac{d\tau_\kappa}{d\sigma} = -\left(\frac{\partial\kappa}{\partial\tau}\right)^{-1}\frac{\partial\kappa}{\partial\sigma}. \tag{106}$$

The Cauchy-Riemann differential equations, Eqs. (75) and (76), in amplitude and phase immediately reduce Eq. (103) to

$$\frac{d\tau_\theta}{d\sigma} = \left(\frac{\partial\kappa}{\partial\sigma}\right)^{-1}\frac{\partial\kappa}{\partial\tau}. \tag{107}$$

A comparison with Eq. (106) leads us to the reciprocity relation

$$\frac{d\tau_\theta}{d\sigma} = -\left(\frac{d\tau_\kappa}{d\sigma}\right)^{-1} \tag{108}$$

for the slopes expressing the well-known fact that lines of constant phase and constant height are orthogonal on each other.

We conclude this appendix by noting that this orthogonality is also a direct consequence of the Cauchy-Riemann differential equations in amplitude and phase.

References

1. D. Bohm, *Quantum Theory* (Dover Publications, New York, 1989)
2. H.M. Edwards, *Riemann's Zeta Function* (Academic Press, New York, 1974)
3. R.P. Feynman, A.R. Hibbs, *Quantum Mechanics and Path Integrals* (McGraw-Hill, New York, 1965)
4. R.A. Hauser, J. Nedić, The continuous Newton–Raphson method can look ahead. SIAM J. Optim. **15**(3), 915–925 (2005)
5. M. Heitel, D. Lebiedz, On analytical and topological properties of separatrices in 1-D holomorphic dynamical systems and complex-time Newton Flows (2019). arXiv:1911.10963
6. J.D. Jackson, *Classical Electrodynamics* (Wiley, New York, 1999)
7. M.B. Kim, J.W. Neuberger, W.P. Schleich, A perfect memory makes the continuous Newton method look ahead. Phys. Script. **92**(8), 085201 (2017)
8. D. Lebiedz, Holomorphic Hamiltonian ξ-flow and Riemann zeros (2020). arXiv:2006.09165
9. P.M.C. Morse, H. Feshbach, *Methods of Theoretical Physics* (McGraw-Hill, New York, 1953)
10. J.W. Neuberger, Continuous Newton's method for polynomials. Math. Intell. **21**, 18 (1999)
11. J.W. Neuberger, C. Feiler, H. Maier, W.P. Schleich, Newton flow of the Riemann zeta function: separatrices control the appearance of zeros. New J. Phys. **16**, 103023 (2014)
12. J.W. Neuberger, C. Feiler, H. Maier, W.P. Schleich, The Riemann hypothesis illuminated by the Newton flow of ζ. Phys. Script. **90**, 108015 (2015)
13. B. Riemann, Ueber die Anzahl der Primzahlen unter einer gegebenen Grösse, in *Monatsberichte der Berliner Akademie* (1859). Transcribed German version and English translation by D. R. Wilkins see http://www.claymath.org/publications/riemanns-1859-manuscript
14. W.P. Schleich, I. Bezděková, M.B. Kim, P.C. Abbott, H. Maier, H.L. Montgomery, J.W. Neuberger, Equivalent formulations of the Riemann hypothesis based on lines of constant phase. Phys. Script. **93**, 065201 (2018)
15. A. Sommerfeld, *Elektrodynamik* (Harri Deutsch, Frankfurt, 2005)
16. E.C. Titchmarsh, *The Theory of the Riemann Zeta-Function* (Clarendon Press, Oxford, 1967)
17. E.T. Whittaker, G.N. Watson, *A Course of Modern Analysis* (Cambridge University Press, Cambridge, 1996)

Exploring the Hottest Atmosphere with the *Parker Solar Probe*

Gary P. Zank, Lingling Zhao, Laxman Adhikari, Daniele Telloni, Justin C. Kasper, and Stuart D. Bale

Contents

Abstract

Identifying the physical process(s) responsible for heating the solar corona and thus driving the solar wind is perhaps the most important open question in heliophysics. This is a primary goal of the *Parker Solar Probe* (*PSP*) Mission. Various heating mechanisms have been suggested but one that is gaining increasing credence is associated with the dissipation of low frequency

G. P. Zank (✉) · L. Zhao · L. Adhikari
Center for Space Plasma and Aeronomic Research (CSPAR) and Department of Space Science,
University of Alabama in Huntsville, Huntsville, AL, USA

D. Telloni
INAF—Astrophysical Observatory of Torino, Torinese, Italy

J. C. Kasper
BWX Technologies, Inc., Washington, DC, USA

Department of Climate and Space Sciences and Engineering, University of Michigan, Ann Arbor, MI, USA

S. D. Bale
Physics Department, University of California, Berkeley, CA, USA

© The Author(s), under exclusive license to Springer Nature Switzerland AG 2023
R. Citro et al. (eds.), *Sketches of Physics*, Lecture Notes in Physics 1000,
https://doi.org/10.1007/978-3-031-32469-7_6

magnetohyrodynamic (MHD) turbulence. We review two current turbulence models, describe the modeling that has been done, and relate *Parker Solar Probe (PSP)* observations to the basic predictions of both models. *PSP* entered a region of sub-Alfvénic solar wind during encounter 8 and we review the low-frequency turbulence properties in this novel region. The observed spectra are well fitted using a spectral theory for nearly incompressible magnetohydrodynamics.

1   Introduction

Illustrated in Fig. 1 is arguably the most significant open question in heliophysics. Fueled by convection and magnetic fields, the temperature of the solar atmosphere, shown in Fig. 1, increases rapidly from the visible surface of the Sun (the photosphere, $\sim$5800 K), through the chromosphere and an abrupt transition region ($\sim$500 km thick, i.e., less than 1000th of the Sun's radius), to eventually reach 1–10M K in the solar corona. The hot corona is unstable, producing a supersonic solar wind that pervades interplanetary space and carves the protective heliospheric bubble out of the interstellar medium. Fifty years have passed since the prediction and detection of the solar wind, and we still do not understand how the corona is heated and how the solar wind is accelerated. Absent any special circumstances, the temperature above the photosphere would be expected to decrease with increasing height, as with the atmosphere of the Earth. What is [are] the physical process [processes] responsible for the high temperature of the solar corona and the continued heating of the solar wind in interplanetary space? This critical question

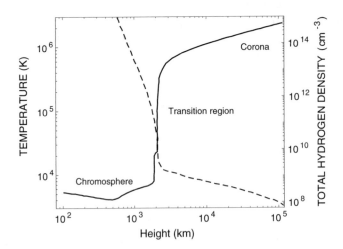

Fig. 1 Model of solar atmospheric temperature (solid line) and density (dotted) versus height above the photosphere. After a reprint from *The Solar Corona*, by Leon Golub and Jay M. Pasachoff, pp. 388. ISBN 0521480825. Cambridge, UK: Cambridge University Press, September 1997 [2], with permissions by Cambridge University Press

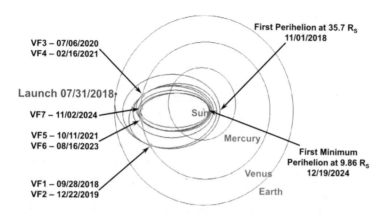

Fig. 2 *PSP* trajectory viewed from above the ecliptic plane with launch date on July 31, 2018. The seven Venus gravity assists (i.e., Venus Flybys [VF]; green dots) with the corresponding dates are shown together with the first perihelion and the first minimum perihelion. Reprinted from [1]

is now being addressed by the *Parker Solar Probe*, a NASA mission designed to carry instruments deep into the hot atmosphere of the Sun in an effort to find an answer. The *Parker Solar Probe (PSP)* (Fig. 3) was launched in August 2018, and was designed to reach a radial distance of $<10R_\odot$ (solar radii) above the solar surface [1]. The seven-year mission will use seven Venus gravity assist maneuvers to gradually reduce the perihelia to $\sim10R_\odot$ (Fig. 2). Already, *PSP* has flown closest to the Sun's surface than any other spacecraft in completing 10 close encounters with the solar atmosphere. *PSP* will spend 937 hours inside 20 $R_\odot$, 440 hours inside 15 $R_\odot$, and 14 hours inside 10 $R_\odot$ [1], and will therefore have sufficient time to sample slow, fast, and transient solar wind as the solar cycle evolves from solar minimum towards solar maximum, i.e., a period of increasing solar activity that results in a complex structured solar corona. The orbit of *PSP* is restricted to the ecliptic plane, ensuring that the fast wind directly above the Sun's polar regions is inaccessible. However, fast solar wind is sampled frequently as it emerges from low latitude or equatorial coronal holes providing considerable insight into what happens in the large, long-lived polar coronal holes that are responsible for the fast solar wind during solar minimum.

Identifying the heating processes in the solar corona is the primary goal of both the NASA *Parker Solar Probe* and the ESA *Solar Orbiter* missions. The *Solar Orbiter* mission was launched about a year after *PSP* on 10 February 2020. In closely approaching the Sun *PSP* is making unprecedented in situ measurements of the plasma and magnetic fields. *PSP* is a once-in-a-lifetime opportunity to explore the atmosphere of our Sun, make unexpected discoveries, and enable a fundamental understanding of the physics of the corona and solar wind. Previous models for the origin of the wind rested on remote observations and in situ measurements no closer than 0.3 AU from the Sun. With the loss of energy in the corona through radiation, heat conduction, waves, and the escaping solar wind, a significant amount of power is required to maintain the observed coronal temperatures. The source of this energy is the convective motion of the surface of the photosphere and its embedded magnetic field, but the mechanism by which the large scale and low frequency motion of the field is able to dissipate sufficient heat in the corona is not yet understood. Clues to the nature of the heating mechanism have been gleaned from spectroscopic measurements of the corona that show that heavy ions have large perpendicular temperature anisotropies. In situ measurements confirm that heating also continues in the solar wind out to at least 1 AU since ion and electron temperatures fall more slowly with distance than expected of adiabatic expansion.

Five major classes of models have been advanced to describe the heating and acceleration mechanisms of the solar corona and wind, with varying degrees of success, and none is yet universally accepted. The two leading candidates, ion cyclotron heating and the turbulence cascade, both rely on the presence of upwardly propagating Alfvénic or magnetized fluctuations. The conversion of magnetic energy due to photospheric-driven motion of the magnetic carpet or even the higher canopy of magnetic field via either direct heating (some form of reconnection) or the launching of waves is generally thought to be the origin of the energy needed to heat the corona. Ion cyclotron heating and the turbulent cascade models exploit the notion that low frequency shear Alfvén waves excited in the chromosphere will survive into the lower corona. Upward propagating Alfvén modes have been observed in the chromosphere and lower corona by Hinode and ground-based detectors. Ion cyclotron resonant heating, one of the most widely accepted mechanisms for coronal heating, results from the energization of particles interacting resonantly with Alfvén waves, thus heating the solar corona. The turbulent cascade, the alternative to ion cyclotron heating, exploits the possibility that upwardly propagating low-frequency Alfvén waves are partially reflected, thereby driving a turbulent cascade through coupling to zero frequency modes. The cascade is expected to be quasi-perpendicular and the dissipation of energy at high wave number may therefore lead to quasi-perpendicular heating of the corona. However, as we discuss below, the turbulence models have undergone significant modification in attempting to describe expectations and theory. In this essay, we provide an overview of the most promising coronal heating mechanisms and their comparison with *PSP* observations.

2   The *Parker Solar Probe* Spacecraft

The detailed description of the *Parker Solar Probe* spacecraft can be found in Fox et al. [1] and is summarized here and illustrated in Fig. 3. The mass of *PSP* is 685 kg (including propellant) at launch, it stands approximately 3 m in height, and has a 2.3 m diameter at the thermal protection system (TPS). I was fortunate enough to see the pre-launch spacecraft at the Applied Physics Laboratory (APL) of The Johns Hopkins University. *PSP* is a little larger than a small car and exuded an air of impressive solidity, size, and technological sophistication. *PSP* is unique in possessing a heat shield to shield the spacecraft and its instrumentation from the very harsh radiation environment of the Sun's atmosphere—this is the thermal protection

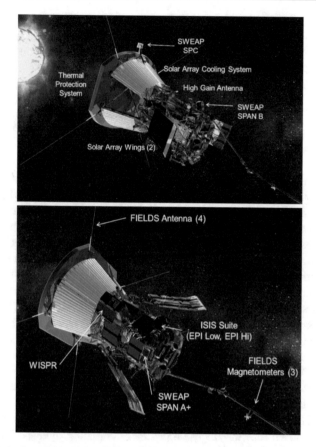

Fig. 3 (Top) Anti-ram side of the *PSP* spacecraft, identifying instruments and significant components. (Bottom) Ram side of the *PSP* spacecraft identifying the location of most science instruments. Reprinted from [1]

system (TPS) that is a carbon-carbon composite/carbon foam sandwich mounted ahead of the spacecraft (Fig. 3), i.e., pointed sunward with only portions of the solar arrays, the SWEAP SPC, and the FIELDS Electric Field antennae outside the umbra.

Four instruments comprise the *PSP* science package: the Electromagnetic Fields Investigation (FIELDS); the Integrated Science Investigation of the Sun, Energetic Particle Instruments (ISI⊙S); the Solar Wind Electrons Alphas and Protons Investigation (SWEAP), and the Wide Field Imager for Solar Probe Plus (WISPR). The two instruments that are most pertinent to this essay are SWEAP and FIELDS. SWEAP (Principal Investigator: Justin Kasper) [3] has two electrostatic analyzers and a Faraday cup and counts solar wind electrons, protons and helium ions, from which the distribution functions and their moments such as velocity, density, and temperature can be derived. The solar wind plasma conditions are measured by the

SWEAP Solar Probe ANalyzer (SPAN) top-hat electrostatic analyzer and the Solar Probe Cup (SPC), a Faraday cup. FIELDS (Principal Investigator: Stuart Bale) [4] comprises two fluxgate magnetometers, a search coil magnetometer and five electric antennas to measure electric and magnetic fields and waves, the spacecraft floating potential, density fluctuations, and radio emissions.

3   Overview of Solar Wind Turbulence and Coronal Heating Models

The solar wind is classified typically into two classes according to the plasma flow velocity observed at 1 astronomical unit (au): fast solar wind with a velocity ≥ 500 km/s and slow solar wind with a velocity ≤ 500 km/s. It is generally accepted that the fast solar wind originates from coronal holes with open magnetic field lines, but the origin of slow solar wind streams is still debated [5, 6]. Central to the coronal heating problem, whether in fast or slow wind, is the transport of waves and turbulence beyond the photosphere. Several models for heating the solar corona and the subsequent driving of the solar wind have been advanced, the two most popular being the dissipation of low frequency magnetohydrodynamic (MHD) turbulence or the dissipation of ion cyclotron waves. The MHD turbulence models can be further distinguished by a two class classification of the physical models as introduced by Cranmer and van Ballegooijen [7]. The first is described as "wave/turbulence-driven" (W/T-)models and the other as "reconnection/loop-opening" (RLO-)models.

Most of the *PSP* measurements plasma and magnetic fields have so far been made above the Alfvén surface but the last few encounters have been in sub-Alfvénic regions of the solar wind [8, 9]. These observations are providing us with our best glimpse so far into the likely properties of coronal turbulence. Observations of low-frequency MHD turbulence by *PSP* allow us to now begin constraining models of the turbulently heated solar corona that drives the supersonic solar wind. These models postulate a source of turbulence at a base location, perhaps just above the photosphere or, more typically for wave-turbulence models, just above the chromosphere. By modeling the evolution of MHD turbulence in the corona, through the Alfvén surface, and into the supersonic solar wind (identifying further possible in situ sources of turbulence), the constraints inferred from *PSP* observations can now be applied to the models. Although preliminary, these new observational constraints on the models already shed considerable light upon the origin of coronal and solar wind turbulence, the heating of coronal plasma, and the origin of the fast and slow solar wind.

In this Essay, we discuss briefly the possible turbulence mechanisms thought to be responsible for the heating of the solar corona and hence the driving of the fast and slow solar wind. An extensive theory has been developed and this is now being tested in detail against observational signatures revealed by *PSP* and that may allow us to distinguish between the various turbulence models of coronal heating and solar wind driving. Two sets of observations identify a central conundrum for the turbulence models that has to be addressed. The first is that in highly magnetic field-aligned solar wind flows near 1 astronomical units (au), Wang et al. [10] and Telloni et al. [11] found that Alfvén waves were propagating uni-directionally (normalized cross-helicity $|\sigma_c| \simeq 1$) and that the spectrum was a power law with spectral index $-5/3$ i.e., $k_\parallel^{-5/3}$, where $k_\parallel$ is the wave vector parallel to the mean magnetic field. Secondly, Zhao et al. [12] reported Parker Solar Probe observations identifying highly magnetic field-aligned coronal flows that exhibited uni-directionally propagating Alfvén waves with a corresponding Kolmogorov-like spectrum $k_\parallel^{-5/3}$. Uni-directionally propagating Alfvén waves do not interact nonlinearly, requiring counter-propagating Alfvén waves for a nonlinear cascade in wave number space to occur, and hence there should not be a Kolmogorov-like spectrum for uni-directionally propagating Alfvén waves nor should there be any related dissipation of Alfvén waves and hence heating of the solar wind or solar corona. Resolving this puzzle is essential if turbulence models are argued to be responsible for the heating of the solar corona and hence driving of the solar wind.

We discuss below the two turbulence models advanced to explain the heating of the solar corona. One can take the view that both the slow wind and fast wind are subject to the same underlying heating mechanism but that the geometry of the large-scale magnetic fields (open coronal holes and large-scale loops) mediates the nature of the wind in important ways. Illustrated in Fig. 4 is a cartoon showing one perspective of the expansion of (left) the fast solar wind from the base of a coronal hole and (right) the slow solar wind from higher up in the corona. In both cases, from the perspective of turbulence models that dissipate their energy via a cascade process, the mechanism for heating the open coronal hole plasma and the

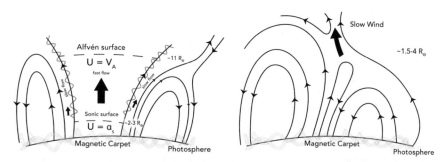

Fig. 4 Left: Cartoon illustrating the possible origin of the fast solar wind from a coronal hole driven by turbulence associated with the magnetic carpet. Additional sources of turbulence in the corona, below and above the Alfvén surface, may be due to Kelvin-Helmholtz instabilities on the edges of coronal regions that separate fast coronal flows from adjacent slow flows associated with closed magnetic field regions as illustrated. The sonic and Alfvén surfaces are identified, i.e., surfaces at which the coronal flow transitions from subsonic to supersonic and sub-Alfvénic to super-Alfvénic respectively. Right: A possible model describing the origin of the slow solar wind. Large-scale loop plasma is heated via turbulence associated with the magnetic carpet, and interchange reconnection at higher altitudes, possibly 2–5 $R_\odot$ above the photosphere, leads to the release of hot loop plasma. The hot plasma expands into higher open magnetic field coronal regions to reach supersonic and super-Alfvénic speeds thereafter

loop plasma is the same but the processes by which the plasma eventually forms the wind are different.

The W/T-model advanced originally by Matthaeus et al. [13] assumes that a broad spectrum of low-frequency MHD Alfvén waves carries kinetic and magnetic energy from the photosphere into the solar corona. The waves are generated by rapid displacements of open magnetic flux tubes that are advected by motion in the photosphere. An important concern is the difficulty that Alfvén waves have in propagating from the chromosphere, across the transition region, into the corona. Since $\ll 1\%$ of the incident Alfvénic flux (perhaps as little as $4 \times 10^{-2}\%$ [14–17]— see below) is thought to be transmitted across the transition region, this requires extremely high Alfvén wave fluxes emanating from the photosphere. Once through the transition region, the upwardly propagating Alfvén waves in W/T models experience partial non-WKB reflection off the large-scale density and magnetic field gradients of the solar corona and the counter-propagating Alfvénic modes interact nonlinearly to generate quasi-2D fluctuations that cascade to smaller scales and eventually dissipate, thus heating the coronal plasma.

An alternative turbulence model, described as quasi-2D turbulence models, identifies the "magnetic carpet" [18], a region of both open and closed magnetic field that is distributed uniformly across the solar surface, as the source of predominantly quasi-2D MHD turbulence that acts to heat the solar corona. The mixed polarity small-scale loops that form the magnetic carpet rise from below the photosphere and interact on a replenishment time scale of about 40 hours [18]. Interspersed with the small-scale loops in the magnetic carpet are large-scale magnetic field lines, which could either be open as in coronal holes or closed as with large coronal loops. The constant stirring and mixing of the magnetic carpet loops generates small-scale quasi-2D turbulence above the photosphere [19] through the interaction of emerging and evolving carpet magnetic field. Zank et al. [20] argue that quasi-2D turbulence [21, 22] generated by the magnetic carpet is advected through the chromosphere, across the transition region without reflection, and into the solar corona. Quasi-2D advected turbulence obviously does not suffer the losses that a corresponding flux of Alfvén waves would experience across the transition region. The dissipation of the quasi-2D turbulence is found to be sufficient to heat the underlying plasma to temperatures in excess of 10^6 K. Zank et al. [20] suggest that the basic mechanism for coronal heating via the dissipation of quasi-2D advected turbulence applies to both the fast and slow solar wind. The fast wind is heated by the dissipation of quasi-2D turbulence originating from the magnetic carpet and is advected by the heated expanding coronal flow. Models invoking a dominant quasi-2D turbulent component [20, 23] are found to accelerate the solar wind to supersonic speeds within 2–4 $R_\odot$, consistent with observations [24,25] and to super-Alfvénic speeds within ~ 10–11 $R_\odot$. These models remain somewhat preliminary and further factors that need to be incorporated into these model are super-radial expansion and the possibility that the supersonic and super-Alfvénic flows in the coronal hole initiate possible Kelvin-Helmholtz instabilities along the boundaries of a coronal hole that separates faster and slower coronal flows. A first model that incorporates Kelvin-Helmholtz instabilities along the boundaries of a coronal hole has been presented by Telloni et al. [26], finding that the already accelerated solar wind may be further accelerated by the dissipation of Kelvin-Helmholtz turbulence. This model has been used to interpret remote spectroscopic measurements of the UV corona acquired with the UltraViolet Coronagraph Spectrometer on board the SOlar and Heliospheric Observatory (SOHO) during the minimum activity of solar cycle 22. High temperature-velocity correlations are found along the fast/slow solar wind interface region and interpreted as manifestations of KH vortices formed by the roll-up of the shear flow, whose dissipation could lead to higher heating and, because of that, higher velocities.

Remote observations offer some general if inconclusive insights into the two classes of turbulence models for the solar corona. The transverse displacement of spicules in the chromosphere with amplitudes of 10–25 km/s and periods 100–500 s was observed by De Pontieu et al. [27], from which they estimated an energy flux of ~ 100 Wm^{-2}. These fluctuations were interpreted as Alfvén waves with energy flux "sufficient to drive the solar wind" [27], but this interpretation was disputed by Erdelyi and Fedun [28] who favored an interpretation as kink modes. The debate has not been settled by either further observations [29–34] or further theory, modeling, and analysis [35, 36]. Indeed, estimates for the energy content and flux of the waves are highly uncertain observationally [37–42]. More generally, waves with periods greater than 30–100 s are thought not to propagate energy efficiently from the coronal base [15, 16, 43].

Zank et al. [20] estimated the observed magnitude of quasi-2D turbulent magnetic field fluctuations to determine whether there is sufficient energy to power coronal heating in polar coronal holes. The unperturbed magnetic field is very nearly radial in the center of a polar coronal hole, and has a strength of about 10 G at $1.03 R_\odot$ (i.e., $0.03 R_\odot$ above the surface). Withbroe [44] estimated that the upward energy flux required at $1.03 R_\odot$ to sustain coronal heating above this height along the radial field in the center of a polar coronal hole is about 3×10^5 erg cm^{-2} s^{-1}. Averaged over several hours, the polar coronal hole outflow speed at $1.03 R_\odot$ is ≤ 10 km s^{-1}. For a 10 km s^{-1} outflow at $1.03 R_\odot$, the constraint of an energy flux of $\sim 3 \times 10^5$ erg cm^{-2} s^{-1} requires that magnetic field fluctuations perpendicular to the average radial field $B_\perp$ be about 1.9 G in strength. This corresponds to angular deviations of about $11°$ around the radial direction for the total field vector at $1.03 R_\odot$. De Pontieu et al. [27] and Moore et al. [45] have identified oscillatory swaying motions of the open field with amplitudes of order $15°$ and periods of about 5 minutes at $1.03 R_\odot$ in polar coronal holes. These observations correspond to $B_\perp \sim 2.7$ G, suggesting that there is sufficient magnetic energy flux corresponding to perpendicular magnetic field components to heat the solar corona and possibly drive the solar wind.

In concluding this discussion, another popular model for heating the solar coronal plasma is the dissipation of ion cyclotron waves created at the base of the solar corona. The spectrum of ion cyclotron waves is treated as a static, linear, non-interacting superposition of waves. Ion cyclotron heating is also sometimes described as a "cyclotron sweep mechanism," and was addressed critically by Leamon et al. [46]. The dissipation rate of the cyclotron sweep mechanism depends on the local proton (or minor ion) cyclotron frequency, which decreases with increasing heliocentric distance. Outward propagating fluctuations are transported into regions for which damping occurs at progressively lower frequencies. As a consequence, the linear dissipation of the fluctuations results in the absorption of energy at the local gyro-frequency. Linear dissipation of the wave spectrum therefore corresponds to a "sweep" through spectral space towards lower frequencies, thereby depleting the spectrum. Such a depletion of the wave spectrum will occur at even lower frequencies if minor or heavy ions are included in the dissipative process, with the consequence that the wave spectrum will already be partially depleted before the protons even experience heating [47]. A further and critical problem with the ion cyclotron heating mechanism [46] is that a large wave energy flux at high frequencies, up to perhaps the kilohertz range, is required to account for rapid cyclotron damping in the lower corona [13]. Whether such a large enhancement of kHz power at the coronal base exists is unclear as this has not been observed. Furthermore, the question of how such high-frequency waves in the lower solar corona could be generated [48] is very unclear. A further problem with the ion cyclotron dissipation mechanism is that the energy cascade in the parallel wavenumber direction is very slow, implying that the resonantly dissipated spectrum is unlikely to be replenished quickly enough to ensure rapid heating of the coronal plasma within 1–4 $R_\odot$. Finally, high-frequency modes experience only weak nonlinear couplings and weak WKB reflections [13]. High-frequency, parallel-propagating Alfvén waves will therefore experience rapid transport through the corona, experiencing only direct kinetic damping [16, 49]. For coronal conditions, the estimated perpendicular cascade rate [46] will lead to a much faster dissipation of energy at the small scales than linear proton cyclotron dissipation. Consequently, the ion cyclotron sweep mechanism does not appear to be a viable mechanism for heating the solar corona.

However, it is important to distinguish the possibility that a cyclotron-resonant dissipation mechanism likely participates in the spectral cascade together, with other possible kinetic non-cyclotron-resonant mechanisms, from the ion cyclotron-sweep mechanism. For example, Bruno and Trenchi [50] showed that the ion spectral break moves to higher and higher frequencies as the Sun is approached. They showed quite clearly that, of the possible correspondences of the ion gyro-frequency and the frequencies corresponding to the ion-inertial length and Larmor radius, the ion-cyclotron resonance frequency is the one that better matches the location of the break point. This suggests that a cyclotron-resonant dissipation mechanism might be one of the processes involved in the spectral cascade. Related ion-cyclotron dissipation mechanisms that may lead to temperature anisotropy have been observed statistically at 1 au by Telloni et al. [51]. Another set of kinetic fluctuations that may be important for dissipation is kinetic Alfvén waves (KAWs), which are ion scale dispersive waves with primarily perpendicular wave vectors [52] and are formed naturally by a turbulent cascade [53]. KAWs have been considered in the context of e.g., solar wind heating [54] and solar electron heating in the corona [55].

4   Magnetohydrodynamic Turbulence Models Applied to the Coronal Heating Problem

The solar wind originating from coronal holes is likely to be highly aligned with the large-scale open radial magnetic field, making it observationally difficult to identify quasi-2D turbulence. Instead, the turbulence is likely to present itself as essentially slab turbulence i.e., turbulence comprised of counter-propagating Alfvén waves. According to the basic W/T model [13] and extensions thereof [56–70], the normalized cross helicity σ_c (the difference between the forward and backward Elsässer energies normalized to the total Elsässer energy) should be small to ensure that nonlinear interactions generate quasi-2D fluctuations [71] that then undergo a rapid 2D ($\mathbf{k}_\perp$, relative to the mean magnetic field) turbulent cascade that dissipates energy and heats the plasma. However, if the normalized cross helicity $|\sigma_c| \sim 1$, i.e., uni-directionally propagating Alfvén waves, then no turbulent interactions are possible within the W/T framework. However, the nearly incompressible (NI) MHD model in the small and $O(1)$ plasma beta regime admits uni-directional Alfvén wave propagation, and the Alfvén waves interact passively with the advected quasi-2D fluctuations to produce a $k_\parallel^{-5/3}$ wave number spectrum [22]. This is a key observational discriminator between the W/T and the quasi-2D turbulence models of solar coronal turbulence when the flow is highly aligned with the magnetic field.

A possible origin of the slow solar wind is illustrated in the right panel of Fig. 4. In this model, the quasi-2D turbulence generated by the magnetic carpet in the footpoints of a large magnetic loop heats the loop plasma in much the same way as coronal hole plasma heating occurs. However, the heating generates counter-propagating flows that cannot escape the loop, allowing loop plasma to be heated to high temperatures. Thereafter, Fig. 4, the loop plasma escapes into the high corona via interchange reconnection, at heights of perhaps ~2–4 $R_\odot$. We suggest that it is the hot escaping loop plasma that forms the slow solar wind as it exits the upper corona through higher open field regions. This is basically the mechanism suggested by Fisk et al. [72, 73] for the origin of the slow solar wind, although here we explicitly incorporate a mechanism for heating the loop plasma that was not addressed in Fisk et al. [72, 73]. Such a model of the slow solar wind predicts a majority quasi-2D component and a minority slab component that should be relatively easily observed since the flows are unlikely to be highly aligned with the mean magnetic field [74]. By contrast, it is possible for the W/T model to heat the loop plasma since counter-propagating Alfvén waves will be generated at either end of a coronal loop. Indeed, Nigro et al. [75, 76], both theoretically and via reduced MHD simulations, found that fluctuations with a large perpendicular wave vector component can be generated in a loop by nonlinear interactions of counter-propagating Alfvénic fluctuations resulting in low cross-helicity values. Under these circumstances, with the opening of the loop via interchange reconnection, the hot loop plasma will expand as described already but the dominant magnetic turbulence component will be slab, with a cross helicity $\sigma_c \simeq 0$, and not the quasi-2D component. As with the fast solar wind, two clearly distinct and testable predictions emerge from the NI MHD quasi-2D model and the W/T model: the former predicts a majority quasi-2D component and a minority slab component and the latter a majority slab component (with σ_c relatively close to although not necessarily 0) and a minority quasi-2D component. We note that if the slow wind happened to be highly field-aligned, the observed turbulence would appear to be slab or highly Alfvénic, and could even have high values ($|\sigma_c| \sim 1$) of the normalized cross helicity (uni-directional Alfvén wave propagation) and have a parallel wave number spectrum of the Kolmogorov form. In support of this perspective, we recall the very interesting results presented by D'Amicis et al. [74, 77] reporting observations of Alfvénic slow wind streams that possess a high cross-helicity value and a Kolmogorov-like spectrum. In their Figure 5 (bottom panel) D'Amicis et al. [74] show a correlation between θ_{BR} and V_{sw} that is almost identical to that in the fast wind (Figure 5, top panel). Essentially, D'Amicis et al. [74] find that the observed slow wind is field-aligned, possesses high cross helicity and yet has a Kolmogorov-like spectrum. Their result corresponds to the results presented by Telloni et al. [25] and Zhao et al. [12] for fast and slow field-aligned flows respectively. Matteini et al. [78, 79] report related results for fast wind. The important point is that both the slow and fast wind were field-aligned during the time these three sets of observations were made and for this reason, only Alfvénic observations can be observed (the 2D component not being easily visible to a single spacecraft in this geometry). The observation of Kolmogorov-like spectra in such flows with a high cross helicity can be explained by NI MHD in the $\beta \ll 1$ or $O(1)$ limits thanks to a passive scalar-like interaction of uni-directionally propagating Alfvén waves with advected quasi-2D turbulence that results in a $k_\parallel^{-5/3}$ spectrum [22], i.e., NI MHD generates a $k_\parallel^{-5/3}$ spectrum dynamically from uni-directionally propagating Alfvén waves. This stands in contrast to the W/T model which cannot do so. Such observations would therefore favor the NI MHD quasi-2D model rather than the W/T model.

Adhikari et al. [80] published an interesting paper entitled "*Does turbulence turn off at the Alfvén critical surface?*" that is germane to the discussion here. They examined two turbulence models in the vicinity of the Alfvén surface, the one underlying the W/T model and the other the NI MHD model in the plasma beta $\beta \ll 1$ or $O(1)$ limits. In essence, the W/T model predicts that at the Alfvén surface [80]

1. only outwardly propagating modes exist;
2. the velocity and magnetic field fluctuations are perfectly correlated, with equal velocity and magnetic field fluctuation correlation lengths; and
3. the nonlinear and dissipation terms are zero. Since these terms describe the transfer of large-scale energy to small scales, no transfer of energy occurs, suggesting that turbulence "turns off" at the Alfvén surface for the W/T model.

Within the W/T model, the Alfvén surface therefore acts to produce a "fossil"-like turbulence state shortly after the solar wind flow crosses the surface, "filtering" out all the backward propagating Alfvén modes, leaving only outwardly propagating modes. An important question is what the spectrum of the transmitted/filtered outwardly propagating Alfvén modes might be. If "fossil" turbulence, then if one follows the argument of Dobrowolny et al. [81, 82] (see also Zank et al. [22, 83]) it is possible that the spectrum transmitted across the Alfvén surface would be Kolmogorov-like for the outwardly propagating modes just below the Alfvén surface, provided the wave-wave couplings are governed by the nonlinear timescale rather than the Alfvén timescale (see Appendix B of Zank et al. [83]). It is unclear how long such a superposition of linear Alfvén modes will retain a Kolmogorov-like spectrum as they propagate further into the super-Alfvénic solar wind in the absence of dynamical nonlinear couplings. Quasi-linear theory would suggest that resonant scattering of ions by Alfvén waves would result in a k^{-2} spectrum [84]. The continued expansion of the solar wind will lead eventually to the generation of (now advected) backward propagating Alfvénic modes that can "restart" nonlinear interactions and hence turbulence.

By contrast, the NI MHD $\beta \ll 1$ or $O(1)$ turbulence model predicts that at the Alfvén surface

1. the majority quasi-2D component is not subject to a critical point, although the Alfvén surface does represent a critical point for the minority slab turbulence component;
2. higher-order slab turbulence, like the W/T model, reduces to only outwardly propagating modes, with the slab turbulence velocity and magnetic field fluctuations perfectly aligned and having equal velocity and magnetic fluctuations correlation lengths;
3. despite only outwardly propagating higher-order slab turbulence modes existing, the nonlinear dissipation term is non-zero because the dissipation of slab turbulence in the NI MHD description is due primarily to mixing with the dominant quasi-2D turbulence component;
4. both outward and inward non-propagating modes of the dominant quasi-2D component are transmitted and the corresponding velocity and magnetic field fluctuations are not perfectly correlated; and
5. the total nonlinear/dissipation terms are nonzero, ensuring that energy is transferred through the inertial range to eventually be dissipated, implying that turbulence in the plasma beta regime $\beta \ll 1$ or $O(1)$, as described by NI MHD, does not turn off at the Alfvén critical surface.

Consequently, NI MHD turbulence in the $\beta \ll 1$ or $O(1)$ regimes remains active and dominated by nonlinear interactions associated with the dominant quasi-2D component at and beyond the Alfvén surface. Furthermore, because of the interaction between quasi-2D structures and uni-directionally propagating Alfvénic modes, the minority slab turbulence also remains active at and beyond the Alfvén surface.

One can list the basic NI quasi-2D and W/T model predictions that can be tested by observations made by both *Parker Solar Probe (PSP)* and *Solar Orbiter*. *PSP* crossed below the Alfvén surface during encounters 8, 9 and 10, and is therefore making observations in both the super-Alfvénic and sub-Alfvénic solar wind [8, 9, 85, 86]. Observationally, we stress that measuring spectral anisotropy and anisotropic structures is difficult when solar wind and coronal flows are highly aligned with the magnetic field, thereby preventing the observation of quasi-2D structures [12, 22]. The NI quasi-2D models predict [21, 22]

1. that the energy-containing range in both the slow and fast wind is a superposition of a majority quasi-2D component and a minority slab component, likely in the ratio of 80:20.
2. that the inertial range is similarly anisotropic in both the slow and fast wind with a corresponding majority quasi-2D component and a minority slab component.
3. an arbitrary (normalized) cross-helicity for both the majority and minority component in both the slow and fast wind. For highly field-aligned flows, whether in the fast wind or even in slow wind, it is possible that uni-directional (i.e., with high normalized cross-helicity values $|\sigma_c| \simeq 1$) Alfvén wave/slab propagation exhibiting a $k_\parallel^{-5/3}$ spectrum can occur.
4. arbitrary values of the (normalized) residual energy σ_r, including evolving toward -1, i.e., magnetic energy dominated, or $+1$, i.e., vortex-dominated, in the majority 2D component.
5. that the density fluctuations are primarily advected entropy fluctuations that behave as a passive scalar (a slow and fast mode wave contribution enters only at the higher order), have an amplitude that is ordered roughly by the turbulent Mach number $M \equiv \delta u / C_s$, where δu is a characteristic velocity of the turbulent fluctuations and C_s is the characteristic sound speed, and that the density variance spectrum is $k^{-5/3}$.

By contrast, the W/T models predict [13, 61]

1. that the turbulence in the energy-containing range is primarily slab with a minority 2D component generated by the interaction of counter-propagating Alfvén waves for both the fast and slow wind.
2. that the inertial range is either isotropic or possibly possesses a Goldreich-Sridhar scaling [87].
3. that the cross helicity for both fast and slow wind is small since counter-propagating Alfvén waves are essential to ensure the turbulent cascade of energy to small scales, and uni-directional Alfvén/slab propagation with a $k_\parallel^{-5/3}$ spectrum is not in general possible unless observations are made close to and just above the Alfvén surface. If the cross helicity is large, then turbulence is no longer operative within the context of the W/T model and can be regarded as "fossil" turbulence with the spectrum corresponding to a superposition of linear modes, possibly described by quasi-linear theory.
4. that the residual energy for both the fast and slow solar wind should be close to zero since the turbulence is primarily slab.

5. nothing about density fluctuations, at least not in the standard W/T models, although Lithwick and Goldreich [88] suggest that slow mode waves may be responsible for density fluctuations.

A more general discussion comparing these predictions to *PSP* observations can be found in Zank et al. [89] with additional details in Adhikari et al. [90–92] and Zhao et al. [93, 94]. Below, we focus on the first encounter of *PSP* with the Alfvén critical surface that separate the super-and sub-Alfvénic coronal regions, the latter of which is frequently described as the "magnetically dominated solar corona [8].

5   First Observations near the Alfvén Critical Surface

For about 5 hours between 09:30 and 14:40 UT on 2021-04-28 at around 0.1 au, the NASA *Parker Solar Probe (PSP)* entered a sub-Alfvénic region of the solar wind [8]. Two further shorter sub-Alfvénic intervals were subsequently sampled during encounter 8. Kasper et al. [8] ascribe the first sub-Alfvénic region to a steady flow in a region of rapidly expanding magnetic field above a pseudostreamer. The discovery of this hitherto in situ unobserved region of the solar wind represents a major accomplishment of the *PSP* mission, particularly for the insight it will provide in our understanding of how the solar corona is heated and the solar wind accelerated.

Zank et al. [9] examine the properties of low-frequency MHD turbulence in the first sub-Alfvénic interval observed by *PSP* and show that these observations admit a natural interpretation in terms of the NI MHD spectral theory [22].

Figure 5 is an overview of the first and longest of three sub-Alfvénic intervals identified by Kasper et al. [8]. The data includes magnetic field measurements from PSP/FIELDS [4], ion moments data from PSP/SWEAP instrument, and electron density derived from quasi-thermal noise (QTN) spectroscopy [3, 8]. The radial magnetic field is extremely steady with relatively small amplitude fluctuations. The radial velocity is also rather steady and the alignment between the flow and magnetic field vectors is very high. The quantity $\Psi \equiv \langle \theta_{B_0 U_0} \rangle$ represents the mean angle between the magnetic $\mathbf{B}_0$ and plasma velocity $\mathbf{U}_0$ mean fields during the intervals of interest. The velocity $\mathbf{U}_0$ is the relative velocity that includes the spacecraft speed, i.e., the spacecraft frame velocity. This is important for a generalized form of Taylor's hypothesis. The flow appears to be mostly sub-Alfvénic, although relatively marginally, across much of the interval, since the Alfvénic Mach number $M_A = V_R/V_A < \sim 1$. That the flow is so highly aligned renders quasi-2D fluctuations essentially invisible to *PSP*, and only fluctuations propagating along or anti-parallel to the inwardly directed (towards the Sun) magnetic field are observable. The plasma beta β_p is well below 1, being close to 10^{-1} for most of the interval. In the NI MHD context, this would imply that the predicted majority quasi-2D component was not observed because of the close alignment of the flow with the magnetic field; instead the observed fluctuations correspond to a minority slab component.

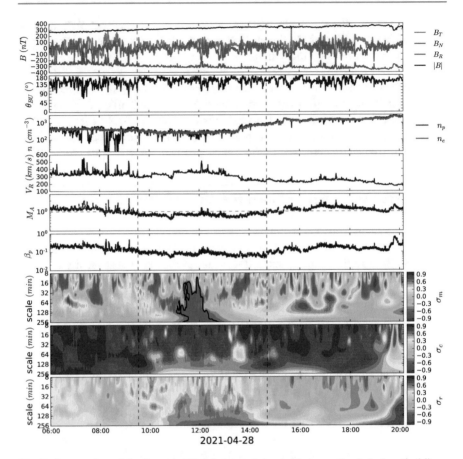

Fig. 5 An overview of the first sub-Alfvénic interval, located between the dashed vertical lines, and the adjacent super-Alfvénic intervals observed by *PSP* during encounter 8 showing from top to bottom the radial B_R, transverse B_T, and normal B_N magnetic fields, and the intensity $|\mathbf{B}|$, the angle θ_{BU} between the relative flow velocity $\mathbf{U}_0$ (i.e., the relative velocity between the solar wind flow and the spacecraft velocity vectors) and magnetic field vectors, the proton n_p (black) and electron number density n_e (red), the radial component of the plasma speed V_R measured in the inertial RTN frame, Alfvénic Mach number of the radial flow $M_A = V_R/V_A$, and the plasma beta β_p. The bottom three colored panels, in descending order, show frequency spectrograms of the normalized magnetic helicity σ_m, normalized cross helicity σ_c, and normalized residual energy σ_r. Reprinted from [9]

Zhao et al. [12, 95] have developed a method to automatically identify magnetic flux ropes and Alfvénic fluctuations using the observed rotation of the magnetic field [96, 97] and the normalized reduced magnetic helicity [98], which is usually high in regions of magnetic flux ropes [99, 100]. Alfvénic structures and flux ropes can be distinguished by evaluating the normalized cross helicity σ_c ($\equiv (\langle z^{+2} \rangle - \langle z^{-2} \rangle)/(\langle z^{+2} \rangle + \langle z^{-2} \rangle)$), where $\mathbf{z}^{\pm} = \mathbf{u} \pm \mathbf{b}/\sqrt{\mu_0 \rho}$ are the Elsässer variables for the fluctuating velocity $\mathbf{u}$ and magnetic $\mathbf{b}$ fields and ρ the mean plasma density, and the normalized residual energy σ_r ($\equiv \langle \mathbf{z}^+ \cdot \mathbf{z}^- \rangle / (\langle z^{+2} \rangle + \langle z^{-2} \rangle) = ((\langle u^2 \rangle - \langle b^2 \rangle / (\mu_0 \rho)) / (\langle u^2 \rangle + \langle b^2 \rangle / (\mu_0 \rho)))$. In most cases, the cross helicity of a magnetic flux rope is low (because of their closed-loop field structure, implying both sunward and anti-sunward Alfvénic fluctuations inside) and the residual energy is negative (indicating the dominance of magnetic fluctuation energy), whereas Alfvénic structures typically exhibit high cross helicity and a small residual energy values. Frequency spectrograms of the normalized magnetic helicity σ_m, normalized cross helicity σ_c, and normalized residual energy σ_r are illustrated in the bottom three panels of Fig. 5. Numerous small and mid-scale magnetic positive and negative rotations are present, including a particularly large structure bounded by black contour lines with $\langle \sigma_m \rangle \simeq -0.7$. Within this high magnetic helicity structure, the averaged σ_c is around 0.55 and the averaged σ_r is about 0.5. The scale size of this structure is around 80 minutes. The plasma outside this large structure has a cross helicity, close to 1, indicating almost exclusively outward propagating anti-parallel to the mean magnetic field, Elsässer fluctuations $\mathbf{z}^+$. The residual energy spectrogram shows large regions with $\sigma_r \simeq 0$, i.e., Alfvénic fluctuations, although the border region near 16:00 hours, 2021-04-28, appears to be comprised of largely magnetic structures with negative σ_r. The left-handed helical structure ($\sigma_m < 0$) at about 12:00 has a positive σ_r (~0.5), indicating the relative dominance of kinetic fluctuating energy. It suggests that *PSP* may have observed a vortical structure with a relatively weak wound-up magnetic field.

The combination of $\sigma_c \sim 0.9$ and $\sigma_r \sim 0.0$ for much of the sub-Alfvénic interval indicates that *PSP* is observing primarily Alfvénic fluctuations. This is typically described as "Alfvénic turbulence" despite the inability of *PSP* to discern non-Alfvénic structures easily in a highly magnetic field-aligned flow. The very high value of σ_c raises considerable problems for the notion of slab turbulence, which relies on counter-propagating Alfvén waves to generate the non-linear interactions that allow for the cascading of energy to ever smaller scales [71, 81, 82]. Such highly field-aligned flows with high σ_c, and hence populated by uni-directionally propagating Alfvén waves, have been observed near 1 au [10, 25] by the *Wind* spacecraft and closer to the Sun by *PSP* [12, 93]. The 1D reduced spectra in both cases exhibited a $k_{\parallel}^{-5/3}$ form in the inertial range, ($k_{\parallel}$ the wave number parallel to the mean magnetic field), raising questions about the validity of critical balance theory [25, 87]. The spectral theory of NI MHD in the $\beta_p \ll 1$ and ~ 1 regimes [22] shows that the interaction of a dominant quasi-2D component with uni-directional Alfvén waves can yield a $k_{\parallel}^{-5/3}$ spectrum. The sub-Alfvénic interval of Fig. 5 exhibits a number of features quite similar to those observed in the highly field-aligned flows discussed by Telloni et al. [25] and Zhao et al. [12, 93].

Zank et al. [9] analyze the turbulent properties of the sub-Alfvénic interval shown in Fig. 5 and interpret the observations based on the NI MHD spectral theory appropriate to the anisotropic superposition of 2D+slab turbulence [22]. This requires the correct application of Taylor's hypothesis for the sub-Alfvénic and modestly super-Alfvénic flows. Such an extension is straightforwardly developed in the context of the NI MHD superposition model, but it does require that the $\mathbf{z}^{\pm}$ Elsässer modes be treated separately for the minority forward and backward propagating slab modes [9] (see also [101]). The composite spectra for super-Alfvénic and sub-Alfvénic flows are given by (1) and (2), and (3) and (4) respectively below [9],

$$P_{\parallel}^{total\pm}(f) = P_{\parallel}^{\infty}(f) + P_{\parallel}^{*\pm}(f)$$

$$= \frac{C^{\infty}}{q^{\infty}+1}\left(\frac{U_0 \sin\Psi}{2\pi}\right)^{q^{\infty}-1} f^{-q^{\infty}} + \frac{C^{*\pm}}{2}\frac{2\pi}{U_0\cos\Psi \mp V_A}G^{*\pm}(k_z);$$

$$(1)$$

$$P_{\perp}^{total\pm}(f) = P_{\perp}^{\infty}(f) + P_{\perp}^{*\pm}(f)$$

$$= \frac{q^{\infty}}{q^{\infty}+1}C^{\infty}\left(\frac{U_0 \sin\Psi}{2\pi}\right)^{q^{\infty}-1} f^{-q^{\infty}}$$

$$+ \frac{C^{*\pm}}{2}\frac{2\pi}{U_0\cos\Psi \mp V_A}G^{*\pm}(k_z),$$

$$(2)$$

and $k_z = 2\pi f/(U_0\cos\Psi \mp V_A)$, C^{∞} the amplitude of the 2D turbulence, and q^{∞} the spectral index of the 2D component; and

$$P_{\parallel}^{total\pm}(f) = \frac{C^{\infty}}{q^{\infty}+1}\left(\frac{U_0 \sin\Psi}{2\pi}\right)^{q^{\infty}-1} f^{-q^{\infty}} + \frac{C^{*\pm}}{2}\frac{2\pi}{V_A \mp U_0\cos\Psi}G^{*\pm}(k_z);$$

$$(3)$$

$$P_{\perp}^{total\pm}(f) = \frac{q^{\infty}}{q^{\infty}+1}C^{\infty}\left(\frac{U_0 \sin\Psi}{2\pi}\right)^{q^{\infty}-1} f^{-q^{\infty}} + \frac{C^{*\pm}}{2}\frac{2\pi}{V_A \mp U_0\cos\Psi}G^{*\pm}(k_z),$$

$$(4)$$

and $k_z = 2\pi f/(V_A \mp U_0\cos\Psi)$. The spectral theory for $\beta_p \ll 1$, ~ 1 NI MHD predicts that the dominant 2D spectrum is a $-5/3$ power law in $k_{\perp}$ i.e., $G^{\infty}(k_{\perp}) \equiv E^{\infty}(k_{\perp})k_{\perp} = C^{\infty}k_{\perp}^{-5/3}$ [21, 22], where $E^{\infty}(k_{\perp})$ is the 1D Elsässer energy spectrum. However, because the flow is so highly magnetically aligned, $P_{\parallel}^{\infty}(f)$ and $P_{\perp}^{\infty}(f)$ are effectively zero. Thus, as described above, the quasi-perpendicular fluctuations are essentially invisible to *PSP* measurements unfortunately.

Figure 6 is a plot of the power spectral densities (PSDs) of the forward and backward Elsässer variables $\mathbf{z}^{\pm}$ for the sub-Alfvénic region (9:33–14:42 UT) and a neighboring super-Alfvénic region (15:00–20:10 UT). The PSD is evaluated over a 5-hour interval for each region using standard Fourier methods to calculate the trace spectra of $\mathbf{z}^{\pm}$ based on the Fourier-transformed autocorrelation function of the three components. The vertical dashed-dotted line in each panel denotes the frequency corresponding to the correlation scale that separates the energy-containing range and inertial range. A f^{-1} spectrum for the energy-containing range is displayed as a reference. The cyan and green curves in each panel are the theoretical predicted spectra for $\mathbf{z}^{\pm}$ in both sub- and super-Alfvén regions. The dominant component in

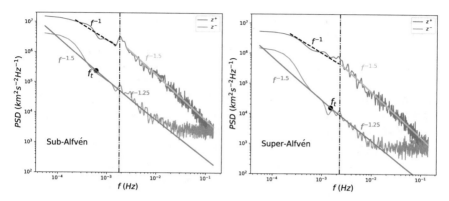

Fig. 6 Trace spectra of the Elsässer variables $z^{\pm}$ calculated in 5-hour intervals for the sub-Alfvénic region (left panel), and a neighboring super-Alfvénic region (right panel). The sub-Alfvénic region is delineated by the two vertical dashed lines shown in Fig. 1. The 5-hour interval just after the sub-Alfvénic region is selected as representative of the super-Alfvénic region. The solid green and cyan lines are predicted theoretical spectra, the dashed f^{-1} curve is to guide the eye, and f_t identifies the transition frequency (see text for details). The dashed vertical line separates the f^{-1} and the $f^{-1.5}$ sections of the z^+ spectra. Reprinted from [9]

both regions is the outward propagating z^+ component, with the PSD being at least an order of magnitude larger than that of the inward z^- component. The flattening of the z^- PSD at frequencies $\geq 10^{-2}$ Hz has no physical significance and is due to the noise floor in the plasma measurements. Careful examination of the z^- PSD shows that the low frequency part of the spectrum is steeper than the high frequency part, whereas the z^+ PSD, other than a bump at the inner scale [8], is a single power law in frequency with $\sim f^{-1.5}$. The modestly super-Alfvénic spectra for $z^{\pm}$ are very similar, and the dominance of z^+ is again evident.

The values of $\Psi \equiv \langle \theta_{B_0 U_0} \rangle$ in the sub- and super-Alfvénic regions analyses in Fig. 6 are 15° and 18° respectively. Both the strong alignment of mean magnetic field and mean velocity and the limited range of $\theta_{B_0 U_0}$ in both intervals of interest make it very difficult to evaluate the ratio of 2D and slab power. From both the values of Ψ and the normalized cross-helicity and residual energy spectrograms, it is clear that the spectra correspond to predominantly forward and minority backward propagating Alfvénic fluctuations since the 2D component is observationally invisible to *PSP*. While one may argue that the $f^{-3/2}$ z^+ spectrum is consistent with the Iroshnikov-Kraichnan theory, the argument is not credible since the needed counter-propagating z^- waves are almost entirely absent. The spectral theory of NI MHD [22] predicts that the general form of the NI/slab turbulence spectrum is given by

$$G^*(k_z) \equiv E^*(k_z, k_\perp) k_\perp^2 = C^* k_z^{-(2a+3)/3} \left(1 + \left(\frac{k_z}{k_t} \right)^{-(2a-3)/3} \right)^{1/2}, \tag{5}$$

where a describes a possible relationship between wave numbers $k_\perp$ and k_z, i.e., wave number anisotropy such that $k_\perp = k_z^a / k_t^{a-1}$ for $a > 0$. Equation (5), has a "transition wave number" k_t or frequency f_t (Fig. 6). The physical interpretation of k_t is that it represents the transition from a wave number regime controlled primarily by nonlinear interactions (with nonlinear timescale τ_∞) to a regime controlled by Alfvénic interactions (with a generalized Alfvénic timescale τ_A); specifically we have formally from the above definitions that $\tau_\infty / \tau_A = (k_z / k_t)^{-(2a-3)/3}$ (detailed discussions of the two time scales are found in [9,22]).

Equation (5) can be applied to the spectra shown in Fig. 6 by choosing $a = 3/4$, i.e., $k_\perp \sim k_z^{3/4}$. For the $\mathbf{z}^+$ spectrum, assume that $\tau_\infty \ll \tau_A^+$, i.e., nonlinear interactions mediated by quasi-2D fluctuations dominate, to obtain

$$G^{*+}(k_z) \simeq C^{*+} k_z^{-3/2}. \tag{6}$$

This is reasonable given the absence of sufficient $\mathbf{z}^-$ modes with which to interact nonlinearly [81,82]. For the $\mathbf{z}^-$ spectrum, suppose that τ_∞ and τ_A are finite since the minor $\mathbf{z}^-$ component can interact with the more numerous counter-propagating $\mathbf{z}^+$ modes. Thus, there exists a transition wave number k_t after which the $\mathbf{z}^-$ spectrum will flatten. Expression (5) becomes

$$G^{*-}(k_z) = C^{*-} k_z^{-3/2} \left(1 + \left(\frac{k_z}{k_t} \right)^{1/2} \right)^{1/2}, \tag{7}$$

showing that at small wave numbers the spectrum is $\sim k_z^{-3/2}$ and at large wave numbers the asymptotic spectrum is $\sim k_z^{-1.25}$. Equations (1)–(4) are used to express the wave number spectra for the 2D component ($C^\infty k_\perp^{-5/3}$) and the slab component (6) and (7) in frequency space. The results are illustrated in Fig. 6. The left panel for the sub-Alfvénic region shows that the observed $\mathbf{z}^\pm$ PSDs are very well fitted by the theory for values of $C^{*\pm} = 0.13$ and 1.5 respectively, and $k_t = 4.3 \times 10^{-4}$ km^{-1}. The predicted flattening of the theoretical frequency spectrum $f^{-3/2}$ at higher frequencies to $f^{-1.25}$ fits the observed $\mathbf{z}^-$ PSD well. The super-Alfvénic interval spectra (right panel) are similarly well fitted with similar parameters $C^{*\pm} = 0.08$ and 0.012 respectively, and $k_t = 1.0 \times 10^{-4}$ km^{-1}. Since Ψ is essentially parallel in both intervals, the contribution from the 2D spectrum is very small and prevents evaluation of C^∞. The full anisotropy cannot therefore be determined because we cannot evaluate the power spectrum $P_\perp^\infty(f)$ (Eqs. (2) and (4)) observationally. The parameter a relating $k_\perp$ and k_z introduces a modest slab anisotropy such that $P_\parallel^*(k_z) > P_\parallel^*(k_\perp)$, at sufficiently small scales. In the context of nearly incompressible MHD, this implies the ordering $P_\perp^\infty(k_\perp) \gg P_\parallel^*(k_z) > P_\parallel^*(k_\perp)$ with $a = 3/4$ [22]. As discussed in the physical interpretation below, where observations of non-slab turbulence can be made by *PSP*, a significant quasi-2D component has been observed [102, 103].

The fitting parameters apply primarily to the spectral slopes and although the spectral indices for the $z^{\pm}$ spectra are similar for both the super- and sub-Alfvénic intervals, there are some obvious differences. For example, the transition frequency f_t shifts to a larger frequency in the super-Alfvénic region, suggesting that nonlinear interactions rather than Alfvénic interactions dominate more of the low-frequency spectrum. In addition, the spectral amplitude in the inertial range for both $z^{\pm}$ in the sub-Alfvénic region is approximately 5 times larger than that in the super-Alfvénic region, and this appears to be true of the f^{-1} energy-containing range for the z^{+} spectra too. However, the important point, discussed above, is that there is no major change in the nature of the low-frequency MHD turbulence in the modestly sub-Alfvénic and modestly super-Alfvénic flows, as expected from a turbulence description dominated by 2D fluctuations and a uni-directionally propagating Alfvén wave component.

Shown in Fig. 7 are plots of the electron number density PSDs in the sub- and super-Alfvénic regions. The more accurate low-frequency receiver (LFR) electron data set from the Radio Frequency Spectrometer (RFS) part of the PSP/FIELDS instrument suite [4] has been used since it is sufficiently well populated over the range shown in Fig. 7. The flattening at higher frequencies beyond some 10^{-2} Hz is likely real despite the LFR data being unreliable at these higher frequencies [9]. The spectra over the interval 4×10^{-4} Hz to 10^{-2} Hz are simple power laws with a spectral index of about -1.59 in the sub-Alfvénic flow and about -1.89 in the super-Alfénic flow. The density spectra for both regions do not correspond to the z^{+} spectra in the inertial range that have a -1.5 spectral index, and are quite

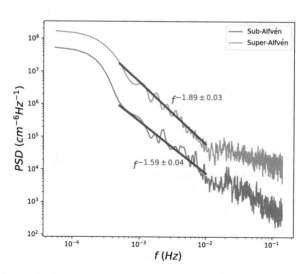

Fig. 7 PSDs for the fluctuating density variance in the sub- and super-Alfvénic regions corresponding to those used for Fig. 3. The LFR electron density from FIELDS has been used. The fitted frequency range for both intervals starts from 5×10^{-4} Hz to 1×10^{-2} Hz. Reprinted from [9]

unlike the convex $\mathbf{z}^-$ spectra (Fig. 6), thus ruling out the possibility that the density spectra are due to the parametric decay instability [104–106]. Density spectra can be determined from the NI MHD theory. Within the NI MHD theory, density fluctuations are entropic modes, i.e., zero frequency fluctuations, that are advected by the dominant quasi-2D turbulent velocity fluctuations and therefore behave as a passive scalar [21]. The relevant time scale is the quasi-2D nonlinear time scale and the underlying spectrum responsible for advection is that associated with the 2D velocity fluctuations. The NI MHD theory explicitly shows that the 2D Elsässer variable spectrum satisfies $E_{2D} \sim k_\perp^{-5/3}$. Under some circumstances [21], this serves as a proxy for the spectrum of 2D velocity fluctuations but more generally, the dominant quasi-2D velocity spectrum $E_{2D,v}$ will have the form $E_{2D,v} \sim k_\perp^{-q}$. This yields a density spectrum of the form $E_\rho(k_\perp) = C_\rho k_\perp^{-q}$, where $q = 5/3$ only if e.g., kinetic energy dominates or the 2D residual energy is zero. Zank et al. [9] show that the frequency spectrum for the density PSD $P_\rho(f)$ is related to the density wave number PSD $P_\rho(k_\perp)$ according to

$$P_\rho(f) = C_\rho \int_{|k_x|}^{\infty} \frac{k_\perp^{-q}}{\sqrt{k_\perp^2 - k_x^2}} dk_\perp = \frac{\sqrt{\pi} C_\rho k_x^{-q} \Gamma\left(\frac{q}{2}\right)}{2\Gamma\left(\frac{q+1}{2}\right)}, \quad k_x = \frac{2\pi f}{U_0 \sin \Psi}, \quad (8)$$

where C_ρ is the amplitude of the density spectrum. The frequency spectrum therefore has the form f^{-q}. That the density spectrum is noticeably distinct from both the observed $\mathbf{z}^\pm$ spectra suggests its origin is unrelated to the slab spectra, whether via the parametric decay instability or passive advection of density fluctuations by slab turbulence. That leaves the possibilities that the density fluctuations are either zero frequency NI MHD entropic modes advected by quasi-2D incompressible turbulence [21] or compressible wave modes, which have been identified in *PSP* data in the presence of dominant incompressible turbulence [93].

The compressibility of the fluctuations is presented in Fig. 8, illustrating that the transverse or incompressible magnetic fluctuations, shown by the orange curve, are clearly dominant compared to the compressible magnetic field aligned fluctuations (blue curve). This is true of both sub-and super-Alfvénic regions and is consistent with observations discussed previously by Zhao et al. [93]. This assures us that the turbulence observed by PSP is largely incompressible, including in the sub-Alfvénic solar wind. The spectral slopes are found to be about -1.52 for incompressible magnetic fluctuations and about -1.48 for the corresponding compressible component in the sub-Alfvénic wind and respectively -1.48 and -1.44 in the super-Alfvénic interval over the frequency range 2×10^{-3} to 0.2 Hz. It is evident that the spectral indices of the observed density PSDs are quite different from those of the compressible spectra shown in Fig. 8. Whereas the compressible fluctuations shown in Fig. 8 can be associated with fast (and possibly slow) magnetosonic modes [93], the very different characteristics of the density PSDs suggest that these fluctuations are not associated with waves but instead are likely to be entropy fluctuations with zero frequency. This result demonstrates post facto the rationale for using Taylor's

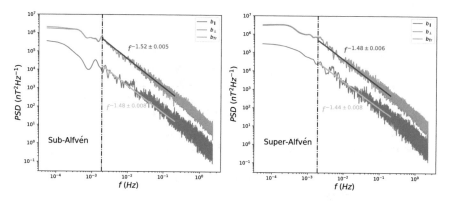

Fig. 8 PSDs for magnetic fluctuations in the parallel (blue) and transverse (orange) directions for the sub-Alfvénic (left panel) and super-Alfvénic (right panel) regions. The total trace spectrum (green) effectively overlays the transverse fluctuations. Pink and blue colored lines are fitted to the data over the frequency range 2×10^{-3} to 0.2 Hz to estimate the spectral indices of the incompressible and compressible magnetic field PSDs. Reprinted from [9]

hypothesis for density fluctuations in the form of Eq. (8), i.e., for advected density fluctuations. Within NI MHD, density fluctuations are advected by the dominant 2D turbulence in the $O(1)$ and $\ll 1$ plasma beta regimes [21, 107], thereby providing insight into the dominant 2D velocity turbulence that *PSP* is unable to observe in the highly field-aligned intervals. In particular, the density spectrum in the sub-Alfvénic region suggest a quasi-2D velocity spectrum very slightly flatter than $-5/3$ and rather steeper than $-5/3$ in the super-Alfvénic interval.

Evidence for the presence of quasi-2D fluctuations has been presented elsewhere. Zhao et al. [12, 93] have identified small-scale flux ropes observed in previous *PSP* encounters. Bandyopadhyay and McComas [102] and Zhao et al. [103] find that turbulence in the inner heliosphere is highly anisotropic with significant contributions from a 2D component, in many cases dominating the slab contribution despite the challenges for *PSP* observing quasi-2D fluctuations when the magnetic field and plasma flow become increasingly highly aligned with decreasing distance above the solar surface. Accordingly, we can interpret the spectra illustrated in Fig. 6 as due to the possibly minority slab component of quasi-2D+slab turbulence. This then yields the following physical interpretation of the observed $z^{\pm}$ turbulent fluctuations in the sub-Alfvénic and modestly super-Alfvénic regions of the solar wind. The z^{+} (outward) fluctuations dominate, with the spectral amplitude of inward propagating slab modes nearly an order of magnitude smaller, meaning the slab component is comprised almost entirely of uni-directionally propagating Alfvén waves. Given the much smaller intensity of inward propagating modes, the z^{+} are obliged to interact nonlinearly almost exclusively with quasi-2D modes to produce the observed power law spectrum. The nonlinear interaction is governed by the nonlinear time scale τ_{∞} with virtually no interaction on the Alfvén time scale τ_A with the counter-propagating minority z^{-} component. The spectral slope of

$-3/2$ for the $\mathbf{z}^+$ (and low frequency part of the $\mathbf{z}^-$ spectrum) indicates modest slab wave number anisotropy with $k_z \sim k_\perp^{4/3}$ in the inertial range. By contrast, the low-frequency $\mathbf{z}^-$ component is dominated by nonlinear rather than Alfvénic interactions, unlike the high frequencies that are governed primarily by interactions with counter-propagating Alfvén modes on the time scale τ_A^-. This is manifest in the concavity of the $\mathbf{z}^-$ spectrum due to the presence of a transition wave number or frequency at which $\tau_\infty = \tau_A$. Nonetheless, to explain the slab observations presented here in the context of nearly incompressible MHD, the 2D component power anisotropy should dominate the power in the slab component.

6   Conclusions

1. *Parker Solar Probe* observed primarily outwardly propagating Alfvénic fluctuations during the first of the sub-Alfvénic intervals observed. This likely reflects the highly magnetic field-aligned flow of the interval that renders quasi-2D fluctuations effectively invisible to observations. Nonetheless, some evidence of magnetic structures is present near the interval boundaries, as well as a large vortex-like structure embedded in the interval.

2. An extended form of Taylor's hypothesis is necessary to relate frequency and wave number spectra in the analysis of turbulence in sub-Alfvénic and the modestly super-Alfvénic flows. Zank et al. [9] introduced a modified form of Taylor's hypothesis based on a decomposition of the turbulence into 2D and forward and backward propagating slab components (an alternative but related approach has been suggested by Bourouaine and Perez [101]).

3. The PSDs for the Elsässer fluctuations in the sub- and super-Alfvénic interval show that the forward $\mathbf{z}^+$ component dominates, having a spectral amplitude much greater than that of the inward component, and a frequency (wave number) spectrum of the form $f^{-3/2}$ $(k_\parallel^{-3/2})$ throughout the inertial range. By contrast, the inward $\mathbf{z}^-$ PSD exhibits a convex spectrum: $f^{-3/2}$ $(k_\parallel^{-3/2})$ at low frequencies that flattens around a transition frequency (wave number) f_t (k_t) to $f^{-1.25}$ $(k_\parallel^{-1.25})$ at higher frequencies. Because *PSP* makes measurements in a highly aligned flow, the observations correspond largely to slab fluctuations.

4. A NI MHD 2D+slab spectral theory [22] provides a physical interpretation of the $\mathbf{z}^\pm$ spectra. The theoretically predicted slab spectra are in excellent agreement with the observed spectra provided there exists a modest slab wave number anisotropy $k_\perp \sim k_\parallel^{3/4}$. The $\mathbf{z}^+$ wave number spectrum is predicted to be $k_\parallel^{-3/2}$ because it interacts primarily with quasi-2D fluctuations on a time scale τ_∞ rather than the significantly smaller $\mathbf{z}^-$ component. By contrast, the minority $\mathbf{z}^-$ fluctuations can interact with both quasi-2D and counter-propagating slab modes, so that both the nonlinear τ_∞ and Alfvén τ_A time scales determine the form of the spectrum. Theoretically, this combination of time scales predicts a convex spectrum with an inflection or transition point determined by the balance of the

time scales, $\tau_\infty = \tau_A$, and the spectrum is predicted to flatten from a $f^{-3/2}$ ($k_\parallel^{-3/2}$) low frequency or nonlinear dominated regime to a $f^{-1.25}$ ($k_\parallel^{-1.25}$) higher frequency or Alfvénic dominated regime.

5. The density fluctuation PSDs exhibit simple power laws with spectral indices of -1.59 and -1.89 for the sub-and super-Alfvénic cases, respectively. The spectra resemble neither the dominant outward or inward Elsässer spectra, suggesting that the density fluctuations are not due to the parametric decay instability. The compressible magnetic field fluctuation spectrum follows the incompressible magnetic field spectrum closely and is distinctly different from the density spectrum, suggesting that the density fluctuations are not associated with compressible magnetosonic wave modes. Instead, they appear to be zero-frequency entropic modes advected by the background turbulent velocity field. This interpretation is consistent with the expectations of NI MHD in which entropic density fluctuations are advected by the dominant quasi-2D velocity fluctuations, indicating that the density spectra offer insight into quasi-2D turbulence.

6. The spectra in the modestly super-Alfvénic regions closely resemble those in the sub-Alfvénic interval, and indeed the three parameters ($C^{*\pm}$, and a) for the two sets of spectra are very similar, indicating that the same basic turbulence physics holds in both regions. Nonetheless, there are some differences in details, such as the transition frequency shifting to a larger frequency in the super-Alfvénic region, and the fluctuating power for both $\mathbf{z}^+$ and $\mathbf{z}^-$ in the sub-Alfvénic region is approximately 5 times larger than that in the super-Alfvénic region. The larger k_t for the super-Alfvénic flow indicates that nonlinear interactions rather than Alfvénic interactions dominate for a larger part of the low-frequency spectrum. Nonetheless, there is no profound change in turbulence properties in the modestly super-Alfvénic and sub-Alfvénic flows, indicating that it is unlikely that the Alfvén critical surface serves to filter dominant W/T fluctuations and instead the turbulence is likely dominated by quasi-2D fluctuations that are advected unchanged across the Alfvén critical surface.

The physical interpretation of the Elsässer forward and backward slab and density spectra reflects a manifestation of dominant quasi-2D turbulent fluctuations in the solar wind. The same parameters explain both the observed forward and backward Elsässer spectra. Since the fitting of the results is predicated on a quasi-2D nonlinear time scale and a quasi-2D spectrum of the Kolmogorov form, the results presented here suggest the presence of a dominant 2D component that, because of the field-aligned sampling in both intervals during this encounter, cannot be observed by *PSP*, but nevertheless controls the evolution of slab and density turbulence in the sub-Alfvénic solar wind. *Parker Solar Probe* is bringing us closer to answering the outstanding 50-year-old question of how the solar corona is heated and how the solar wind is driven.

Acknowledgments GPZ, LLZ, and LA acknowledge the partial support of a NASA Parker Solar Probe contract SV4-84017, an NSF EPSCoR RII-Track-1 Cooperative Agreement OIA-1655280, a NASA IMAP subaward under NASA contract 80GSFC19C0027, and a NASA award 80NSSC20K1783. DT was partially supported by the Italian Space Agency (ASI) under contract 2018-30-HH.0. *Parker Solar Probe* was designed, built, and is now operated by the Johns Hopkins Applied Physics Laboratory as part of NASA's Living with a Star (LWS) program (contract NNN06AA01C). Support from the LWS management and technical team has played a critical role in the success of the *Parker Solar Probe* mission.

References

1. N.J. Fox, M.C. Velli, S.D. Bale, R. Decker, A. Driesman, R.A. Howard, J.C. Kasper, J. Kinnison, M. Kusterer, D. Lario, M.K. Lockwood, D.J. McComas, N.E. Raouafi, A. Szabo, Space Sci. Rev. **204**, 7–48 (2016)
2. L. Golub, J.M. Pasachoff, *The Solar Corona* (Cambridge University Press, 1997)
3. J.C. Kasper, R. Abiad, G. Austin, M. Balat-Pichelin, S.D. Bale, J.W. Belcher, P. Berg, H. Bergner, M. Berthomier, J. Bookbinder, E. Brodu, D. Caldwell, A.W. Case, B.D.G. Chandran, P. Cheimets, J.W. Cirtain, S.R. Cranmer, D.W. Curtis, P. Daigneau, G. Dalton, B. Dasgupta, D. DeTomaso, M. Diaz-Aguado, B. Djordjevic, B. Donaskowski, M. Effinger, V. Florinski, N. Fox, M. Freeman, D. Gallagher, S.P. Gary, T. Gauron, R. Gates, M. Goldstein, L. Golub, D.A. Gordon, R. Gurnee, G. Guth, J. Halekas, K. Hatch, J. Heerikuisen, G. Ho, Q. Hu, G. Johnson, S.P. Jordan, K.E. Korreck, D. Larson, A.J. Lazarus, G. Li, R. Livi, M. Ludlam, M. Maksimovic, J.P. McFadden, W. Marchant, B.A. Maruca, D.J. McComas, L. Messina, T. Mercer, S. Park, A.M. Peddie, N. Pogorelov, M.J. Reinhart, J.D. Richardson, M. Robinson, I. Rosen, R.M. Skoug, A. Slagle, J.T. Steinberg, M.L. Stevens, A. Szabo, E.R. Taylor, C. Tiu, P. Turin, M. Velli, G. Webb, P. Whittlesey, K. Wright, S.T. Wu, G. Zank Space Sci. Rev. **204**, 131–186 (2016)
4. S.D. Bale, K. Goetz, P.R. Harvey, P. Turin, J.W. Bonnell, T. Dudok de Wit, R.E. Ergun, R.J. MacDowall, M. Pulupa, M. Andre, M. Bolton, J.L. Bougeret, T.A. Bowen, D. Burgess, C.A. Cattell, B.D.G. Chandran, C.C. Chaston, C.H.K. Chen, M.K. Choi, J.E. Connerney, S. Cranmer, M. Diaz-Aguado, W. Donakowski, J.F. Drake, W.M. Farrell, P. Fergeau, J. Fermin, J. Fischer, N. Fox, D. Glaser, M. Goldstein, D. Gordon, E. Hanson, S.E. Harris, L.M. Hayes, J.J. Hinze, J.V. Hollweg, T.S. Horbury, R.A. Howard, V. Hoxie, G. Jannet, M. Karlsson, J.C. Kasper, P.J. Kellogg, M. Kien, J.A. Klimchuk, V.V. Krasnoselskikh, S. Krucker, J.J. Lynch, M. Maksimovic, D.M. Malaspina, S. Marker, P. Martin, J. Martinez-Oliveros, J. McCauley, D.J. McComas, T. McDonald, N. Meyer-Vernet, M. Moncuquet, S.J. Monson, F.S. Mozer, S.D. Murphy, J. Odom, R. Oliverson, J. Olson, E.N. Parker, D. Pankow, T. Phan, E. Quataert, T. Quinn, S.W. Ruplin, C. Salem, D. Seitz, D.A. Sheppard, A. Siy, K. Stevens, D. Summers, A. Szabo, M. Timofeeva, A. Vaivads, M. Velli, A. Yehle, D. Werthimer, J.R. Wygant, Space Sci. Rev. **204**, 49–82 (2016)
5. S.D. Bale, S.T. Badman, J.W. Bonnell, T.A. Bowen, D. Burgess, A.W. Case, C.A. Cattell, B.D.G. Chandran, C.C. Chaston, C.H.K. Chen, J.F. Drake, T.D. de Wit, J.P. Eastwood, R.E. Ergun, W.M. Farrell, C. Fong, K. Goetz, M. Goldstein, K.A. Goodrich, P.R. Harvey, T.S. Horbury, G.G. Howes, J.C. Kasper, P.J. Kellogg, J.A. Klimchuk, K.E. Korreck, V.V. Krasnoselskikh, S. Krucker, R. Laker, D.E. Larson, R.J. MacDowall, M. Maksimovic, D.M. Malaspina, J. Martinez-Oliveros, D.J. McComas, N. Meyer-Vernet, M. Moncuquet, F.S. Mozer, T.D. Phan, M. Pulupa, N.E. Raouafi, C. Salem, D. Stansby, M. Stevens, A. Szabo, M. Velli, T. Woolley, J.R. Wygant, Nature **576**, 237–242 (2019)
6. J.C. Kasper, S.D. Bale, J.W. Belcher, M. Berthomier, A.W. Case, B.D.G. Chandran, D.W. Curtis, D. Gallagher, S.P. Gary, L. Golub, J.S. Halekas, G.C. Ho, T.S. Horbury, Q. Hu, J. Huang, K.G. Klein, K.E. Korreck, D.E. Larson, R. Livi, B. Maruca, B. Lavraud, P. Louarn, M. Maksimovic, M. Martinovic, D. McGinnis, N.V. Pogorelov, J.D. Richardson, R.M. Skoug,

J.T. Steinberg, M.L. Stevens, A. Szabo, M. Velli, P.L. Whittlesey, K.H. Wright, G.P. Zank, R.J. MacDowall, D.J. McComas, R.L. McNutt, M. Pulupa, N.E. Raouafi, N.A. Schwadron Nature **576**, 228–231 (2019)

7. S.R. Cranmer, A.A. van Ballegooijen, Astrophys. J. **720**, 824–847 (2010). (Preprint 1007.2383)
8. J.C. Kasper, K.G. Klein, E. Lichko, J. Huang, C.H.K. Chen, S.T. Badman, J. Bonnell, P.L. Whittlesey, R. Livi, D. Larson, M. Pulupa, A. Rahmati, D. Stansby, K.E. Korreck, M. Stevens, A.W. Case, S.D. Bale, M. Maksimovic, M. Moncuquet, K. Goetz, J.S. Halekas, D. Malaspina, N.E. Raouafi, A. Szabo, R. MacDowall, M. Velli, T. Dudok de Wit, G.P. Zank, Phys. Rev. Lett. **127**, 255101 (2021)
9. G.P. Zank, L.L. Zhao, L. Adhikari, D. Telloni, J.C. Kasper, M. Stevens, A. Rahmati, S.D. Bale, Astrophys. J. Lett. **926**, L16 (2022)
10. X. Wang, C. Tu, J. He, E. Marsch, L. Wang, C. Salem, Astrophys. J. Lett. **810**, L21 (2015)
11. D. Telloni, F. Carbone, R. Bruno, L. Sorriso-Valvo, G.P. Zank, L. Adhikari, P. Hunana, Astrophys. J. **887**, 160 (2019)
12. L.L. Zhao, G.P. Zank, L. Adhikari, M. Nakanotani, D. Telloni, F. Carbone, Astrophys. J. **898**, 113 (2020)
13. W.H. Matthaeus, G.P. Zank, S. Oughton, D.J. Mullan, P. Dmitruk, Astrophys. J. Lett. **523**, L93–L96 (1999)
14. H. Alfvén, C.G. Fält hammer, *Cosmical Electrodynamics*, 2nd edn., vol. 877 (Clarendon, Oxford, 1963)
15. C.A. Ferraro, C. Plumpton, Astrophys. J. **127**, 459 (1958)
16. J.F. McKenzie, M. Banaszkiewicz, W.I. Axford, Astron. Astrophys. **303**, L45 (1995)
17. D. Mullan, *Physics of the Sun* (CRC Press, 2009)
18. A.M. Title, C.J. Schrijver, The Sun's magnetic carpet, in *Cool Stars, Stellar Systems, and the Sun* (*Astronomical Society of the Pacific Conference Series*, vol. 154), ed. by R.A. Donahue and J.A. Bookbinder (1998), p. 345
19. A.F. Rappazzo, E.N. Parker, Astrophys. J. Lett. **773**, L2 (2013) (*Preprint* 1306.6634)
20. G.P. Zank, L. Adhikari, P. Hunana, S.K. Tiwari, R. Moore, D. Shiota, R. Bruno, D. Telloni, Astrophys. J. **854**, 32 (2018)
21. G.P. Zank, L. Adhikari, P. Hunana, D. Shiota, R. Bruno, D. Telloni, Astrophys. J. **835**, 147 (2017)
22. G.P. Zank, M. Nakanotani, L.L. Zhao, L. Adhikari, D. Telloni, Astrophys. J. **900**, 115 (2020)
23. L. Adhikari, G.P. Zank, L.L. Zhao, Astrophys. J. **901**, 102 (2020)
24. D. Telloni, E. Antonucci, M.A. Dodero, Astron. Astrophys. **476**, 1341–1346 (2007)
25. D. Telloni, S. Giordano, E. Antonucci, Astrophys. J. Lett. **881**, L36 (2019)
26. D. Telloni, L. Adhikari, G.P. Zank, L. Zhao, L. Sorriso-Valvo, E. Antonucci, S. Giordano, S. Mancuso, Astrophys. J. **929**, 98 (2022)
27. B. De Pontieu, S.W. McIntosh, M. Carlsson, V.H. Hansteen, T.D. Tarbell, C.J. Schrijver, A.M. Title, R.A. Shine, S. Tsuneta, Y. Katsukawa, K. Ichimoto, Y. Suematsu, T. Shimizu, S. Nagata, Science **318**, 1574 (2007)
28. R. Erdélyi, V. Fedun, Science **318**, 1572 (2007)
29. S. Tomczyk, S.W. McIntosh, S.L. Keil, P.G. Judge, T. Schad, D.H. Seeley, J. Edmondson, Science **317**, 1192 (2007)
30. J.W. Cirtain, L. Golub, L. Lundquist, A. van Ballegooijen, A. Savcheva, M. Shimojo, E. DeLuca, S. Tsuneta, T. Sakao, K. Reeves, M. Weber, R. Kano, N. Narukage, K. Shibasaki, Science **318**, 1580 (2007)
31. S. Tomczyk, S.W. McIntosh, Astrophys. J. **697**, 1384–1391 (2009) (Preprint 0903.2002)
32. G. Verth, M. Goossens, J.S. He, Astrophys. J. Lett. **733**, L15 (2011)
33. T.J. Okamoto, B. De Pontieu, Astrophys. J. Lett. **736**, L24 (2011) (Preprint 1106.4270)
34. D. Kuridze, R.J. Morton, R. Erdélyi, G.D. Dorrian, M. Mathioudakis, D.B. Jess, F.P. Keenan, Astrophys. J. **750**, 51 (2012) (Preprint 1202.5697)
35. T.V. Zaqarashvili, R. Erdélyi, Space Sci. Rev. **149**, 355–388 (2009) (Preprint 0906.1783)

36. M. Mathioudakis, D.B. Jess, R. Erdélyi, Space Sci. Rev. **175**, 1–27 (2013) (Preprint 1210.3625)
37. T. Van Doorsselaere, V.M. Nakariakov, P.R. Young, E. Verwichte, Astron. Astrophys. **487**, L17–L20 (2008)
38. S.W. McIntosh, B. de Pontieu, M. Carlsson, V. Hansteen, P. Boerner, M. Goossens, Nature **475**, 477–480 (2011)
39. R.J. Morton, J.A. McLaughlin, Astron. Astrophys. **553**, L10 (2013) (Preprint 1305.0140)
40. G. Nisticò, V.M. Nakariakov, E. Verwichte, Astron. Astrophys. **552**, A57 (2013)
41. J.O. Thurgood, R.J. Morton, J.A. McLaughlin, Astrophys. J. Lett. **790**, L2 (2014) (Preprint 1406.5348)
42. R.J. Morton, S. Tomczyk, R.F. Pinto, Astrophys. J. **828**, 89 (2016) (Preprint 1608.01831)
43. M. Velli, Astron. Astrophys. **270**, 304–314 (1993)
44. G.L. Withbroe, Astrophys. J. **325**, 442 (1988)
45. R.L. Moore, A.C. Sterling, D.A. Falconer, Astrophys. J. **806**, 11 (2015) (Preprint 1504.03700)
46. R.J. Leamon, W.H. Matthaeus, C.W. Smith, G.P. Zank, D.J. Mullan, S. Oughton, Astrophys. J. **537**, 1054–1062 (2000)
47. S.R. Cranmer, Astrophys. J. **532**, 1197–1208 (2000)
48. J.L. Kohl, G. Noci, S.R. Cranmer, J.C. Raymond, Astron. Astrophys. Rev. **948**, 31–66 (2006)
49. W.I. Axford, J.F. McKenzie, The solar wind, in *Cosmic Winds and the Heliosphere*, ed. by J.R. Jokipii, C.P. Sonett, M.S. Giampapa (1997), p. 31
50. R. Bruno, L. Trenchi, Astrophys. J. Lett. **787**, L24 (2014) (Preprint 1404.2191)
51. D. Telloni, F. Carbone, R. Bruno, G.P. Zank, L. Sorriso-Valvo, S. Mancuso, Astrophys. J. Lett. **885**, L5 (2019)
52. J.V. Hollweg, J. Geophys. Res. **104**, 14811–14820 (1999)
53. O. Pezzi, F. Malara, S. Servidio, F. Valentini, T.N. Parashar, W.H. Matthaeus, P. Veltri, Phys. Rev. E **96**, 023201 (2017)
54. S.D. Bale, P.J. Kellogg, F.S. Mozer, T.S. Horbury, H. Reme, Phys. Rev. Lett. **94**, 215002 (2005) (Preprint physics/0503103)
55. F. Malara, G. Nigro, F. Valentini, L. Sorriso-Valvo, Astrophys. J. **871**, 66 (2019)
56. S. Oughton, W.H. Matthaeus, P. Dmitruk, L.J. Milano, G.P. Zank, D.J. Mullan, Astrophys. J. **551**, 565–575 (2001)
57. P. Dmitruk, L.J. Milano, W.H. Matthaeus, Astrophys. J. **548**, 482–491 (2001)
58. P. Dmitruk, W.H. Matthaeus, L.J. Milano, S. Oughton, G.P. Zank, D.J. Mullan, Astrophys. J. **575**, 571–577 (2002) (Preprint astro-ph/0204347)
59. T.K. Suzuki, S.I. Inutsuka, Astrophys. J. Lett. **632**, L49–L52 (2005) (Preprint astro-ph/0506639)
60. S.R. Cranmer, A.A. van Ballegooijen, R.J. Edgar, Astrophys. J. Suppl. **171**, 520–551 (2007) (Preprint astro-ph/0703333)
61. S.R. Cranmer, A.A. van Ballegooijen, L.N. Woolsey, Astrophys. J. **767**, 125 (2013) (Preprint 1303.0563)
62. Y.M. Wang, Y.K. Ko, R. Grappin, Astrophys. J. **691**, 760–769 (2009)
63. B.D.G. Chandran, J.V. Hollweg, Astrophys. J. **707**, 1659–1667 (2009) (Preprint 0911.1068)
64. A. Verdini, M. Velli, W.H. Matthaeus, S. Oughton, P. Dmitruk, Astrophys. J. Lett. **708**, L116–L120 (2010) (Preprint 0911.5221)
65. T. Matsumoto, K. Shibata, Astrophys. J. **710**, 1857–1867 (2010) (Preprint 1001.4307)
66. B.D.G. Chandran, T.J. Dennis, E. Quataert, S.D. Bale, Astrophys. J. **743**, 197 (2011) (Preprint 1110.3029)
67. A.V. Usmanov, W.H. Matthaeus, B.A. Breech, M.L. Goldstein, Astrophys. J. **727**, 84–+ (2011)
68. R. Lionello, M. Velli, C. Downs, J.A. Linker, Z. Mikić, A. Verdini, Astrophys. J. **784**, 120 (2014) (Preprint 1402.4188)
69. A.V. Usmanov, M.L. Goldstein, W.H. Matthaeus, Astrophys. J. **788**, 43 (2014)

70. L.N. Woolsey, S.R. Cranmer, Astrophys. J. **787**, 160 (2014) (Preprint 1404.5998)
71. J.V. Shebalin, W.H. Matthaeus, D. Montgomery, J. Plasma Phys. **29**, 525–547 (1983)
72. L.A. Fisk, N.A. Schwadron, T.H. Zurbuchen, J. Geophys. Res. **104**, 19765–19772 (1999)
73. L.A. Fisk, J. Geophys. Res. (Space Phys.) **108**, 1157 (2003)
74. R. D'Amicis, L. Matteini, R. Bruno, Mon. Not. R. Astron. Soc. **483**, 4665–4677 (2019) (Preprint 1812.01899)
75. G. Nigro, F. Malara, V. Carbone, P. Veltri, Phys. Rev. Lett. **92**, 194501 (2004)
76. G. Nigro, F. Malara, P. Veltri, Astrophys. J. **685**, 606–621 (2008)
77. R. D'Amicis, R. Bruno, B. Bavassano, J. Atmos. Solar-Terrestrial Phys. **73**, 653–657 (2011)
78. L. Matteini, T.S. Horbury, M. Neugebauer, B.E. Goldstein, Geophys. Res. Lett. **41**, 259–265 (2014)
79. L. Matteini, T.S. Horbury, F. Pantellini, M. Velli, S.J. Schwartz, Astrophys. J. **802**, 11 (2015) (Preprint 1501.00702)
80. L. Adhikari, G.P. Zank, L.L. Zhao, Astrophys. J. **876**, 26 (2019)
81. M. Dobrowolny, A. Mangeney, P. Veltri, Phys. Rev. Lett. **45**, 144–147 (1980)
82. M. Dobrowolny, A. Mangeney, P. Veltri, Astron. Astrophys. **83**, 26–32 (1980)
83. G.P. Zank, A. Dosch, P. Hunana, V. Florinski, W.H. Matthaeus, G.M. Webb, Astrophys. J. **745**, 35 (2012)
84. L.L. Williams, G.P. Zank, J. Geophys. Res. **99**, 19229–+ (1994)
85. R. Bandyopadhyay, W.H. Matthaeus, D.J. McComas, R. Chhiber, A.V. Usmanov, J. Huang, R. Livi, D.E. Larson, J.C. Kasper, A.W. Case, M. Stevens, P. Whittlesey, O.M. Romeo, S.D. Bale, J.W. Bonnell, T. Dudok de Wit, K. Goetz, P.R. Harvey, R.J. MacDowall, D.M. Malaspina, M. Pulupa, Astrophys. J. Lett. **926**, L1 (2022) (Preprint 2201.10718)
86. L.L. Zhao, G.P. Zank, D. Telloni, M. Stevens, J.C. Kasper, S.D. Bale, Astrophys. J. Lett. **928**, L15 (2022)
87. P. Goldreich, S. Sridhar, Astrophys. J. **438**, 763–775 (1995)
88. Y. Lithwick, P. Goldreich, Astrophys. J. **562**, 279–296 (2001) (Preprint astro-ph/0106425)
89. G.P. Zank, L.L. Zhao, L. Adhikari, D. Telloni, J.C. Kasper, S.D. Bale, Phys. Plasmas **28**, 080501 (2021)
90. L. Adhikari, G.P. Zank, L.L. Zhao, J.C. Kasper, K.E. Korreck, M. Stevens, A.W. Case, P. Whittlesey, D. Larson, R. Livi, K.G. Klein, Astrophys. J. Suppl. **246**, 38 (2020) (Preprint 1912.02372)
91. L. Adhikari, G.P. Zank, L. Zhao, Fluidika **6**, 368 (2021)
92. L. Adhikari, G.P. Zank, L.L. Zhao, M. Nakanotani, S. Tasnim, Astron. Astrophys. **650**, A16 (2021)
93. L.L. Zhao, G.P. Zank, Q. Hu, D. Telloni, Y. Chen, L. Adhikari, M. Nakanotani, J.C. Kasper, J. Huang, S.D. Bale, K.E. Korreck, A.W. Case, M. Stevens, J.W. Bonnell, T. Dudok de Wit, K. Goetz, P.R. Harvey, R.J. MacDowall, D.M. Malaspina, M. Pulupa, D.E. Larson, R. Livi, P. Whittlesey, K.G. Klein, N.E. Raouafi, Astron. Astrophys. **650**, A12 (2021) (Preprint 2010.04664)
94. L.L. Zhao, G.P. Zank, J.S. He, D. Telloni, L. Adhikari, M. Nakanotani, J.C. Kasper, S.D. Bale, Astrophys. J. **922**, 188 (2021)
95. L.L. Zhao, G.P. Zank, Q. Hu, Y. Chen, L. Adhikari, J.A. leRoux, A. Cummings, E. Stone, L.F. Burlaga, Astrophys. J. **886**, 144 (2019)
96. L. Burlaga, E. Sittler, F. Mariani, R. Schwenn, J. Geophys. Res. **86**, 6673–6684 (1981)
97. M.B. Moldwin, J.L. Phillips, J.T. Gosling, E.E. Scime, D.J. McComas, S.J. Bame, A. Balogh, R.J. Forsyth, J. Geophys. Res. **100**, 19903–19910 (1995)
98. W.H. Matthaeus, M.L. Goldstein, C. Smith, Phys. Rev. Lett. **48**, 1256–1259 (1982)
99. D. Telloni, R. Bruno, R. D'Amicis, E. Pietropaolo, V. Carbone, Astrophys. J. **751**, 19 (2012)
100. D. Telloni, S. Perri, R. Bruno, V. Carbone, R.D. Amicis, Astrophys. J. **776**, 3 (2013)
101. S. Bourouaine, J.C. Perez, Astrophys. J. Lett. **879**, L16 (2019) (Preprint 1906.05644)
102. R. Bandyopadhyay, D.J. McComas, Astrophys. J. **923**, 193 (2021) (Preprint 2110.14756)

103. L.L. Zhao, G.P. Zank, L. Adhikari, M. Nakanotani, Astrophys. J. Lett. **924**, L5 (2022) (Preprint 2112.01711)
104. M.L. Goldstein, Astrophys. J. **219**, 700–704 (1978)
105. D. Telloni, R. Bruno, V. Carbone, E. Antonucci, R. D'Amicis, Astrophys. J. **706**, 238–243 (2009)
106. R. Bruno, D. Telloni, L. Primavera, E. Pietropaolo, R. D'Amicis, L. Sorriso-Valvo, V. Carbone, F. Malara, P. Veltri, Astrophys. J. **786**, 53 (2014) (Preprint 1411.3473)
107. P. Hunana, G.P. Zank, Astrophys. J. **718**, 148–167 (2010)

A Primer on the Riemann Hypothesis

Michael E. N. Tschaffon, Iva Tkáčová, Helmut Maier,
and Wolfgang P. Schleich

Contents

M. E. N. Tschaffon (✉)
Institut für Quantenphysik and Center for Integrated Quantum Science and Technology (IQST),
Universität Ulm, Ulm, Germany
e-mail: michael.tschaffon@uni-ulm.de

I. Tkáčová
Department of Physics, Faculty of Electrical Engineering and Computer Science, VSB-Technical
University of Ostrava, Ostrava, Poruba, Czech Republic

H. Maier
Institut für Reine Mathematik, Universität Ulm, Ulm, Germany

W. P. Schleich
Institut für Quantenphysik and Center for Integrated Quantum Science and Technology (IQST),
Universität Ulm, Ulm, Germany

Hagler Institute for Advanced Study, Institute for Quantum Science and Engineering (IQSE), and
Texas A&M AgriLife Research, Texas A&M University, College Station, TX, USA
e-mail: wolfgang.schleich@uni-ulm.de

© The Author(s), under exclusive license to Springer Nature Switzerland AG 2023   191
R. Citro et al. (eds.), *Sketches of Physics*, Lecture Notes in Physics 1000,
https://doi.org/10.1007/978-3-031-32469-7_7

Abstract

We provide an introduction for physicists into the Riemann Hypothesis. For this purpose, we first introduce, and then compare and contrast the Riemann function and the Dirichlet L-functions, with the Titchmarsh counterexample. Whereas the first two classes of functions are expected to satisfy the Riemann Hypothesis, the Titchmarsh counterexample is known to violate it. Throughout our article we employ elementary mathematical techniques known to every physicist. Needless to say, we do not verify the Riemann Hypothesis but suggest heuristic arguments in favor of it. We also build a bridge to quantum mechanics by interpreting the Dirichlet series central to this field as a superposition of probability amplitudes leading us to an unusual potential with a logarithmic energy spectrum opening the possibility of factoring numbers.

1 Introduction

In his talk at the *International Congress of Mathematicians* on August 8, 1900 in Paris, David Hilbert presented ten at that time unsolved problems [19] which proved to be very influential for the mathematics of the twentieth century. Since then, many of them have been solved. However, the 8th problem, that is, the Riemann Hypothesis [7, 27, 28] has resisted vehemently.

In order to appreciate the importance of the Riemann Hypothesis, we need to recall that the Riemann zeta function is intimately connected to the distribution of primes. This feature stands out most clearly in the prime-counting function, defined by the number of primes below a given real number x. Obviously, this function consists of jumps by one at primes and plateaus of different lengths in between.

Whereas the average behavior of this curve is given by $x/\ln x$ which dates back to Carl Friedrich Gauss, the locations of the steps are determined by the imaginary parts of the non-trivial zeros of the Riemann zeta function, provided they all have identical real part 1/2. The conjecture that *all* these zeros have this property constitutes the Riemann Hypothesis. If it were wrong, it would have catastrophic consequences. The steps would not appear at the locations of primes.

The importance that Hilbert attributed to this enigma can be judged from the fact that he once said that if he could come back in 100 years from now, his first question would be: Has anybody presented a proof of the Riemann Hypothesis? In this article, we summarize this longstanding problem of number theory from the perspective of theoretical physics and connect it to quantum mechanics.

Indeed, the Riemann zeta function ζ plays not only a central role in mathematics but also in physics. Three examples may suffice to illustrate this point. The distribution of energy levels in nuclei can be described [23] by random matrices, and their statistics of eigenvalues is similar to that of the non-trivial zeros of ζ. There also exists [35] an intimate connection between the inverted harmonic oscillator [36] and the Mellin transform, which emerges in the analytic continuation of ζ. The longstanding Polya-Hilbert Hypothesis of finding a Hamiltonian whose eigenvalues are given by the zeros of ζ has recently been verified [2]. For a detailed discussion of the physics of the Riemann Hypothesis, we refer to Refs. [31, 39].

1.1   Riemann Hypothesis

During the last 30 years, scientific publishing has undergone a dramatic change. Manuscripts prepared by the author in LATEX, open access publishing in electronic journals, and arXiv servers are only a few of the novel tools enabling instant exchange of knowledge between scientists using the internet.

It is worthwhile to compare and contrast the present day situation to the one of the year 1859 when Bernhard Riemann submitted his seminal article [26] entitled *On the Number of Prime Numbers less than a Given Quantity*, obviously not in LATEX but in his beautiful handwriting. In Fig. 1 we show the first page of his manuscript.

In the introduction, which we amplify in Fig. 2, he thanks the *Berlin Academy* for electing him a member, which gives him the privilege to publish articles in its journal. In particular, he states:

> I believe that I can best convey my thanks for the honour which the Academy has to some degree conferred on me, through my admission as one of its correspondents, if I speedily make use of the permission thereby received to communicate an investigation into the accumulation of the prime numbers; a topic which perhaps seems not wholly unworthy of such a communication, given the interest which *Gauss* and *Dirichlet* have themselves shown in it over a lengthy period.

Despite the lack of today's technology, his pioneering article left a lasting impression on number theory.

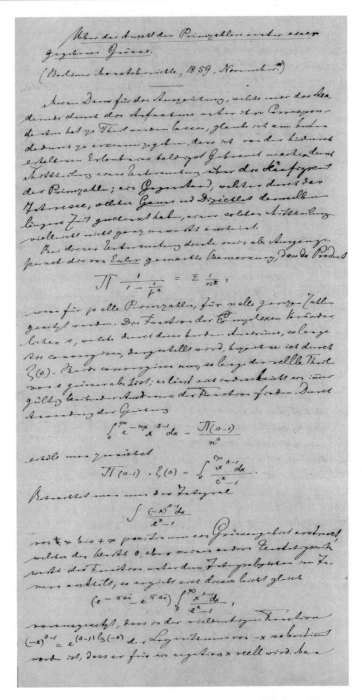

Fig. 1 The first page of Bernhard Riemann's seminal article [26] (With kind permission from the Niedersächsische Staats- und Universitätsbibliothek (SUB) Göttingen)

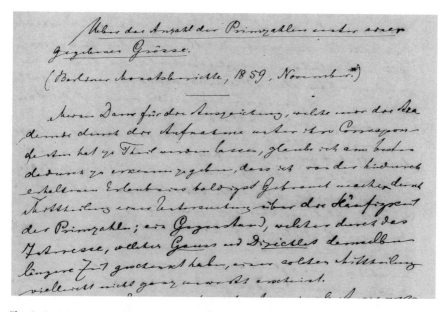

Fig. 2 Introduction of Riemann's article (Meinen Dank für die Auszeichnung, welche mir die Akademie durch die Aufnahme unter ihre Correspondenten hat zu Theil werden lassen, glaube ich am besten dadurch zu erkennen zu geben, dass ich von der hierdurch erhaltenen Erlaubniss baldigst Gebrauch mache durch Mittheilung einer Untersuchung über die Häufigkeit der Primzahlen; ein Gegenstand, welcher durch das Interesse, welches *Gauss* und *Dirichlet* demselben längere Zeit geschenkt haben, einer solchen Mittheilung vielleicht nicht ganz unwerth erscheint.) (With kind permission from the SUB Göttingen)

In the second paragraph, enlarged in Fig. 3, he formulates the problem.

For this investigation my point of departure is provided by the observation of *Euler* that the product

$$\prod \frac{1}{1 - \frac{1}{p^s}} = \sum \frac{1}{n^s}$$

if one substitutes for p all prime numbers, and for n all whole numbers. The function of the complex variable s which is represented by these two expressions, wherever they converge, I denote by $\zeta(s)$. Both expressions converge only when the real part of s is greater than 1;

He then proposes an analytic continuation of ζ and notes that all zeros of the so-defined function are located on a line in the complex plane. He states, as shown in Fig. 4, that

... it is very probable that all roots are real. Certainly one would wish for a stricter proof here; I have meanwhile temporarily put aside the search for this after some fleeting futile attempts, as it appears unnecessary for the next objective of my investigation.

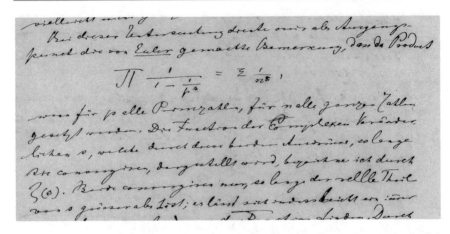

Fig. 3 Formulation of the problem of Riemann's article (Bei dieser Untersuchung diente mir als Ausgangspunkt die von *Euler* gemachte Bemerkung, dass das Product

$$\prod \frac{1}{1 - \frac{1}{p^s}} = \sum \frac{1}{n^s}$$

wenn für *p* alle Primzahlen, für *n* alle ganzen Zahlen gesetzt werden. Die Function der complexen Veränderlichen *s*, welche durch diese beiden Ausdrücke, so lange sie convergiren, dargestellt wird, bezeichne ich durch $\zeta(s)$. Beide convergiren nur, so lange der reelle Theil von *s* grösser als 1 ist;) (With kind permission from the SUB Göttingen)

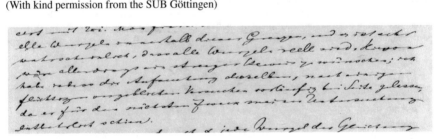

Fig. 4 Riemann Hypothesis (... und es ist sehr wahrscheinlich, dass alle Wurzeln reell sind. Hiervon wäre allerdings ein strenger Beweis zu wünschen; ich habe indess die Aufsuchung desselben nach einigen flüchtigen vergeblichen Versuchen vorläufig bei Seite gelassen, da er für den nächsten Zweck meiner Untersuchung entbehrlich schien.) (With kind permission from the SUB Göttingen)

Since Riemann had performed a rotation in the complex plane, these zeros are real. However, in the non-rotated variables these so-called *non-trivial* zeros are conjectured to *all* have the real part 0.5 defining the critical line. This statement constitutes the famous Riemann Hypothesis. Although Riemann's claim has crucial consequences for the distribution of prime numbers, it has still not been verified.

In this article, we provide an introduction into the Riemann Hypothesis [5]. Here we not only focus on the Riemann zeta function but also address the *generalized* Riemann Hypothesis which is supposed to hold for the Dirichlet L-functions [9,

20]. Moreover, we introduce the Titchmarsh counterexample [34] which satisfies the same functional equation as the Riemann function but is known to have zeros off the critical line.

Needless to say, we do not solve this great conundrum either but compare and contrast the three functions. Our article was composed under the watchful eyes of a professional number theorist by theoretical physicists for theoretical physicists. It is therefore *not* structured in the language of mathematics as a sequence of definition, lemma, and theorem.

1.2 Outline

Our article is organized as follows: We start by introducing in Sect. 2 the Riemann zeta function ζ, and then recall Riemann's analytic continuation of ζ into the complete complex plane. This approach leads us in Sect. 3 to the Riemann function ξ, defined as a product of several functions resulting from the analytic continuation. We then derive two different but equivalent formulations of ξ and conclude by presenting an asymptotic representation of ξ.

Sections 4 and 5 lay the foundations for our discussion of the Dirichlet-L functions. Here, we first introduce and then summarize properties of Dirichlet characters χ as well as Gauss sums.

In Sect. 6 we define the Dirichlet L-functions $\Lambda = \Lambda(s, \chi)$ of complex-valued argument s and character χ, and review the essential ingredients of their analytic continuation, such as the associated functional equation and the symmetry relations of the absolute value as well as the phase of Λ with respect to the critical line. In particular, we emphasize that in contrast to ξ, where the phase is anti-symmetric, a phase shift can occur for Λ. This distinction is the result of the functional equation for Λ which contains a phase factor originating from the normalized Gauss sum.

In Sect. 7 we then discuss the Titchmarsh counterexample. Here, we emphasize the consequence of the lack of the existence of an Euler product resulting in zeros off the axis. We conclude in Sect. 8 by a brief summary.

In order to keep our article self-contained while concise, we have included several appendices. For example, in Appendix 1 we recall the derivation of the functional equation of the Jacobi theta function. This quantity is of utmost importance in determining the fundamental building block of ξ which is at the very heart of the corresponding functional equation.

We then present in Appendix 2 an exact alternative expression for the integral determining the imaginary parts of the non-trivial zeros. Indeed, they are governed by the identity of this integral and a Lorentzian, which decays with the square of the imaginary part τ of s. This new formula for the integral simplifies this condition considerably, since it brings out explicitly that the integral also decays as $1/\tau^2$.

In Appendix 3 we obtain the asymptotic expansion of the prefactor of the Riemann function, defined by an exponential and the Gamma function. Since this exponential product also appears in the Dirichlet L-functions, we derive an approximate but analytic expressions for it, and apply this result to the Riemann function.

In Appendix 4 we use the asymptotic expansion of ξ given solely by the asymptotic limits of the exponential product to investigate the dependence of the amplitude and phase of ξ in the complex plane. We test these expressions by applying the Cauchy-Riemann differential equations in amplitude and phase.

We then in Appendix 5 evaluate explicitly the normalized Gauss sums of three characteristic Dirichlet sums. In order to bring out most clearly the similarities and differences between the analytic continuations of ξ and Λ we also re-derive in Appendix 6 the functional equation of the *generalized* Jacobi theta function. Here, we emphasize especially the fact that the Dirichlet character $\chi(0)$ vanishes. Indeed, this feature ensures a functional equation that is different from the one corresponding to the Jacobi theta function of ξ, discussed in Appendix 1.

2   Riemann Zeta Function

The Dirichlet representation of the Riemann zeta function [8]

$$\zeta(s) \equiv \sum_{n=1}^{\infty} \frac{1}{n^s} \tag{1}$$

with complex argument $s \equiv \sigma + i\tau$ involves in its sum all natural numbers n.

In this section, we first show that its convergence is restricted to the domain of the complex plane with $1 < \sigma$ and provide a physical argument [22] based on the normalization condition of a quantum state, which is intimately related to the Born interpretation of quantum mechanics. We then recall the arguments of Riemann in his celebrated article [26] to derive the analytic continuation and thus a definition of ζ in the *complete* complex plane. Although this derivation is well-known and contained in many text books [8] we find it useful to include it here in order to be able to point to the origin of the individual terms compared to the analogous analysis for Dirichlet L-functions in Sect. 6.1.

Here we first consider the so-derived expression as an alternative representation of ζ and obtain properties of ζ not immediately obvious from the Dirichlet sum. Then we provide reasons for its validity in the complete complex plane.

We conclude this overview by emphasizing that the interpretation of the expression for ζ provided by the analytic continuation is not just interference [11] which could be realized even by a classical wave theory but also involves [10] entanglement, the earmark of quantum mechanics.

2.1   Dirichlet Series as Interfering Probability Amplitudes

In this section we build a bridge between the central ingredient of analytic number theory, that is, a Dirichlet series such as the one of the Riemann zeta function and quantum theory. In particular, we show that ζ can be interpreted [11,22] as the auto-correlation function of a time-dependent wave packet moving in a potential with a logarithmic energy spectrum. We briefly address the question as to the existence of such a potential, which represents an example of an inverse problem [37]. Finally, we outline the use of a logarithmic energy spectrum to factor numbers [13–16].

2.1.1   Riemann Zeta Function and Wave Packet Dynamics

In order to gain insight into ζ, and make a connection to quantum mechanics, we first use the relation

$$n^{-s} = e^{-s \ln n} = e^{-\sigma \ln n} e^{-i\tau \ln n} \tag{2}$$

to cast the Dirichlet series, Eq. (1), in the form

$$\zeta(s) = \sum_{n=1}^{\infty} e^{-\sigma \ln n} e^{-i\tau \ln n} \tag{3}$$

which is reminiscent of the auto-correlation function of a time-dependent quantum state.

Indeed, when we start from a quantum state

$$|\psi(t=0)\rangle \equiv \sum_{m=0}^{\infty} \psi_m |m\rangle \tag{4}$$

expanded into energy eigenstates $|m\rangle$ with the eigenvalue equation

$$\hat{H}|m\rangle = E_m|m\rangle \tag{5}$$

provided by a time-independent Hamiltonian $\hat{H}$, the time evolution of $|\psi(t=0)\rangle$ governed by the Schrödinger equation

$$i\hbar \frac{d}{dt}|\psi(t)\rangle = \hat{H}|\psi(t)\rangle \tag{6}$$

leads us immediately to the state

$$|\psi(t)\rangle = \sum_{m=0}^{\infty} \psi_m e^{-i E_m t/\hbar}|m\rangle . \tag{7}$$

Hence, the complex-valued auto-correlation function

$$C(t) \equiv \langle \psi(t=0) | \psi(t) \rangle \,, \tag{8}$$

that is, the time-dependent overlap between the initial and the time-evolved state takes the form

$$C(t) = \sum_{m=0}^{\infty} |\psi_m|^2 \exp\left(-i \frac{E_m t}{\hbar}\right) . \tag{9}$$

A comparison between the expressions, Eqs. (3) and (9), for ζ and C yields immediately the identifications

$$\frac{1}{\hbar} E_m t \equiv \tau \ln(m+1) \tag{10}$$

and

$$|\psi_m|^2 \equiv e^{-\sigma \ln(m+1)} . \tag{11}$$

Hence, the Dirichlet series, Eq. (1), of ζ represents [10, 22] the auto-correlation function of a wave packet in a potential with a logarithmic energy spectrum

$$E_m \equiv \hbar\omega \ln(m+1) . \tag{12}$$

Here $\hbar\omega$ has the units of an energy and $E_0 = 0$.

The real part σ of the complex-valued argument s of ζ determines the occupation probability $|\psi_m|^2$ of the m-th level with energy E_m, and the imaginary part τ of s is the dimensional time $\tau \equiv \omega t$.

Due to the unitary time evolution the state $|\psi(t)\rangle$ remains normalized at all times provided the initial state $|\psi(0)\rangle$ is normalized, which leads us to the condition

$$1 = \sum_{n=0}^{\infty} |\psi_m|^2 \,, \tag{13}$$

and translates with Eq. (11) into the requirement

$$1 = \sum_{n=1}^{\infty} e^{-\sigma \ln n} = \sum_{n=1}^{\infty} \frac{1}{n^\sigma} . \tag{14}$$

We can employ the triangle inequality to obtain the estimate

$$|\zeta| \leq \sum_{n=1}^{\infty} \left| \frac{1}{n^s} \right| = \sum_{n=1}^{\infty} \frac{1}{n^\sigma} \,, \tag{15}$$

and to convince ourselves that the convergence of this sum is guaranteed, provided $1 < \sigma$.

The physical interpretation of $n^{-\sigma}$ as occupation probabilities provides us with a vivid picture of the normalization of the state, or the convergence condition of the Dirichlet series. Indeed, as σ approaches unity from above, the function $|\psi_m|^2$ reaches ever higher values of m, that is, the occupation goes into higher and higher states.

We conclude that this interpretation of the Dirichlet series is not restricted to the one of ζ but holds true for all Dirichlet series. For a more detailed discussion, we refer to Ref. [11].

2.1.2   Inverse Problem

One crucial question remains: Do quantum systems exist that display a logarithmic energy spectrum? The answer is yes, but we have to construct them by solving the inverse problem [37].

Indeed, in a typical quantum scenario the potential is given, and we search for the energy eigenvalue by solving the time-independent Schrödinger equation. However, now we face the inverse situation, that is, the spectrum is given and we need to find the potential.

Many techniques offer themselves to solve the inverse problem. In semi-classical quantum mechanics, the Bohr-Sommerfeld quantization rule [29] allows us to obtain the approximate energy spectrum for a given potential. However, we can also employ the Abel integral transform [22] to find the potential for a given semi-classical energy distribution as pursued by the Rydberg-Klein-Rees technique; Moreover, it is always possible to solve for the potential numerically.

Since this question is not central to our article, we refer to the extensive literature [12, 21, 37] on this topic. It suffices to say that when we assume that the desired potential is one-dimensional and symmetric, for every spectrum there exists a potential and it is unique.

2.1.3   Factorization

We conclude this interpretation of the Dirichlet series of ζ, Eq. (1), as interfering probability amplitudes [11] by noting that a logarithmic energy spectrum allows [13–16] us to also factor a composite natural number

$$N \equiv p \cdot q \tag{16}$$

consisting of two primes p and q due to the elementary functional equation

$$\ln N = \ln(p \cdot q) = \ln p + \ln q \tag{17}$$

of the logarithm.

Indeed, when we excite a quantum system consisting of two degrees of freedom by an energy proportional to $\ln(p \cdot q)$, the two subsystems must contain the energies proportional to $\ln p$ and $\ln q$. This fact is the direct consequence of the additivity of energies in quantum mechanics.

For this reason, it is also interesting to create a potential whose energy spectrum contains solely the logarithms of prime numbers. For a detailed discussion of this technique, we refer to Ref. [16]. We emphasize that a potential with a spectrum solely of primes has recently been realized in an experiment reported in Ref. [6].

Experiments of this type rely on the fact that atoms, whose internal energy levels are far detuned from the frequency of a light field, feel [18] in their center-of-mass motion a potential governed by the spatial intensity distribution. Hence, by creating the intensity in the shape of the desired potential leads to the potential for the atom.

We emphasize that experiments to realize such a factorization technique based on a specially designed potential are presently on their way in the group of Los Alamos (private communication, October 2022).

2.2 Sum Representation

We now turn to an equivalent formulation of ζ in three terms rather than an infinite sum. This expression is also the basis for the analytic continuation discussed in the next section.

At the very heart of Riemann's derivation [26] of the analytic continuation of ζ is the identity

$$n^{-s} = \frac{\pi^{s/2}}{\Gamma(s/2)} \int_0^\infty dx\, e^{-\pi n^2 x} x^{s/2-1} \tag{18}$$

which follows immediately from the integral representation [38]

$$\Gamma(s) \equiv \int_0^\infty dt\, e^{-t} t^{s-1} \tag{19}$$

of the Gamma function with the substitution $t \equiv \pi n^2 x$.

When we substitute the integral representation, Eq. (18), of n^{-s} into the expression, Eq. (1) for the Dirichlet series of ζ, and interchange summation and integration we arrive at the formula

$$\zeta(s) = \frac{\pi^{s/2}}{\Gamma(s/2)} \int_0^\infty dx\, \omega(x) x^{s/2-1} \tag{20}$$

where we have introduced the Jacobi theta function [38]

$$\omega(x) \equiv \sum_{n=1}^\infty e^{-\pi n^2 x} . \tag{21}$$

Next we decompose the integration over x in Eq. (20) into the two intervals $0 \leq x \leq 1$ and $1 \leq x \leq \infty$ which leads us to the representation

$$\zeta(s) = \frac{\pi^{s/2}}{\Gamma(s/2)}[\gamma(s) + I(s)] \tag{22}$$

where we have introduced the abbreviation

$$\gamma(s) \equiv \int_1^\infty dx\, \omega(x) x^{s/2-1} \tag{23}$$

and the integral

$$I(s) \equiv \int_0^1 dx\, \omega(x) x^{s/2-1} . \tag{24}$$

When we recall the definition

$$\tilde{a}(s) \equiv \int_0^\infty dx\, a(x) x^{s-1} \tag{25}$$

of the Mellin transform $\tilde{a}$ of the function a, we realize that γ is a *truncated* Mellin transform of the Jacobi theta function ω, since the integration variable x only extends from 1 to infinity rather than from 0.

We note in passing that the Gamma function, Eq. (19), is the Mellin transform of the exponential function. Moreover, we point out that also a *quantum* Mellin transform [35] associated with the quantum mechanics of the inverted harmonic oscillator [36] has been proposed.

In Appendix 1 we recall the functional equation [38]

$$\omega(x) = \frac{1}{\sqrt{x}}\omega\left(\frac{1}{x}\right) + \frac{1}{2}\left[\frac{1}{\sqrt{x}} - 1\right] \tag{26}$$

of ω, and substitute it now into the integral I given by Eq. (24) which yields the expression

$$I = \int_0^1 dx\, \omega\left(\frac{1}{x}\right) x^{s/2-3/2} + \frac{1}{2}\int_0^1 dx\, \left(x^{(s-1)/2-1} - x^{s/2-1}\right) . \tag{27}$$

With the new integration variable $y \equiv 1/x$ in the first integral, and performing the integration in the second one we arrive at the result

$$I = \gamma(1-s) + \frac{1}{s-1} - \frac{1}{s} , \tag{28}$$

where we have recalled the definition, Eq. (23), of γ.

As a result, we find the representation

$$\zeta(s) = \frac{\pi^{s/2}}{\Gamma(s/2)} \left[\gamma(s) + \gamma(1-s) + \frac{1}{\wp(s)} \right] \tag{29}$$

of the Riemann zeta function ζ as a sum of three terms given by the truncated Mellin transform γ of ω evaluated at s and $1-s$ as well as the quadratic polynomial

$$\wp(s) \equiv s(s-1) \tag{30}$$

which also involves the product of s and $1-s$.

2.3   Analytic Continuation

We emphasize that in Riemann's treatment summarized in the preceding section, the sum over n has *not* been eliminated, nor has it been resummed. It is still there in the Jacobi theta function ω except that now the coefficients n^{-s} have been replaced due to the integral identity, Eq. (18), by $\exp\left(-\pi n^2 x\right)$ where according to the truncated Mellin transform γ of ω, given by Eq. (23), the real-valued integration variable x is restricted to the interval $1 < x$.

This restriction guarantees that the sum representing ω is rapidly converging. Moreover, again due to the condition $1 < x$ of the integration variable x in γ, the term

$$x^{s/2-1} = e^{(s/2-1)\ln x} \tag{31}$$

is free of any singularity. Hence, γ is free of any singularity as well.

It is the functional equation, Eq. (26), of ω which ensures the mapping of the integral I onto γ. Indeed, if the integration domain contained the point $x = 0$ where all terms in the sum would be unity the resulting sum would be divergent.

We also note that as a result of the functional equation, Eq. (26), γ appears in the representation, Eq. (29), of ζ evaluated at s *and* at $1-s$. Moreover, the terms independent of ω in the functional equation, Eq. (26) are the origin of the polynomial $\wp$ in Eq. (29).

The appearance of the ratio $\pi^{s/2}/\Gamma(s/2)$ in Eq. (29), is a result of the integral identity, Eq. (18), and has important consequences. Indeed, the relation

$$\pi^{s/2} = \exp\left(\frac{s}{2}\ln \pi\right) \tag{32}$$

immediately shows us that this term is also defined in the complete complex plane, as is the Gamma function. Hence, Eq. (29) represents the analytic continuation of ζ.

2.4   A First Glimpse at the Zeros

The analytic continuation, Eq. (29), of ζ allows us to also obtain elementary properties of ζ and provides us with some insight into the location of the zeros. In the present section, we show that there exist zeros at even negative integers, which are called the *trivial* zeros of ζ. We also address the so-called *non-trivial* zeros, which are the subject of the Riemann Hypothesis.

2.4.1   Trivial Zeros

We start by recalling that the Gamma function has simple poles [38] at the negative integers, including zero. Due to the appearance of $s/2$ in the argument of Γ these poles are now at the negative *even* integers including zero.

Since $\Gamma(s/2)$ appears in the *denominator* of ζ, the simple *poles* of $\Gamma(s/2)$ translate into simple *zeros* of ζ at the negative even integers. Since we can easily understand the origin of these zeros of ζ as well as their locations, they carry the name *trivial zeros*.

However, there is also the pole of Γ at $s = 0$. Since $\wp$ is also in the denominator of ζ, and according to the definition, Eq. (30) of $\wp$, is *linear* in s, the product of s and $\Gamma(s/2)\,s$ is finite. This fact follows immediately from the functional equation [38]

$$s\Gamma(s) = \Gamma(s + 1) \tag{33}$$

of the Gamma function. Indeed, due to the shift of the argument by unity, there is no pole of Γ at $s = 0$ anymore.

Since $\Gamma(1/2) = \sqrt{\pi}$ is finite, the pole of ζ due to the term $1 - s$ in $\wp$ in the denominator in Eq. (29) is not eliminated. We emphasize that it is already present in the Dirichlet representation, Eq. (1), of ζ.

Indeed, we find that for $s = 1$ the Dirichlet series reduces to the harmonic sum

$$\zeta(1) = \sum_{n=1}^{\infty} \frac{1}{n} \tag{34}$$

which is well-known not to converge.

These considerations show us that ζ as determined by Eq. (29) is now defined everywhere in the complex plane. It has simple zeros at the even negative integers and a simple pole at $s = 1$.

2.4.2   Non-trivial Zeros

However, the representation, Eq. (29), of ζ also allows a first glimpse at the Riemann Hypothesis. Additional zeros of ζ might arise when the terms $\gamma(s)$, $\gamma(1 - s)$ and $1/\wp(s)$ compensate each other, leading us to the condition

$$\gamma(s_0) + \gamma(1 - s_0) + \frac{1}{\wp(s_0)} = 0 \tag{35}$$

which with the definition, Eq. (30), of $\wp$ and $s \equiv \sigma + i\tau$ translates into

$$\gamma(\sigma_0 + i\tau_0) + \gamma(1 - \sigma_0 - i\tau_0) - \frac{1}{(\sigma_0 + i\tau_0)(1 - \sigma_0 - i\tau_0)} = 0. \tag{36}$$

Due to the appearance of σ_0 and $\sigma_0 - 1$ it is tempting to set $\sigma_0 = 1/2$, that is, we consider the condition for a zero on the critical line which leads us to the requirement

$$\gamma\left(\frac{1}{2} + i\tau_0\right) + \gamma\left(\frac{1}{2} - i\tau_0\right) - \frac{1}{\frac{1}{4} + \tau_0^2} = 0, \tag{37}$$

which with the identity $\gamma(s^*) = \gamma^*(s)$ reduces to

$$2\,\mathrm{Re}\left(\gamma\left(\frac{1}{2} + i\tau_0\right)\right) = \frac{1}{\frac{1}{4} + \tau_0^2}. \tag{38}$$

From the definition, Eq. (23), of γ we arrive at the condition

$$2\int_1^\infty dx\, \omega(x) x^{-3/4} \cos\left(\frac{\tau_0}{2}\ln x\right) = \frac{1}{\frac{1}{4} + \tau_0^2} \tag{39}$$

for zeros on the critical line, equating the Lorentzian on the right-hand side of the equation with a type of a cosine transform which involves the logarithm of the integration variable due to the fact that we have started from the Mellin transform.

In order to gain some insight into this condition of a zero on the critical line, we show in Fig. 5 in its dependence of the integration variable x the integrand of the integral in Eq. (39) for increasing values of τ which increases the frequency of the oscillations of the cosine. Since the product of the Jacobi theta function and the fractional power $x^{-3/4}$ is positive and decays for increasing x, there should be a value of τ where the first two oscillations of the cosine lead to an almost cancellation of the integral, with a small value identical to the right-hand side of Eq. (39).

For increasing τ, more oscillations of the cosine emerge as shown in Fig. 5 by the waves moving to the left with increasing frequency, and lead to an oscillatory behavior of the integrand. Additional near-cancellations are thus possible as to achieve a value identical to the $1/\tau^2$ decay of the Lorentzian on the right-hand side of Eq. (39).

Based on this argument one might suspect that eventually the frequency of the cosine is so high that the decaying function $\omega(x)x^{-3/4}$ appears locally to be constant and the integral vanishes. It is at this point that the special form of the Jacobi theta function enters.

Indeed, according to its definition, Eq. (21) it consists of a sum of exponentials with ever-increasing steepness. As a result, no matter how large the frequency of the

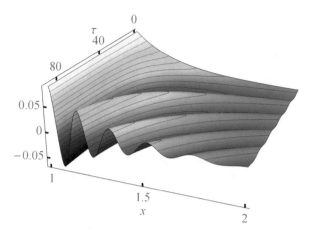

Fig. 5 Heuristic argument supporting the claim that there exist infinitely many zeros of ξ on the critical line. The integrand of the integral in the condition, Eq. (39), determining these zeros and shown in its dependence on the integration variable x on the horizontal axis, develops oscillations with ever increasing frequency for increasing τ. Due to these oscillations the integral can assume values identical to the decaying Lorentzian on the right-hand side of the condition

oscillation, there will always be an exponential that decays faster leading to a zero of ξ. Therefore, we expect an infinite amount of zeros on the critical line.

In Appendix 2 we derive an exact, alternative expression for the integral in Eq. (39), which brings out most clearly a $1/\tau^2$-decay identical to the one of the Lorentzian on the right-hand side of Eq. (39). In the new representation of the integral, the frequency of the oscillations is constant for increasing values of τ, but the decay of the individual exponentials forming the Jacobi theta function ω is slowed down. In complete agreement with the considerations summarized in Fig. 5, the zeros are determined by those exponentials which decay faster than the period 2π of the oscillation.

We conclude by emphasizing that this analysis of the integrand explains that there are infinitely many zeros *on* the critical line, but of course, it does not provide any insight into the question of zeros *off* the critical line.

3   Riemann Function

In the preceding section we have derived the expression, Eq. (29), of ζ in terms of the product of two terms: (i) the ratio $\pi^{s/2}/\Gamma(s/2)$ originating from the integral representation, Eq. (18) of n^{-s}, and (ii) a *sum* of *three* expressions involving the truncated Mellin transform γ at s and $1 - s$, as well as the quadratic polynomial $\wp$.

In the present section, we use this representation to introduce the Riemann function ξ by a product of these ingredients. As a result, the so-defined ξ is governed by a single function f_R evaluated at s *and* at $1 - s$, and an elementary functional equation.

3.1   Product Representation

The analytic continuation of ζ, given by Eq. (29), suggests introducing the Riemann function ξ by the product

$$\xi(s) \equiv \frac{1}{2} \wp(s) \left(\frac{1}{\pi}\right)^{s/2} \Gamma\left(\frac{s}{2}\right) \zeta(s) . \tag{40}$$

which emerges in a natural way when we multiply ζ, given by Eq. (29), by the factor $\pi^{-s/2} \Gamma(s/2)$ as to cancel the prefactor in the analytic continuation, and by $\wp$ to eliminate its inverse.

We have also introduced the factor $1/2$, which provides us with an equivalent compact expression for ξ. Indeed, when we recall the definition, Eq. (30), of the polynomial $\wp$ of second degree, we arrived the explicit form

$$\xi(s) \equiv \pi^{-s/2}(s-1)\frac{s}{2}\Gamma\left(\frac{s}{2}\right)\zeta(s) , \tag{41}$$

which with the help of the functional equation, Eq. (33), of the Gamma function reads

$$\xi(s) = \pi^{-s/2}(s-1)\Gamma\left(\frac{s}{2}+1\right)\zeta(s) . \tag{42}$$

Hence, the multiplication of ζ, given by Eq. (29), by $\pi^{-s/2}\Gamma(s/2)$, that is the inverse of the factor in front of the term in the square bracket, and by $\wp$, has created a function ξ that is free of the trivial zeros as well as the simple pole at $s = 1$. However, ξ still enjoys any other zero that ζ might have.

3.2   Functional Equation

When we recall the analytic continuation Eq. (29) of ζ, we find from Eq. (40) the expression

$$\xi(s) = \frac{1}{2}\wp(s)\left[\gamma(s) + \gamma(1-s)\right] + \frac{1}{2} \tag{43}$$

for ξ. Here the truncated Mellin transform γ, introduced by Eq. (23), of the Jacobi theta function ω, given by Eq. (21), enters with the arguments s and $1-s$. Moreover, the polynomial $\wp$, defined by Eq. (30), also consists of the product of s and $1 - s$, and therefore satisfies the functional equation

$$\wp(1 - s) = \wp(s).$$ (44)

Hence, we can express ξ as the superposition

$$\xi(s) = f_R(s) + f_R(1 - s)$$ (45)

of a *single* function

$$f_R(s) \equiv \frac{1}{2}\wp(s)\gamma(s) + \frac{1}{4}$$ (46)

evaluated at s *and* $1 - s$ as shown in Fig. 6.

When we introduce the mirror image $\bar{s} \equiv 1 - s^*$ of s with respect to the critical line we can also represent ξ in the form

$$\xi(s) = f_R^*(\bar{s}) + f_R(s).$$ (47)

Here we have used the symmetry relation

$$f_R^*(s) = f_R(s^*)$$ (48)

which indicates that the absolute value $|f_R|$ of f_R is *symmetric* with respect to the real axis, whereas the phase μ of f_R is *anti-symmetric*.

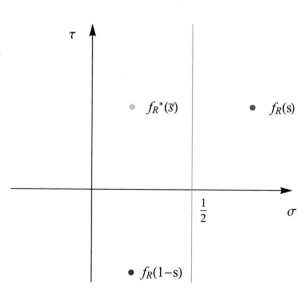

Fig. 6 Determination of ξ from f_R according to Eq. (47). The value of ξ at the argument s exemplified by the blue dot in the complex plane is the sum of f_R at s *and* f_R at $1 - s$ marked by the violet dot with a point symmetry with respect to $s = 1/2$. Due to the identity $f_R^*(s) = f_R(s^*)$ the contribution of f_R at $1 - s$ can also be interpreted as arising from the mirror image $\bar{s}$ of s with respect to the critical line, and indicated by the green dot, but with the complex conjugate of f_R

According to Eq. (45), ξ obviously satisfies the functional equation

$$\xi(s) = \xi(1-s).$$ (49)

We emphasize the importance of the factor $\pi^{-s/2}\Gamma(s/2)$ in ξ for the emergence of this functional equation. Indeed, in the analytic continuation, Eq. (29), the expressions in the square bracket satisfy the symmetry relation of s being interchanged by $1-s$. However, the elimination of the prefactor in ζ which destroys this symmetry in ζ recovers it again in ξ.

We conclude by noting that the multiplication of ζ by $\wp$ in Eq. (40) has lead in ξ to the off-set of $1/2$ and its appearance in front of $\gamma(1-s)$ and of $\gamma(s)$.

3.3   Symmetry Relations

The functional equation, Eq. (49), of ξ shows immediately elementary symmetries of ξ with respect to the critical line in its absolute value $|\xi|$ and its phase Φ when we represent ξ in the form

$$\xi\left[\frac{1}{2} + \left(\sigma - \frac{1}{2}\right) + i\tau\right] = \xi\left[\frac{1}{2} - \left(\sigma - \frac{1}{2}\right) - i\tau\right] = \xi^*\left[\frac{1}{2} - \left(\sigma - \frac{1}{2}\right) + i\tau\right]$$ (50)

Indeed, we find the relation

$$\left|\xi\left[\frac{1}{2} + \left(\sigma - \frac{1}{2}\right) + i\tau\right]\right| = \left|\xi\left[\frac{1}{2} - \left(\sigma - \frac{1}{2}\right) + i\tau\right]\right|$$ (51)

which demonstrates that $|\xi|$ is *symmetric* with respect to the critical line.

In contrast, the phase Φ of ξ is *anti-symmetric* with respect to the critical line due to the complex conjugate of the right-hand side of Eq. (50).

3.4   Elementary Building Block

According to Eq. (45) we can construct ξ at s from the values of the elementary building block f_R at s *and* $1-s$. Due to its importance, we discuss in this section the function f_R defined by Eq. (46) in its absolute value and phase. Moreover, we connect the zeros of ξ with the behavior of f_R, and provide a different point of view on the Riemann Hypothesis.

3.4.1   General Properties
We start by presenting in Fig. 7, the absolute value $|f_R|$ of f_R which assumes the shape of a bowl with two holes. Indeed, $|f_R|$ increases as s approaches the edges of

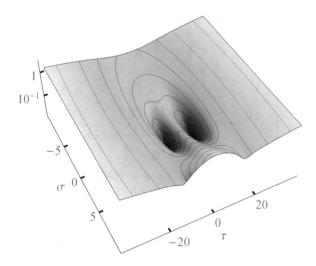

Fig. 7 Absolute value $|f_R|$ of the elementary building block f_R of the Riemann function ξ in the complex plane. The function $|f_R|$ is in the shape of a bowl with two holes representing zeros of f_R. Along the real axis $|f_R|$ displays a ridge or a valley of decaying values for positive or negative decreasing σ, with a minimum located at $\sigma = 1/2$. Due to the presence of the zeros on one side of the critical line, $|f_R|$ is *not* symmetric with respect to this axis. In order to bring out most clearly the characteristic features, we use a logarithmic scale for $|f_R|$.

the complex plane, and the zeros are located symmetrically with respect to the real axis.

Their existence and location is a consequence of three properties of f_R: (i) The real axis is a line of constant phase. (ii) There exists a saddle point at $s = 1/2$, and (iii) all lines of constant phase terminate in a zero of f_R. We now address each point separately.

According to Eq. (46), f_R is real on the real axis with

$$f_R(0) = f_R(1) = \frac{1}{4}$$

following from the identity $\wp(0) = \wp(1) = 0$ dictated by the definition Eq. (30) of the quadratic polynomial and its functional equation, Eq. (44). Whereas $\wp$ is positive along the real axis, except in the domain $0 < \sigma < 1$, γ is always positive. Hence, f_R is smaller than $1/4$ for $0 < \sigma < 1$, but larger everywhere else, leading to a minimum of f_R in this interval of the real axis.

This argument does not tell us that the minimum is at $s = 1/2$. However, it is supported by our numerical computation of f_R.

Since f_R is real along the real axis, it corresponds to a line of constant phase with a value zero or an integer multiple of 2π. As shown in the Chapter "Insights into Complex Functions" of this volume, lines of constant phase must terminate in a zero, and the absolute value of the function of interest along the line has to decay.

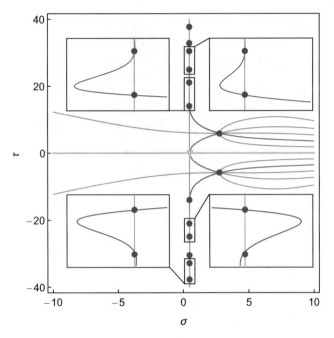

Fig. 8 Lines of constant phase of the building block f_R of ξ in the complex plane. All phase lines (gray curves) originate from infinity, that is, from the edges of the complex plane and end in the two zeros (red dots). The minimum of f_R at $s = 1/2$ corresponds to a saddle point (green triangle) where the two phase lines on the real axis (green line) coming from $-\infty$ and $+\infty$ collide, and break off to eventually terminate in the two zeros. These curves represent a separatrix which separates the flows. The phase lines corresponding to $\pm\pi/2$ (blue curves) approach the center of the complex plane from the top and the bottom by oscillating along the critical line (orange line). These oscillations exemplified by the insets give rise to an infinite number of zeros of ξ (blue dots) located on the critical line. In this visualization, we have used different scales for the horizontal axes of the four insets to overcome the rapid decay of these oscillations as τ increases

For this reason, the minimum of f_R at $s = 1/2$ corresponds to a saddle point where the phase lines from $-\infty$ and from $+\infty$ collide, and eventually go into the two zeros shown in Fig. 8.

Our numerical work also suggests that we deal with double zeros, each of which requires phase lines of an interval of 4π originating from infinity in the upper part of the complex plane, and moving towards the zero *above* the real axis. Likewise, the phase lines of another 4π-interval approach the zero *below* the real axis. The real axis together with the two curves emerging from the saddle point and ending in them separate the flows approaching the two zeros.

3.4.2 Two Special Phase Lines and the Non-trivial Zeros

We note that the phase lines corresponding to $+\pi/2$ and $-\pi/2$ play a very special role. The one corresponding to $-\pi/2$ emerges from infinity at the top, and the one for $+\pi/2$ from the bottom of the complex plane. Both oscillate around the critical line.

Most importantly, they cross it at the non-trivial zeros of ξ and only move towards the zero of f_R after having passed the first zero of ξ. Hence, the imaginary parts of the non-trivial zeros of ξ are defined by the crossings of the phase lines $\pm\pi/2$ with the critical line.

This property stands out most clearly when we recall from Eq. (47) that a zero of ξ requires the cancellation of f_R^* at $\bar{s}$ by f_R at s, which translates into the condition

$$f_R^*(\bar{s}) + f_R(s) = 0, \tag{52}$$

or

$$|f_R(\bar{s})|e^{-i\mu(\bar{s})} + |f_R(s)|e^{i\mu(s)} = 0. \tag{53}$$

Hence, for a zero in ξ to occur, we need to satisfy two conditions: (i) The absolute values of f_R at $\bar{s}$ and at s have to be identical, that is,

$$|f_R(\bar{s})| = |f_R(s)|, \tag{54}$$

and (ii) the sum of the phases μ at $\bar{s}$ and s must be an odd integer multiple of π, that is,

$$\mu(\bar{s}) = (2k + 1)\pi - \mu(s), \tag{55}$$

where k is an integer.

Although f_R enjoys the symmetry relation Eq. (48) with respect to the real axis, we emphasize that f_R does *not* satisfy a general symmetry with respect to the critical line. This feature makes it very unlikely that symmetrically located points $\bar{s}$ and s exist in the complex plane where the two conditions Eqs. (54) and (55) are satisfied.

However, when s is located on the critical line, $\bar{s}$ is obviously identical to s. In this case, the absolute values of f_R and the phases μ at s and $\bar{s}$ are identical, and we arrive at the condition

$$2\left|f_R\left(\frac{1}{2} + i\tau\right)\right|\cos\mu = 0$$

for a zero of ξ which can be now be easily satisfied when μ is an odd integer multiple of $\pi/2$, in complete agreement with the phase condition Eq. (55).

In Fig. 8 the phase lines corresponding to $\pm\pi/2$ cross the critical line. However, the ones with $\pm3\pi/2$, $5\pi/2$ and $\pm7\pi/2$ go straight from infinity into the zeros without crossing the critical line. It is for this geometry that these phase lines do not lead to additional zeros on the critical axis in addition to the ones at $\pm\pi/2$.

3.4.3   Riemann Hypothesis in Terms of Lines of Constant Height and Phase

The behavior of f_R, illustrated in Figs. 7 and 8, suggests a much-needed plausibility argument supporting the Riemann Hypothesis, expressed so pointedly in the book [8] by Harold Mortimer Edwards:

> One of the things which makes the Riemann hypothesis so difficult is the fact that there is no plausibility argument, no hint of a reason, however unrigorous, why it should be true.

In the upper half of the complex plane, phase lines of an interval of 4π approach a zero located to the right of the critical line and, similarly, in the lower half. The lines of constant phase $\pm\pi/2$ divide the two halves into domains of dramatically different densities of phase lines. Indeed, on the left side of the critical line, there are phase lines from an interval from zero to approximately $\pi/2$, whereas the right side contains the remaining part from approximately $\pi/2$ to full 4π.

Therefore, it is difficult to imagine that two phase lines, crossing the points s and $\bar{s}$ located symmetrically with respect to the critical line, would satisfy the condition, Eq. (55). Likewise, it seems inconceivable that in addition to this rather stringent phase condition also the absolute values of f_R at these points are identical, especially in view of the asymmetry of f_R with respect to the critical line.

We emphasize that this argument is by no means rigorous but could be what Edwards had in mind. Moreover, it could serve as a guide for a more detailed study.

3.5   The Riemann Hypothesis at the Edge of the Complex Plane

In the preceding section, we have studied properties of the building block f_R of ξ in order to gain insight into the Riemann Hypothesis. Here we have employed the overall structure of the lines of constant phase of f_R. In the present section we present another approach [24, 25, 30] based on the lines of constant phase towards the Riemann Hypothesis which requires the knowledge of the asymptotic behavior of ξ, a topic which we pursue in more detail in the following sections.

In order to illustrate this idea, we show in Fig. 9 the lines of constant phase of ξ over a range of values of σ extending from the critical line at $\sigma = 1/2$ to the right edge of the complex plane. Here, we have chosen a non-linear scale for the horizontal axis to bring to light the characteristic features. Due to the anti-symmetry of the phase lines dictated by the functional equation, Eq. (49), of ξ in the form of Eq. (50), we only display the region $-1 < \sigma$.

We recognize three distinct domains: (i) Close to the critical line the network of phase lines is dominated by their approach towards the zeros which for the domain of τ-values shown in the figure are indeed located at $\sigma = 1/2$. (ii) For intermediate values of σ but large ones of τ, that is, $0 < \sigma \ll \tau$, the phase lines leave the neighborhoods of their zeros at large values of τ and descend to smaller values of τ, and (iii) in the asymptotic limit of $0 < \tau \ll \sigma \to \infty$ all phase lines align with, and approach the real axis leading to an ever-increasing amount of lines

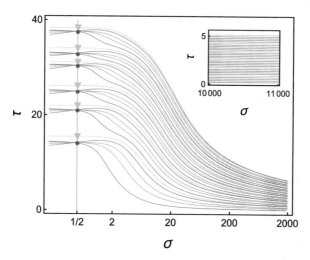

Fig. 9 Mapping of the neighborhood of the critical line (orange curve) of ξ onto the right edge of the complex plane close to the real axis by means of the lines of constant phase (thin black curves). The flows to the individual zeros (red dots) located for the depicted set of τ-values on the critical line are guided by the separatrices (green lines) to approach asymptotically the real axis. Since they get more and more compressed as they tend towards infinity, the flows from ever higher laying zeros are stacked on top of each other. Provided the phase difference between two consecutive separatrices is π, that is, the flow to each zero contains phase lines only from a complete π-interval all zeros must [30] be located on the critical line. The missing π-interval to complete the 2π-interval of phase lines necessary for a simple zero emerges from the left edge of the complex plane due to the anti-symmetry of the phase dictated by the functional equation Eq. (49) of ξ. In order to bring out this behavior most clearly, we have used a non-linear scale for the real axis

being compressed into an identical τ-interval. Here the flow of phase lines is rather orderly.

Figure 9 expresses also an intriguing mapping of the behavior of the phase flow in the neighborhood of the critical line, that is, $\sigma \cong 1/2$ and large values of τ, onto the right edge of the complex plane close to the real axis corresponding to large values of σ but small ones of τ. Indeed, Fig. 9 draws attention to an interesting connection between the flows of ξ in the domains $1/2 \lesssim \sigma \ll \tau$ and $0 < \tau \ll \sigma \to \infty$.

It is well-known that the zeros of ξ are distributed along the critical line in a rather irregular way. Nevertheless, at the edges of the complex plane the flows associated with each zero are almost parallel and stacked on top of each other as indicated by the inset of Fig. 9.

Moreover, due to their continuous approach towards the real axis for increasing values of σ, the flows of ever higher laying zeros get compressed into the same interval of τ-values. It is for this reason that we can investigate the flows associated with zeros of arbitrarily large τ-values by considering them close to the real axis but for large values of σ.

Needless to say, this mapping cannot provide us with any information about the imaginary parts of the zeros. However, it does predict the real parts to be always $1/2$

provided [30] each flow terminating in a zero contains phase lines from a complete π-interval separated from each other by the separatrices shown in Fig. 9 by the green lines.

This approach towards verifying the Riemann Hypothesis requires the determination of the separatrices, and thus a detailed knowledge of the asymptotic behavior of ξ. We dedicate the next sections to this task but admit already at this point that we have not yet been able to develop a criterion to identify separatrices in the sea of phase lines.

3.6   Exponential Representation

It is conventional wisdom that the existence of an Euler product is a key ingredient of the Riemann Hypothesis. Although not proven yet, the Riemann function ξ as well as the Dirichlet L-functions Λ which enjoy an Euler product are expected to have *all* their zeros *on* the critical line.

In contrast, the Titchmarsh counterexample ξ_T discussed in Sect. 7 which is a linear combination of *two* Dirichlet L-functions, and therefore does not obey an Euler product, has not only zeros *on* but also *off* [33] the critical line. In Sect. 7.2, we illuminate this intriguing connection between the location of the zeros and the existence of an Euler product, and trace it back to the well-known difference between a sum of logarithms, and the logarithm of a sum.

In order to compare and contrast these two classes of functions, we first introduce the Euler product of ζ, and then apply it to obtain the exponential representation of ξ. We shall proceed in the next sections in a similar way with the Dirichlet L-functions Λ and the Titchmarsh counterexample ξ_T. We conclude the present section by deriving asymptotic expressions for ξ in two different domains of the complex plane.

3.6.1   Euler Product
In the introduction of this article, we have noted that Riemann in his seminal treatise refers to the Euler product

$$\zeta(s) = \prod_p \left(1 - p^{-s}\right)^{-1} \tag{56}$$

of ζ involving all primes p.

In this section we rederive this formula using two ingredients: (i) The Fundamental Theorem of Arithmetic, which states that every natural number n can be uniquely represented by a product of powers of primes, and (ii) the geometric series

$$\sum_{k=0}^{\infty} x^k = \frac{1}{1-x} \tag{57}$$

valid for $|x| < 1$.

We start by noting that by definition $n = 1$ is *not* a prime number. Hence, the smallest prime number in the infinite product of Eq. (56) is 2, which by the way, is the only even prime number.

Moreover, since $1 < \sigma$ we find with the relation

$$\frac{1}{p^s} = e^{-s \ln p} = e^{-\sigma \ln p} e^{-i\tau \ln p} \tag{58}$$

the inequality

$$\left| \frac{1}{p^s} \right| = e^{-\sigma \ln p} < 1 \tag{59}$$

which allows the application of the geometric series, Eq. (57), leading us to the identity

$$\mathcal{P} \equiv \prod_p \frac{1}{1 - \frac{1}{p^s}} = \prod_p \sum_{k=0}^{\infty} \frac{1}{p^{ks}}, \tag{60}$$

that is,

$$\mathcal{P} = \prod_p \left(1 + \frac{1}{p^s} + \frac{1}{p^{2s}} + \frac{1}{p^{3s}} + \cdots \right). \tag{61}$$

Hence, we arrive the product

$$\mathcal{P} = \left(1 + \frac{1}{p_1^s} + \frac{1}{p_1^{2s}} + \cdots \right) \left(1 + \frac{1}{p_2^s} + \frac{1}{p_2^{2s}} + \cdots \right) \left(1 + \frac{1}{p_3^s} + \frac{1}{p_3^{2s}} + \cdots \right) \cdots \tag{62}$$

of infinite sums containing all powers $k \cdot s$ of all primes $p_1, p_2, \ldots$. Here k is a natural number.

When we perform the multiplication, the infinite product turns into a sum

$$\mathcal{P} = 1 + \cdots \frac{1}{\left(p_{k_1}^{r_1} p_{k_2}^{r_2} \cdots p_{k_\ell}^{r_\ell} \right)^s} + \cdots \tag{63}$$

of terms to the power s where each term is the product of primes $p_{k_1}, p_{k_2}, \ldots, p_{k_\ell}$ to the powers $r_1, r_2 \ldots r_\ell$, and represents a natural number n.

Since all primes and all powers appear, the Fundamental Theorem of Arithmetic guarantees that all natural numbers occur. Thus, we arrive at the Dirichlet representation, Eq. (1), of ζ, verifying the Euler product formula, Eq. (56).

We conclude by considering the limit $\sigma \to \infty$ of ζ and note from the exponential representation, Eq. (58), of p^{-s} that these terms decay exponentially with increasing σ. For this reason, it suffices to reduce the Euler product of ζ to the factor corresponding to $p = 2$. Hence, we arrive at the approximation

$$\zeta(s) \cong \frac{1}{1 - 2^{-s}} \cong 1 + \frac{1}{2^s} ,$$
(64)

in complete agreement with the result following from the Dirichlet series, Eq. (1), of ζ when we restrict the sum to the first two terms.

3.6.2 Exact Expression

We start from the definition, Eq. (41), of ξ, and express the product of functions in terms of products of exponentials. For this purpose we first note the identity

$$\frac{1}{2} \wp(s) = \frac{1}{2} s^2 \left(1 - \frac{1}{s} \right) = e^{-\ln 2} \, e^{2 \ln s} \, e^{\ln\left(1 - \frac{1}{s}\right)} ,$$
(65)

where we have recalled the definition, Eq. (30), of $\wp$.

Moreover, in Appendix 3 we verify the identity

$$\left(\frac{1}{\pi} \right)^{s/2} \Gamma\left(\frac{s}{2} \right) = \sqrt{\pi} \exp\left\{ \frac{s}{2} \left[\ln\left(\frac{s}{2\pi} \right) - 1 \right] - \frac{1}{2} \ln s + \mathcal{R} \right\}$$
(66)

with

$$\mathcal{R} = \ln 2 + \bar{R}\left(\frac{s}{2} \right) ,$$
(67)

and the remainder reads

$$\bar{R}(s) \equiv \int_0^\infty dt \left[\frac{1}{2} - \frac{1}{t} + \frac{1}{e^t - 1} \right] \frac{e^{-ts}}{t} .$$
(68)

When we combine Eqs. (65) and (66), and recall the definition Eq. (41) of ξ we arrive at the final expression

$$\xi = \sqrt{\pi} \, e^{\alpha_\xi} e^{l_\xi}$$
(69)

for the exponential representation of ξ where we have introduced the abbreviations

$$\alpha_\xi(s) \equiv \frac{s}{2} \left[\ln\left(\frac{s}{2\pi} \right) - 1 \right] + \frac{3}{2} \ln s + \mathcal{R}_\xi(s)$$
(70)

with the correction

$$\mathcal{R}_\xi(s) \equiv \ln\left(1 - \frac{1}{s}\right) + \bar{R}\left(\frac{s}{2}\right) \tag{71}$$

containing the remainder $\bar{R}$.

The term

$$l_\xi(s) \equiv -\sum_p \ln\left(1 - p^{-s}\right). \tag{72}$$

follows from the Euler product formula, Eq. (56) of ζ.

3.6.3 Asymptotic Expression

The exponential representation, Eq. (69), of ξ is exact. However, the asymptotic limit $s \to \infty$ simplifies $\mathcal{R}_\xi$, and thus this formula considerably.

In order to bring this fact out most clearly, we evaluate $\mathcal{R}_\xi$ given by Eq. (71) in this limit which according to Appendix 3 yields

$$\bar{R}\left(\frac{s}{2}\right) = \frac{1}{6}\frac{1}{s} + \mathcal{O}\left(\frac{1}{s^3}\right), \tag{73}$$

and with the expansion, Eq. (71), of the logarithm to the lowest order in $1/s$ reduces to

$$\mathcal{R}_\xi(s) = -\frac{5}{6}\frac{1}{s} + \mathcal{O}\left(\frac{1}{s^2}\right). \tag{74}$$

Hence, in the asymptotic limit $s \to \infty$ the correction term $\mathcal{R}_\xi$ in α_ξ in the exponential representation of ξ given by Eq. (69) decays as $1/s$, and can be neglected.

When we recall from Eq. (64) the approximation

$$\zeta(s) \cong 1 \tag{75}$$

for $s \to \infty$, we arrive at the asymptotic limit

$$\xi \cong \sqrt{\pi}\, e^{\alpha_\xi} \tag{76}$$

of the exponential representation of ξ with

$$\alpha_\xi(s) \equiv \frac{s}{2}\left[\ln\left(\frac{s}{2\pi}\right) - 1\right] + \frac{3}{2}\ln s, \tag{77}$$

which is solely determined by the asymptotic behavior of $\wp$ and the product given by Eq. (66).

3.6.4   Amplitude and Phase at the Right Edge of the Complex Plane

In Appendix 4 we use the asymptotic result, Eq. (77), to express ξ in amplitude and phase where

$$\alpha_\xi \equiv \Sigma + i\Phi. \tag{78}$$

According to the formulae derived there, the real part Σ of α_ξ which determines

$$|\xi| = \sqrt{\pi}e^\Sigma \tag{79}$$

takes in its most elementary version the form

$$\Sigma(\sigma, \tau) \cong \frac{\sigma}{2}\left[\ln\left(\frac{\sigma}{2\pi}\right) - 1\right] \tag{80}$$

and is independent of τ. As a result, $|\xi|$ increases exponentially for increasing σ.

Moreover, the phase

$$\Phi(\sigma, \tau) \cong \nu(\sigma)\tau \tag{81}$$

of ξ increases for large but fixed positive values of σ linearly with τ at a rate

$$\nu(\sigma) \cong \frac{1}{2}\ln\left(\frac{\sigma}{2\pi}\right) \tag{82}$$

which is also independent of τ.

Hence, we conclude that at the right edge of the complex plane, that is, for $0 < \tau \ll \sigma \to \infty$ the lines

$$\tau_\Phi(\sigma) \cong \frac{2\Phi}{\ln(\sigma/(2\pi))}, \tag{83}$$

of constant phase Φ of ξ approach with a rate of $1/\ln(\sigma/(2\pi))$ the real axis when we increase σ.

Since the logarithm changes slowly with σ, the lines of constant phase are locally almost parallel to each other. When viewed on a larger scale they get more and more compressed as $\sigma \to \infty$, that is, as σ increases, more and more phase lines get squeezed into a fixed interval of τ, in complete agreement with Fig. 9.

3.7   Representation in Terms of a Switching Function

In Sect. 3.1 we have defined the Riemann function ξ as a product of several functions which emerge in a natural way from the analytic continuation of ζ defined by the Dirichlet series. In this section, we present an alternative representation of ξ in terms of the sum of two infinite sums, closely related to the Dirichlet sum. Indeed, the terms consist of the product of the familiar contribution n^{-s} and a switching function.

3.7.1 Motivation

In Sect. 2.3 we have already emphasized that the analytic continuation of ζ given by Eq. (29) does not rely on a resummation of the Dirichlet series. However, it would be interesting to truncate the Dirichlet sum at an appropriate n_0 and resum the remainder to gain insight into the behaviour of ζ in the complete complex plane.

Indeed, Carl Ludwig Siegel [32] has followed this path, giving rise to the Riemann-Siegel formula for ζ. Motivated by the notes of Riemann, Siegel could reconstruct the key ideas of Riemann that led him to his hypothesis as to the location of the zeros.

Michael Victor Berry and John Keating [3, 4] improved the Riemann-Siegel formula by eliminating the discrete cut-off n_0 of the sum and by introducing an Error function as a switching function. Our approach summarized in this section is similar in spirit but addresses a different domain of the complex plane. Indeed, the Berry-Keating technique focuses on the behavior of ζ along the critical axis whereas we aim at the right edge of the complex plane, that is, for $\sigma \to \infty$.

3.7.2 Formula

We start by considering the expression

$$S \equiv \frac{1}{2}\wp\,(d + \varepsilon) \tag{84}$$

formed by the product of the polynomial $\wp$ and the sum of the terms

$$d \equiv e^{\alpha} \sum_{n=1}^{\infty} \mathcal{S}_n \frac{1}{n^s} \tag{85}$$

and

$$\varepsilon \equiv e^{\alpha} \sum_{n=1}^{\infty} (\mathcal{S}_0 - \mathcal{S}_n) \frac{1}{n^s} . \tag{86}$$

At this moment the functions α and $\mathcal{S}_n$ are not determined yet.

Obviously S is reminiscent of ξ in the form of Eq. (43) where d or ε are then related to $\gamma(s)$ or $\gamma(1 - s)$ and $1/\wp\,(s)$. The task is now to determine α and $\mathcal{S}_n$ in such a way as to achieve the identity

$$S = \xi . \tag{87}$$

We first note that we have a lot of freedom in our choice of the coefficients $\mathcal{S}_n$. Indeed, when we substitute the expressions Eqs. (85) and (86) for d and ε into the definition Eq. (84) of $\mathcal{S}$, the terms $\mathcal{S}_n n^{-s}$ in d and ε cancel each other since they appear in d and ε with opposite signs, and we arrive at the result

$$S = \frac{1}{2} \wp \, e^{\alpha} \sum_{n=1}^{\infty} \mathcal{S}_0 \frac{1}{n^s} \tag{88}$$

or

$$S = \frac{1}{2} \wp \, e^{\alpha} \mathcal{S}_0 \, \zeta(s) . \tag{89}$$

Here we have factored the coefficient $\mathcal{S}_0$ which is independent of n, out of the sum and have recalled the Dirichlet representation, Eq. (1), of ζ.

Next we compare the right-hand side of this equation with the product formula, Eq. (40), of ξ and enforce the identity Eq. (87) by postulating

$$e^{\alpha} \, \mathcal{S}_0 = \pi^{-s/2} \, \Gamma \left(\frac{s}{2} \right) . \tag{90}$$

With the help of the integral representation, Eq. (19), of the Gamma function we now derive explicit expressions for α and $\mathcal{S}_0$ which reduce to the Error function in the asymptotic limit $\sigma \to \infty$, as we show later.

Indeed, when we start from the identity

$$\pi^{-s/2} \, \Gamma \left(\frac{s}{2} \right) = \int_0^{\infty} dt \, \frac{1}{t} \, \exp \left[-t + \frac{s}{2} \ln \left(\frac{t}{\pi} \right) \right] \tag{91}$$

following from Eq. (19) together with the exponential representation, Eq. (32), of $\pi^{-s/2}$, and introduce the integration variable

$$y \equiv \frac{t - \sigma/2}{\sigma/2} \tag{92}$$

for $s \equiv \sigma + i\tau$, we determine the functions

$$\alpha(\sigma, \tau) \equiv \frac{\sigma}{2} \left[\ln \left(\frac{\sigma}{2\pi} \right) - 1 \right] + i \frac{\tau}{2} \ln \left(\frac{\sigma}{2\pi} \right) \tag{93}$$

and

$$\mathcal{S}_0 \equiv \int_{-1}^{\infty} dy \, \mathcal{G} \tag{94}$$

where the complex-valued integrand

$$G(y; \sigma, \tau) \equiv \frac{\exp\left\{-\frac{\sigma}{2}\left[y - \ln(1+y)\right]\right\}}{1+y} \exp\left[i\frac{\tau}{2}\ln(1+y)\right] \tag{95}$$

is rather complicated.

3.7.3 Connection to Sum Representation of ξ

So far we have been able to define the coefficient S_0 but not the terms S_n. This feature is not surprising since they cancel in the *sum*, Eq. (84), of d and ε defining S. Hence, they cannot be determined by comparing S to the product formula, Eq. (40), of ξ and enforcing their identity. For this reason, we now postulate an expression for S_n and show that the so-defined formulae for d and ε are identical to $\gamma(s)$ and $\gamma(1-s) + 1/\wp(s)$, respectively.

Indeed, we define the coefficients

$$S_n \equiv \int_{-1+2\pi n^2/\sigma}^{\infty} dy\, G, \tag{96}$$

and note that for $n = 0$ this expression reduces to the one of Eq. (94) obtained by comparing the product formula, Eq. (40), for ξ with the sum, Eq. (84), of d and ε. However, the lower boundary of the integral now depends quadratically on n, and inversely on the real part σ of the argument s.

This special choice for the functional dependence on n stands out most clearly when we make the substitution

$$x \equiv \frac{\sigma}{2\pi} \frac{1}{n^2}(1+y) \tag{97}$$

which leads us to the expression

$$S_n = e^{-\alpha} n^s \int_1^{\infty} dx\, e^{-\pi n^2 x} x^{s/2-1}, \tag{98}$$

where we have used the definition, Eq. (93), of α.

Thus, for this choice of S_n the contribution d given by Eq. (85) reduces to

$$d = \sum_{n=1}^{\infty} \int_1^{\infty} dx\, e^{-\pi n^2 x} x^{s/2-1} \tag{99}$$

or

$$d = \gamma(s), \tag{100}$$

where we have interchanged the summation and integration, and have recalled the definitions, Eqs. (21) and (23), of ω and γ.

Likewise, we find with the definition, Eq. (96), for S_n the expression

$$\varepsilon = \sum_{n=1}^{\infty} e^{\alpha} \int_{-1}^{-1+2\pi n^2/\sigma} dy \; \mathcal{G} \frac{1}{n^s} \,, \tag{101}$$

which with the substitution, Eq. (97), yields

$$\varepsilon = \int_{0}^{1} dx \; \omega(x) x^{s/2-1} = I \,. \tag{102}$$

Here we have recalled the definition, Eq. (24), of the integral I which also appears in the analytic continuation of the Riemann zeta function.

In this context we have already found with the help of the functional equation of the Jacobi theta function the representation Eq. (28), and with the definition, Eq. (30), of $\wp$ we arrive at the connection

$$\varepsilon(s) = \gamma(1-s) + \frac{1}{\wp(s)} \,. \tag{103}$$

Hence, when we substitute Eqs. (100) and (103) into Eq. (84), we arrive indeed at the sum representation, Eq. (43), of ξ. Thus, the term d in Eq. (84) is identical to γ, whereas ε consists of the sum of $\gamma(1-s)$ and $1/\wp$.

3.7.4 Asymptotic Expansion

So far, our analysis is exact. However, it is not clear yet why the representation of $\gamma(s)$ and $\gamma(1-s)$ in terms of the sums d and ε is useful. We now show that in the limit $\sigma \to \infty$ these infinite sums turn essentially into finite sums where the coefficients S_n play the role of a switching function.

For this purpose, we proceed in two steps: (i) We first perform the asymptotic limit of the integrand $\mathcal{G}$ of the integral defining the coefficients S_n, and (ii) then discuss the resulting Error function in its dependence on n.

We start our analysis by casting $\mathcal{G}$ defined by Eq. (95) into the form

$$\mathcal{G} = e^{-\sigma y^2/4} \frac{\exp\left\{-\frac{\sigma}{2}\left[y - \frac{1}{2}y^2 - \ln(1+y)\right]\right\}}{1+y} \exp\left[i\frac{\tau}{2}\ln(1+y)\right] \tag{104}$$

and employ the expansion

$$\ln(1+y) = y - \frac{1}{2}y^2 + \frac{1}{3}y^3 + \dots \tag{105}$$

of the logarithm which yields

$$\mathcal{G} \cong e^{-\sigma y^2/4} R \tag{106}$$

with the remainder

$$R(y) \equiv \frac{e^{\sigma y^3/6}}{1+y} e^{i\tau(y/2-y^2/4)}. \tag{107}$$

Due to the width $\delta y \equiv (4/\sigma)^{1/2}$ of the Gaussian, we find the estimate

$$R(\delta y) = \frac{e^{4/(3\sqrt{\sigma})} e^{i\tau(1/\sqrt{\sigma}-1/\sigma)}}{1+2/\sqrt{\sigma}} \cong 1. \tag{108}$$

for $\sigma \to \infty$.

Hence, in this limit, the integrand of the integral defining $\mathcal{S}_n$ takes the form

$$\mathcal{G} \cong e^{-\sigma y^2/4} \tag{109}$$

and is essentially a narrow Gaussian located at $y = 0$.

As a result, the coefficients

$$\mathcal{S}_n \cong \int_{-1+2\pi n^2/\sigma}^{\infty} dy \, e^{-\sigma y^2/4} \tag{110}$$

reduce to an Error function as a function of n.

Indeed, for a fixed but large value of σ the lower boundary of the integral in Eq. (110) is approximately independent of n and given by -1. Since the width δy of the Gaussian is proportional to $1/\sqrt{\sigma} \ll 1$, the integral is constant, that is,

$$\mathcal{S}_n \cong \int_{-1}^{\infty} dy \, e^{-\sigma y^2/4} \cong \int_{-\infty}^{\infty} dy \, e^{-\sigma y^2/4} = 2\sqrt{\frac{\pi}{\sigma}}. \tag{111}$$

However, a transition takes place when n is of the order of $n_s \equiv \sqrt{\sigma/(2\pi)}$ and compensates the -1 in the lower boundary of the integral. In this case, we find the approximation

$$\mathcal{S}_n \cong \int_0^{\infty} dy \, e^{-\sigma y^2/4} = \sqrt{\frac{\pi}{\sigma}}. \tag{112}$$

When n is larger than n_s such that

$$\delta y < -1 + \frac{2\pi n^2}{\sigma}, \tag{113}$$

the lower domain of integration does not cover the Gaussian anymore, and $\mathcal{S}_n$ vanishes, that is,

$$\mathcal{S}_n \cong 0. \tag{114}$$

Hence, in the limit of $\sigma \to \infty$ the coefficients $\mathcal{S}_n$ define a switching function. Indeed, $\mathcal{S}_n$ is a constant as long as n is smaller than n_s. In this case, only the factors n^{-s} of the Dirichlet series are relevant. However, when n gets of the order of, or is larger than n_s, the coefficients $\mathcal{S}_n$ become negligible transforming d, and in particular, the Dirichlet series into a finite sum.

In contrast, the terms $\mathcal{S}_0 - \mathcal{S}_n$ in the sum determining ε are approximately vanishing for n smaller than n_s but become relevant for n values larger than n_s. In this case, we regain the terms of the Dirichlet series that were truncated in d.

3.7.5 Connection to the Riemann Hypothesis

Motivated by Fig. 9, we have derived in the preceding section the asymptotic limit of ξ for $\sigma \to \infty$, starting from the exponential representation of ξ. We have found rather elementary expressions for the phase lines at the right edge of the complex plane. However, these expressions do not allow us to identify the separatrices since they do not carry the information of the zeros. Indeed, they arise solely from the exponential product.

In order to remedy this deficiency, we have obtained in this section an exact representation of ξ as a sum of two terms, each of which consists of a Dirichlet sum reminiscent of ζ but with a weight function. One sum corresponds to the truncated Mellin transform at s and the other one to the truncated Mellin transform at $1 - s$ combined with the inverse of $\wp$ which apart from the inverse of the exponential product represents the analytic continuation of ζ.

This representation is tailored to reproduce in a direct way the asymptotic behavior of ξ at the right edge of the complex plane, derived before from the exponential product. In this limit, the weight function turns into a switching function approximated by an Error function. It is constant up to a given large integer n and then rapidly decays creating essentially a truncated Dirichlet sum of ζ. In the second sum of this representation of ξ the weight function only turns on for values larger than n but is again constant.

The zeros of ξ arise from the mutual cancellation of the two sums. In the asymptotic limit, one of them is identical to the asymptotic expression for ξ given by the exponential product and $\wp$. The other one carries the information about the zeros. From this combination of these two contributions we should be able to identify the separatrices and thereby verify the Riemann Hypothesis. However, this task goes beyond the scope of the present article.

4   Dirichlet Characters

We start our review of Dirichlet L-functions by first defining the Dirichlet character, and then deriving elementary properties. Here we focus on features essential for the material discussed in the next sections.

The character χ mod q is a function which maps all integers n to the complex plane and satisfies the following three properties: The function χ is

(i) completely multiplicative, that is,

$$\chi(m \cdot n) = \chi(m)\chi(n), \tag{115}$$

(ii) periodic with period q, that is,

$$\chi(n + q) = \chi(n), \tag{116}$$

and
(iii) only nonzero if n and q do not share a common divisor except 1, that is,

$$\chi(n) = \begin{cases} 0 & \gcd(n, q) > 1 \\ \neq 0 & \gcd(n, q) = 1 \end{cases} \quad \text{with} \quad |\chi(n)| = 1. \tag{117}$$

Equation (117) immediately implies that Dirichlet characters are the $\varphi(q)$-th roots of unity, and consequently we have the identity

$$\chi^*(n)\chi(n) = 1, \tag{118}$$

unless

$$\chi(0) = 0. \tag{119}$$

Here, φ denotes Euler's function, that is, the number of relatively prime residue classes.

The so seemingly innocent identity, Eq. (119), follows immediately from Eq. (117) and will be crucial in the derivation of the functional equation of the generalized Jacobi theta function. Indeed, it is the reason that the expression for the analytic continuation of Λ is less sophisticated than the corresponding one of ξ as discussed in Sect. 6.

Moreover, due to the periodicity property, Eq. (116), we find the relation

$$\chi(q) = \chi(0 + q) = \chi(0), \tag{120}$$

and with the help of Eq. (119) the property

$$\chi(q) = 0. \tag{121}$$

Hence, χ vanishes not only for $n = 0$, but also at the period q and, of course, at integer multiples of it.

Next we calculate $\chi(1)$ and find from the multiplicativity, Eq. (115), of χ the identity

$$\chi(1) = \chi(1 \cdot 1) = (\chi(1))^2 \tag{122}$$

which yields

$$\chi(1) = 1. \tag{123}$$

Here we have used the fact that according to Eq. (117), $\chi(1) \neq 0$ which allows us to divide Eq. (122) by $\chi(1)$. Indeed, $n = 1$ and q only share 1 as a common divisor.

Moreover, in a similar way we establish the relation

$$(\chi(-1))^2 = \chi(-1)\chi(-1) = \chi((-1)(-1)) = \chi(1) = 1 \tag{124}$$

leading us to

$$\chi(-1) = \pm 1. \tag{125}$$

Hence, the parameter

$$\kappa \equiv \frac{1}{2}[1 - \chi(-1)] \tag{126}$$

which appears in the definition of the Dirichlet L-functions Λ introduced in Sect. 6 and contains $\chi(-1)$ can only assume the values $\kappa = 0$ and $\kappa = 1$, that is,

$$\kappa = \frac{1}{2}(1 \mp 1) = \begin{cases} 0 & \text{for } \chi(-1) = 1 \\ 1 & \text{for } \chi(-1) = -1. \end{cases} \tag{127}$$

We conclude by relating the values of $\chi(-n)$ to $\chi(n)$. Indeed, since χ is multiplicative, Eq. (115), we find

$$\chi(-n) = \chi((-1) \cdot n) = \chi(-1)\chi(n). \tag{128}$$

Thus, $\chi(-1)$ connects $\chi(-n)$ and $\chi(n)$.

5 Gauss Sums

In the preceding section, we have introduced the concept of a Dirichlet character and have derived several elementary properties used throughout the remainder of our article. We now turn to Gauss sums, which are sums of products of characters and Fourier phase factors over one period. Gauss sums play a central role in the analytic continuation of Dirichlet L-functions, made possible in part by the functional equation of the generalized Jacobi theta function.

In order to keep the discussion rather elementary, we assume that the module q of the character χ is a prime number. The results are true in greater generality, namely for primitive characters. However, this treatment would require a more elaborate theory.

We first derive a formula connecting two types of Gauss sums, and then calculate the absolute value of the elementary one. This result allows us to express the normalized Gauss sum by a phase factor. We conclude by deriving a symmetry relation for this phase, crucial for the derivation of the functional equation of Dirichlet L-functions.

5.1   Reduction Formula

We start by establishing the identity

$$\tilde{G}(\chi, m) = \chi^*(m)\tilde{G}(\chi, 1) \tag{129}$$

which reduces the Gauss sum

$$\tilde{G}(\chi, m) \equiv \sum_{n=1}^{q} \chi(n)e^{2\pi imn/q} \tag{130}$$

consisting of the sum of products of the characters $\chi(n)$ over a single period and the m-th Fourier component, to the product of the Gauss sum at $m = 1$ and the complex conjugate $\chi^*(m)$ of the character χ at m.

The values of $\chi(n)$ and $\exp(2\pi in/q)$ both depend only on the residue class of $n \bmod q$. Hence, when we represent the product

$$m \cdot n = k + rq \tag{131}$$

by an integer multiple r of q plus an integer correction k with $1 \leq k < q$ then for a fixed m with $1 \leq m < q$ the integer k assumes each value with $1 \leq k < q$ exactly once.

Moreover, we find the relation

$$\chi(n) = \chi\left(\frac{k + rq}{m}\right) = \chi^*(m)\chi(k + rq) = \chi^*(m)\chi(k) \tag{132}$$

where we have used the fact that χ is a root of unity, Eq. (118), together with the periodicity Eq. (116).

Finally, with the help of the identity

$$e^{2\pi imn/q} = e^{2\pi i(k+rq)/q} = e^{2\pi ik/q}e^{2\pi ir} = e^{2\pi ik/q} , \tag{133}$$

we arrive at the representation

$$\tilde{G}(\chi, m) = \chi^*(m)\tilde{G}(\chi) \tag{134}$$

where we have introduced the abbreviation

$$\tilde{G}(\chi) \equiv \tilde{G}(\chi, 1) = \sum_{n=1}^{q} \chi(n)e^{2\pi in/q} \tag{135}$$

for $\tilde{G}(\chi, 1)$.

5.2   Normalized Gauss Sums

Next we verify the identity

$$\left|\tilde{G}(\chi)\right|^2 = q \tag{136}$$

which allows us to introduce the *normalized* Gauss sum G, and represent it by a phase factor. For this purpose, we take advantage of the reduction formula, Eq. (129), and can cast the product

$$\tilde{G}^*(\chi)\tilde{G}(\chi) = \sum_{m=1}^{q} \chi^*(m)e^{-2\pi im/q}\tilde{G}(\chi, 1) \tag{137}$$

following from the definition, Eq. (135), of $\tilde{G}(\chi)$ into the form

$$\left|\tilde{G}\right|^2 = \sum_{m=1}^{q} e^{-2\pi im/q}\tilde{G}(\chi, m) \tag{138}$$

or

$$|G|^2 = \sum_{n=1}^{q} \chi(n) \sum_{m=1}^{q} \left[e^{2\pi i(n-1)/q}\right]^m , \tag{139}$$

where we have recalled the definition, Eq. (130), of $\tilde{G}(\chi, m)$.

Next, we decompose the sum over n into the term $n = 1$ and the remaining part, that is,

$$\left|\tilde{G}\right|^2 = \chi(1) \cdot q + \sum_{n=2}^{q} \chi(n) e^{2\pi i (n-1)/q} \frac{e^{2\pi i (n-1)} - 1}{e^{2\pi i (n-1)/q} - 1}, \tag{140}$$

where the last term is a consequence of the familiar summation formula

$$\sum_{m=1}^{q} x^{m-1} = \frac{x^q - 1}{x - 1}$$

valid for $x \neq 1$.

Since

$$e^{2\pi i (n-1)} = 1,$$

the second contribution in Eq. (140) vanishes, and we arrive with Eq. (123), that is, $\chi(1) = 1$, at the desired result, Eq. (136).

Due to the identity, Eq. (136), the *normalized* Gauss sum

$$G(\chi) \equiv i^{-\kappa} \frac{1}{\sqrt{q}} \tilde{G}(\chi), \tag{141}$$

that is,

$$G(\chi) \equiv i^{-\kappa} \frac{1}{\sqrt{q}} \sum_{n=1}^{q} \chi(n) e^{2\pi i n / q} \tag{142}$$

satisfies the relation

$$|G(\chi)|^2 = 1.$$

Thus, G is just a phase factor, that is,

$$G(\chi) = e^{i\beta(\chi)}, \tag{143}$$

where the phase $\beta = \beta(\chi)$ depends on the character χ.

In Appendix 5 we explicitly evaluate the Gauss sums for the Dirichlet L-functions forming the Titchmarsh counterexample discussed in Sect. 7, and show that the associated phases are non-vanishing. However, we also calculate the Gauss sum for a Dirichlet L-function for which the phase vanishes.

5.3   Symmetry Relation for Phase

In the definition, Eq. (142), of the normalized Gauss sum G we have also included the phase factor $i^{-\kappa}$ without any reasoning. We now show that this term ensures the symmetry relation

$$\beta(\chi^*) = -\beta(\chi) \tag{144}$$

of the phase of the Gauss sum.

For this purpose we first derive an equivalent representation of G and then take the complex conjugate of it. Indeed, when we introduce the new summation index $j \equiv q - n$, the normalized Gauss sum defined by Eq. (142) takes the form

$$G = i^{-\kappa} \frac{1}{\sqrt{q}} \sum_{j=q-1}^{0} \chi(q - j) e^{2\pi i(q-j)/q} , \tag{145}$$

or

$$G = i^{-\kappa} \chi(-1) \frac{1}{\sqrt{q}} \sum_{j=1}^{q} \chi(j) e^{-2\pi ij/q} . \tag{146}$$

Here we have (i) changed the order of the individual terms in the sum and used the periodicity of χ to replace $\chi(0)$ by $\chi(q)$ which both vanish, (ii) employed the periodicity of χ together with Eq. (128), and (iii) taken advantage of the periodicity of the Fourier phase factor.

Next we note the relation

$$i^{-\kappa} \chi(-1) = i^{\kappa} \tag{147}$$

following from Eq. (127), and obtain the equivalent representation

$$G(\chi) = i^{\kappa} \frac{1}{\sqrt{q}} \sum_{j=1}^{q} \chi(j) e^{-2\pi ij/q} . \tag{148}$$

When we now take the complex conjugate of this relation, we arrive with the help of the identity

$$\left(i^{\kappa}\right)^* = (-i)^{\kappa} = i^{-\kappa} \tag{149}$$

at the formula

$$G(\chi)^* = i^{-\kappa} \frac{1}{\sqrt{q}} \sum_{j=1}^{q} \chi^*(j) e^{2\pi i j/q} = G(\chi^*), \tag{150}$$

which with Eq. (143) leads us to the desired symmetry relation Eq. (144).

6 Dirichlet L-functions

We are now in the position of defining the Dirichlet L-functions Λ and deriving their analytic continuations. Indeed, the definition of a Dirichlet L-function $\Lambda = \Lambda(s, \chi)$ of the complex-valued argument $s \equiv \sigma + i\tau$ and Dirichlet character $\chi = \chi(n)$ reads

$$\Lambda(s, \chi) \equiv \left(\frac{q}{\pi}\right)^{s/2} \Gamma\left(\frac{s+\kappa}{2}\right) L(s, \chi), \tag{151}$$

where

$$L(s, \chi) \equiv \sum_{n=1}^{\infty} \chi(n) n^{-s}$$

denotes the Dirichlet L-series.

In analogy to the Riemann function ξ, it is possible to obtain for Λ an analytic continuation into the complete complex plane. In this section we present this derivation and verify the functional equation as well as discuss the ensuing symmetry relations of Λ in amplitude and phase with respect to the critical line.

6.1 Analytic Continuation

We start by noting that with the help of the integral representation

$$\Gamma\left(\frac{s+\kappa}{2}\right) \equiv \int_{0}^{\infty} dt \; e^{-t} t^{(s+\kappa)/2-1}$$

of the Gamma function at the argument $(s + \kappa)/2$, and the integration variable

$$t \equiv \frac{\pi}{q} n^2 x, \tag{152}$$

we can cast Λ given by Eq. (151) into the form

$$\Lambda(s, \chi) = \left(\frac{q}{\pi}\right)^{-\kappa/2} \int_{0}^{\infty} dx \; \omega(x, \chi) x^{(s+\kappa)/2-1}, \tag{153}$$

where we have introduced the generalized Jacobi theta function

$$\omega(x, \chi) \equiv \sum_{n=1}^{\infty} \chi(n) n^{\kappa} e^{-\pi n^2 x/q}. \tag{154}$$

Next we decompose the integration over the variable x in the Mellin transform, Eq. (153), of ω into two integrals: one extending from zero to unity, and one from unity to infinity.

When we apply the functional equation

$$\omega(x, \chi) = G(\chi) x^{-\kappa-1/2} \omega\left(\frac{1}{x}, \chi^*\right) \tag{155}$$

of ω derived in Appendix 6 to the first integral together with the change of variable $y = 1/x$, we arrive at the analytic continuation

$$\Lambda(s, \chi) = e^{i\beta(\chi)} \gamma(1 - s, \chi^*) + \gamma(s, \chi), \tag{156}$$

of Λ where

$$\gamma(s, \chi) \equiv \left(\frac{q}{\pi}\right)^{-\kappa/2} \int_1^{\infty} dx \, \omega(x, \chi) x^{(s+\kappa)/2-1}. \tag{157}$$

Here we have made use of the fact that the normalized Gauss sum G reduces to a phase factor as expressed by Eq. (143).

It is instructive to compare and contrast the analytic continuation of Λ given by Eq. (156) to the corresponding one, Eq. (43), of ξ. Indeed, we find several similarities and differences: (i) Both involve the superposition of a function γ which is an integral, Eqs. (23) and (157), of a Jacobi theta function given by Eqs. (21) and (154), evaluated at s and $1 - s$. (ii) Compared to the one of ξ, the Jacobi theta function of Λ also involves the character χ, its period q and the parameter κ. (iii) The analytic continuation of ξ brings in the quadratic polynomial $s(s - 1)/2$ as well as the constant off-set $1/2$, which are absent in the one of Λ. (iv) One term in the superposition defining Λ is multiplied by the phase factor representing the Gauss sum.

In Appendix 6 we point out the deeper origin of these differences. We show that they are almost exclusively rooted in the simplicity of the functional equation, Eq. (155), of the generalized theta function.

6.2   Functional Equation

Next, we derive the functional equation of Λ. For this purpose we substitute $1 - s$ and χ^* into Eq. (156) and find

$$\Lambda(1 - s, \chi^*) = e^{i\beta(\chi^*)} \gamma(s, \chi) + \gamma(1 - s, \chi^*),$$

which after multiplication by $\exp(i\beta(\chi))$ yields the relation

$$e^{i\beta(\chi)}\Lambda(1 - s, \chi^*) = e^{i\beta(\chi)}\gamma(1 - s, \chi^*) + \gamma(s, \chi),$$

and with the analytic continuation, Eq. (156), of Λ the functional equation

$$\Lambda(s, \chi) = e^{i\beta(\chi)}\Lambda(1 - s, \chi^*). \tag{158}$$

Here we have made use of the symmetry relation, Eq. (144), for the phase β of the Gauss sum, Eq. (142).

6.3   Symmetry Relations

We conclude this brief summary of the analytic continuation of Λ by addressing the symmetry relations for the absolute value $|\Lambda|$ and the phase ϕ of Λ with respect to the critical line following from the functional equation of Λ.

Indeed, Eq. (158) in the form

$$\Lambda\left(\frac{1}{2} + \left(\sigma - \frac{1}{2}\right) + i\tau, \chi\right) = e^{i\beta}\Lambda\left(\frac{1}{2} - \left(\sigma - \frac{1}{2}\right) + i\tau, \chi\right)^*$$

shows that $|\Lambda|$ is symmetric with respect to the critical line, but ϕ experiences a phase shift β resulting from the formula

$$\phi\left(\frac{1}{2} + \left(\sigma - \frac{1}{2}\right) + i\tau, \chi\right) = \beta - \phi\left(\frac{1}{2} - \left(\sigma - \frac{1}{2}\right) + i\tau, \chi\right) + 2\pi k. \tag{159}$$

Here k is an integer.

On the critical line, that is, for $\sigma = 1/2$, we obtain from this identity the relation

$$\phi\left(\frac{1}{2} + i\tau, \chi\right) = \frac{1}{2}\beta(\chi) + k\pi, \tag{160}$$

which states that there, apart from integer multiples of π, ϕ is governed by *half* of the phase β of the Gauss sum G, Eq. (142).

6.4   Special Examples

Finally, we illustrate the Dirichlet L-functions by presenting three examples. The first two constitute the building blocks of the Titchmarsh counterexample. Here the normalized Gauss sums have a non-vanishing phase, which is crucial for their functional equation. The third example has a vanishing phase.

6.4.1 Building Blocks of Titchmarsh Counterexample

We now consider the characters χ_1 and χ_2 defined mod 5 with the values

$$\chi_1(1) = 1, \quad \chi_1(2) = i, \quad \chi_1(3) = -i, \quad \chi_1(4) = -1, \quad \chi_1(5) = 0 \qquad (161)$$

and

$$\chi_2(1) = 1, \quad \chi_2(2) = -i, \quad \chi_2(3) = i, \quad \chi_2(4) = -1, \quad \chi_2(5) = 0 \qquad (162)$$

which enjoy the symmetry relation

$$\chi_1^* = \chi_2. \qquad (163)$$

Since in both functions $q = 5$, we find with $\chi_j(-1) = \chi_j(4 - 5) = \chi_j(4)$ for $j = 1, 2$, and the definitions Eqs. (161) and (162) of χ_1 and χ_2, the result $\chi_1(-1) = \chi_2(-1) = -1$, which yields with Eq. (127) the parameter

$$\kappa = 1.$$

Hence, the associated Dirichlet L-functions following from Eq. (151) take the form

$$\Lambda_j(s) \equiv \Lambda(s, \chi_j) = \left(\frac{5}{\pi}\right)^{s/2} \Gamma\left(\frac{s+1}{2}\right) L(s, \chi_j),$$

where the normalized Gauss sums defined by Eq. (142) read

$$G_j \equiv G(s, \chi_j) \equiv i^{-1} \frac{1}{\sqrt{5}} \sum_{n=1}^{5} \chi_j(n) e^{2\pi i n/5}. \qquad (164)$$

In Appendix 5 we calculate G_1 and G_2 explicitly and find the expressions

$$G_1 = e^{2i\theta}, \qquad (165)$$

and

$$G_2 = e^{-2i\theta} \qquad (166)$$

where

$$\tan(2\theta) = \frac{\sqrt{5} - 1}{2}. \qquad (167)$$

As a result, we are now in the position to present the functional equations of Λ_1 and Λ_2. Indeed, from the functional equation, Eq. (158), for Λ together with the explicit expressions Eqs. (165) and (166) for the normalized Gauss sums, we obtain the identities

$$\Lambda_1(s) \equiv \Lambda(s, \chi_1) = e^{2i\theta} \Lambda(1 - s, \chi_1^*) = e^{2i\theta} \Lambda(1 - s, \chi_2) = e^{2i\theta} \Lambda_2(1 - s) \tag{168}$$

and

$$\Lambda_2(s) \equiv \Lambda(s, \chi_2) = e^{-2i\theta} \Lambda(1 - s, \chi_2^*) = e^{-2i\theta} \Lambda(1 - s, \chi_1) = e^{-2i\theta} \Lambda_1(1 - s), \tag{169}$$

where we have recalled the symmetry relation, Eq. (163), between χ_1 and χ_2.

Hence, multiplication of Eq. (169) by $\exp(i\theta)$ yields the relation

$$e^{i\theta} \Lambda(s, \chi_2) = e^{-i\theta} \Lambda(1 - s, \chi_1).$$

6.4.2 Dirichlet L-function with Vanishing Phase

Next we present an example of a Dirichlet L-function where the phase in the functional equation vanishes and consider the real-valued character

$$\chi(n) \equiv \begin{cases} 1 & \text{for } n = 4k + 1 \\ -1 & \text{for } n = 4k + 3 \\ 0 & \text{for } n \text{ even} \end{cases} \tag{170}$$

with k integer.

Since χ is mod 4, we find $q = 4$ and with $\chi(-1) = \chi(3 - 4) = \chi(3) = -1$ following from the definition, Eq. (170), of χ, we obtain with the help of Eq. (127) the value $\kappa = 1$.

As a result, the Dirichlet L-function corresponding to the character χ defined by Eq. (170) reads

$$\Lambda(s, \chi) \equiv \left(\frac{4}{\pi}\right)^{s/2} \Gamma\left(\frac{s + 1}{2}\right) L(s, \chi)$$

with

$$L(s, \chi) \equiv 1 - 3^{-s} + 5^{-s} - 7^{-s} + \dots .$$

In Appendix 5 we calculate the corresponding Gauss sum and show that the phase β vanishes. As a result, the functional equation, Eq. (158), reduces to the form

$$\Lambda(s, \chi) = \Lambda(1 - s, \chi),$$

familiar from ξ.

Due to the fact that χ is real and $\beta = 0$ the representation

$$\Lambda(s, \chi) = \gamma(s, \chi) + \gamma(1 - s, \chi).$$

following from Eq. (156) involves no phase factors, polynomials or off-sets.

This property is in sharp contrast to the corresponding one for ξ, given by Eq. (43). Hence, the zeros of this Dirichlet L-function Λ arise from the destructive interference of just the generalized truncated Mellin transform at s and $1 - s$.

6.5   Exponential Representation

In Sect. 3.6 we have derived the exponential representation of the Riemann function ξ. Moreover, we have obtained an asymptotic expression for it. In the present section, we pursue the analogous approach for the Dirichlet L-function.

6.5.1   Euler Product

One essential element of the derivation of the exponential representation of ξ is the Euler product, Eq. (56), of ζ. A similar relation holds true for the Dirichlet L-function.

Indeed, we find with the help of the geometric sum, Eq. (57), for the expression

$$\mathcal{P}_L \equiv \prod_p [1 - \chi(p)p^{-s}]^{-1} = \prod_p \sum_{k=0}^{\infty} \frac{\chi(p)^k}{p^{ks}} \tag{171}$$

the formula

$$\mathcal{P}_L = \prod_p \left(1 + \frac{\chi(p)}{p^s} + \frac{\chi(p)^2}{p^{2s}} + \frac{\chi(p)^3}{p^{3s}} + \cdots \right). \tag{172}$$

Due to the multiplicative nature of the Dirichlet character, Eq. (115), we arrive at

$$\chi(p)^j = \chi(p^j) \tag{173}$$

which leads us to

$$\mathcal{P}_L = \prod_p \left(1 + \frac{\chi(p)}{p^s} + \frac{\chi(p^2)}{p^{2s}} + \frac{\chi(p^3)}{p^{3s}} + \cdots \right). \tag{174}$$

When we perform the multiplication, we arrive at

$$\mathcal{P}_L = 1 + \cdots \frac{\chi\left(p_{k_1}^{r_1}\right) \chi\left(p_{k_2}^{r_2}\right) \cdots \chi\left(p_{k_\ell}^{r_\ell}\right)}{\left[p_{k_1}^{r_1} p_{k_2}^{r_2} \cdots p_{k_\ell}^{r_\ell}\right]^s} + \cdots , \tag{175}$$

which with the help of the Fundamental Theorem of Arithmetic, and the multiplicity Eq. (115), of the Dirichlet character in the form

$$\chi\left(p_{k_1}^{r_1}\right) \cdot \chi\left(p_{k_2}^{r_2}\right) \cdots \chi\left(p_{k_\ell}^{r_\ell}\right) = \chi\left(p_{k_1}^{r_1} \cdot p_{k_2}^{r_2} \cdots p_{k_\ell}^{r_\ell}\right) \tag{176}$$

yields the Euler product formula

$$\prod_p \left[1 - \chi(p)p^{-s}\right]^{-1} = \sum_{n=1}^{\infty} \chi(n)n^{-s} = L(s, \chi) \,. \tag{177}$$

When we compare this expression to the one of ζ, Eq. (56), we note that the coefficient unity in front of p^{-s} is replaced by the character $\chi(p)$.

6.5.2 Final Expression

We are now in the position to obtain the exponential representation of the Dirichlet L-functions. For this purpose, we recall from Appendix 3 the expression, Eq. (214), for the exponential product

$$\lambda(s; q, \kappa) \equiv \left(\frac{q}{\pi}\right)^{s/2} \Gamma\left(\frac{s+\kappa}{2}\right) \tag{178}$$

appearing in the definition, Eq. (151), of the Dirichlet L-function Λ, and the Euler product formula (177).

Thus, we arrive at the exponential representation

$$\Lambda = \sqrt{\pi} \, e^{\alpha_\Lambda} e^{l_\Lambda} \tag{179}$$

of the Dirichlet L-function Λ where we have introduced the abbreviation

$$\alpha_\Lambda(s; q, \kappa) \equiv \frac{s}{2}\left[\ln\left(\frac{qs}{2\pi}\right) - 1\right] + \frac{\kappa - 1}{2}\ln(s) + \mathcal{R}_\Lambda(s; \kappa) \tag{180}$$

which contains the correction term

$$\mathcal{R}_\Lambda(s; \kappa) \equiv \frac{s+\kappa-1}{2}\ln\left(1 + \frac{\kappa}{s}\right) - \frac{\kappa}{2} + \left(1 - \frac{\kappa}{2}\right)\ln 2 + \bar{R}\left(\frac{s+\kappa}{2}\right) \,, \tag{181}$$

and the contribution

$$l_\Lambda(s, \chi) \equiv \ln L(s, \chi) = -\sum_p \ln\left[1 - \chi(p)p^{-s}\right] \tag{182}$$

arises from the Euler product, Eq. (177).

We conclude by briefly discussing the asymptotic limit $s \to \infty$ of $\mathcal{R}_\Lambda$. For this purpose we recall from Appendix 3 the expression

$$\mathcal{R}_\Lambda = \left(1 - \frac{\kappa}{2}\right) \ln 2 + \frac{\kappa(\kappa - 2)}{4s} + \frac{1}{6}\frac{1}{s + \kappa} \tag{183}$$

which apart from the constant term $(1 - \kappa/2) \ln 2$ decays as $1/s$.

7 Titchmarsh Counterexample

In this section we discuss the Titchmarsh counterexample which is known [33, 34] to have zeros off the critical line. This function is a superposition of two Dirichlet L-functions discussed already in Sect. 6.4, and satisfies the functional equation of ξ.

We first define the Titchmarsh counterexample and verify the corresponding functional equation. Then we perform an asymptotic expansion and demonstrate that the difference between the Riemann function as well as the Dirichlet L-functions and the Titchmarsh counterexample result from the fact that the latter has no Euler product. This lack leads in the exponential representation to the logarithm of sums rather than the sum of logarithms, which is the deeper reason for the zeros off the critical line.

7.1 Definition and Functional Equation

We start from the definition

$$\xi_T(s) \equiv \frac{1}{2\cos\theta} \left[e^{-i\theta} \Lambda(s, \chi_1) + e^{i\theta} \Lambda(s, \chi_2) \right], \tag{184}$$

of the Titchmarsh counterexample, which is a superposition

$$\xi_T(s) = \frac{1}{2\cos\theta} \left[e^{-i\theta} \Lambda_1(s) + e^{i\theta} \Lambda_2(s) \right] \tag{185}$$

of two Dirichlet L-functions $\Lambda_1 = \Lambda(s, \chi_1)$ and $\Lambda_2 = \Lambda(s, \chi_2)$ of complex-valued character χ_1 mod 5 and χ_2 mod 5 together with the parameter θ defined by Eq. (167).

When we apply the functional equations Eqs. (168) and (169) to the Titchmarsh counterexample, we arrive at

$$\xi_T(s) = \frac{1}{2\cos\theta} [e^{-i\theta} e^{2i\theta} \Lambda_2(1 - s) + e^{i\theta} e^{-2i\theta} \Lambda_1(1 - s)],$$

or

$$\xi_T(s) = \xi_T(1 - s),$$

where we have recalled the representation, Eq. (185), of ξ_T.

We identify three crucial ingredients determining this functional relation: (*i*) The appearance of the phase $+2\theta$ in the functional equation, Eq. (168), for Λ_1, and -2θ in the one, Eq. (169), for Λ_2. (*ii*) The property that the characters χ_1 and χ_2 are the complex conjugates of each other as expressed by Eq. (163), and (*iii*) a superposition of the two Dirichlet L-functions where the phase factors multiplying them contain *half* of the phase appearing in the functional equation, and are of opposite signs as to create the phase factor associated with the other Dirichlet L-function in the Titchmarsh counterexample.

7.2   Exponential Representation

Next we derive the exponential representation of the Titchmarsh counterexample analogous to the one of ξ obtained in Sect. 3.6. For this purpose we recall the definition, Eq. (151), of Λ and find the expression

$$\xi_T(s) \equiv \left(\frac{5}{\pi}\right)^{s/2} \Gamma\left(\frac{s+1}{2}\right)[1 + T(s)] \tag{186}$$

where

$$T(s) \equiv \sum_{n=2}^{\infty} \chi_T(n) n^{-s} \tag{187}$$

with

$$\chi_T(1) = 1, \ \chi_T(2) = \tan\theta, \ \chi_T(3) = -\tan\theta, \ \chi_T(4) = -1 \text{ and } \chi_T(5) = 0, \ \text{mod } 5.$$

Since

$$\chi_T(2)^2 = \tan^2\theta \neq -1 = \chi_T(4),$$

the coefficients $\chi_T(n)$ are not multiplicative. As a result, there does not exist an Euler product.

Nevertheless, we can use the familiar formula

$$1 + T = e^{\ln(1+T)},$$

and employ the representation, Eq. (180), of the exponential product

$$\lambda \equiv \left(\frac{5}{\pi}\right)^{s/2} \Gamma\left(\frac{s+1}{2}\right)$$
(188)

with $q = 5$ and $\kappa = 1$ to cast Eq. (186) into the exponential form

$$\xi_T = \sqrt{\pi}\, e^{\alpha_T} e^{l_T},$$
(189)

where

$$\alpha_T(s) \equiv \frac{s}{2}\left[\ln\left(\frac{5s}{2\pi}\right) - 1\right] + \mathcal{R}_T(s)$$
(190)

with the correction term

$$\mathcal{R}_T \equiv \frac{s}{2}\ln\left(1 + \frac{1}{s}\right) - \frac{1}{2} + \bar{R}\left(\frac{s+1}{2}\right)$$
(191)

and the contribution

$$l_T(s) \equiv \ln[1 + T(s)].$$
(192)

When we compare the expressions, Eqs. (182) and (192) for l_Λ and l_T, we note that l_Λ is a *sum of logarithms* whereas l_T is a *logarithm of a sum*.

We conclude by considering the limit $s \to \infty$ of $\mathcal{R}_T$. For this purpose we recall the expansions Eq. (105) and Eq. (73) of the logarithm as well as the remainder to arrive at

$$\mathcal{R}_T(s) \cong -1/(3s)$$
(193)

which decays as $1/s$.

7.3   Summary

In Table 1 we summarize the exponential representations of the Riemann function ξ, the Dirichlet L-function Λ, and the Titchmarsh counterexample ξ_T. From the second row we note that the three expressions are similar in form. They are the product of $\sqrt{\pi}$, an exponential function of argument α, which arises from the exponential product λ, and an exponential of argument l emerging either from the Euler product of the Riemann zeta function and the Dirichlet L-series, or from the application of the inverse function of the exponential in the case of the Titchmarsh counterexample.

Table 1 Comparison of the exponential representations of the Riemann function ξ, the Dirichlet L-functions Λ and the Titchmarsh counterexample ξ_T. In the limit of $s \to \infty$ the correction term $\bar{R}$ either vanishes or is a constant

	ξ	Λ	ξ_T
ξ, Λ, ξ_T	$\sqrt{\pi}\, e^{\alpha_\xi}\, e^{l_\xi}$	$\sqrt{\pi}\, e^{\alpha_\Lambda}\, e^{l_\Lambda}$	$\sqrt{\pi}\, e^{\alpha_T}\, e^{l_T}$
α	$\frac{s}{2}\left[\ln\left(\frac{s}{2\pi}\right)-1\right]$ $+\frac{3}{2}\ln s + \mathcal{R}_\xi$	$\frac{s}{2}\left[\ln\left(\frac{qs}{2\pi}\right)-1\right]$ $+\frac{\kappa-1}{2}\ln(s) + \mathcal{R}_\Lambda$	$\frac{s}{2}\left[\ln\left(\frac{5s}{2\pi}\right)-1\right] + \mathcal{R}_T$
$\mathcal{R}$	$\ln\left(1-\frac{1}{s}\right) + \bar{R}\left(\frac{s}{2}\right)$	$\frac{s+\kappa-1}{2}\ln\left(1+\frac{\kappa}{s}\right)$ $-\frac{\kappa}{2}+\left(1-\frac{\kappa}{2}\right)\ln 2 + \bar{\mathcal{R}}\left(\frac{s+1}{2}\right)$	$\frac{s}{2}\ln\left(1+\frac{1}{s}\right)-\frac{1}{2}+\bar{\mathcal{R}}\left(\frac{s+1}{2}\right)$
$\mathcal{R}$ $s \to \infty$	$-\frac{5}{6}\frac{1}{s}$	$\left(1-\frac{\kappa}{2}\right)\ln 2$ $+\frac{\kappa(\kappa-2)}{4s}+\frac{1}{6}\frac{1}{s+\kappa}$	$-\frac{1}{3s}$
l	$-\sum_p \ln\left(1-p^{-s}\right)$	$-\sum_p \ln\left[1-\chi(p)p^{-s}\right]$	$\ln\left[1+\sum_{n=2}^{\infty}\chi_T(n)n^{-s}\right]$

The three expressions for α shown in the third row are also rather similar. In particular, the dominant terms, that is, the contributions associated with the square brackets, are almost identical. In the Dirichlet L-function the argument of the logarithm corresponding to ξ is modified by the period q of the character, which for the Titchmarsh counterexample assumes the value $q = 5$.

The second contribution to α given by $\ln s$ has different prefactors in the case of the Riemann and the Dirichlet L-function. Since κ can only assume the values one and zero, the term $\ln s$ has either the weight $-1/2$ corresponding to the value $\kappa = 0$, or is absent for $\kappa = 1$ as is the case for the Titchmarsh counterexample. The appearance of the quadratic polynomial $\wp$ in the definition of ξ is the origin of the increase of the prefactor of $\ln s$ from $-1/2$ to $3/2$.

One, on first sight dramatic difference between the three expressions, seems to be the correction term $\mathcal{R}$, shown in the fourth row. However, in the asymptotic limit $s \to \infty$ the s-dependence of $\mathcal{R}$ disappears due to the decays as $1/s$, but with different prefactors. Hence, even this difference between the three functions is not crucial.

However, the difference stands out most clearly in the last row, where the expressions in the first two columns are a result of the application of the Euler product to express the Riemann zeta function or the Dirichlet L-series as exponentials. They only differ due to the appearance of the character χ. In both cases, we find a sum over logarithms.

Since in the case of the Titchmarsh counterexample no Euler product exists because it is defined as a superposition of two Dirichlet L-functions, we arrive at the logarithm of a sum. We suspect that this property is the origin of the zeros off the critical line.

8   Conclusions and Outlook

In the present article, we have provided an introduction into the longstanding problem of the Riemann Hypothesis and have concentrated first on the Riemann zeta function ζ, and the Riemann function ξ, emerging from the analytic continuation of ζ. Here we have provided a plausibility argument in favor of the Riemann Hypothesis based on the asymmetric distribution of phase lines of the building block f_R of ξ.

Moreover, we have illustrated an alternative approach towards the Riemann Hypothesis by considering the separatrices of ξ on the right edge of the complex plane. A phase difference of π between consecutive separatrices will ensure the validity of the Riemann Hypothesis, whereas a phase difference of 2π or more indicates a zero off the critical line. Hence, this technique requires a detailed understanding of the asymptotic behavior of ξ to identify the separatrices in the sea of phase lines.

For this reason, we have derived several asymptotic expansions of ξ valid in different regions of the complex plane, and analyzed the phase lines in their dependence on the real and imaginary parts of s. Unfortunately, we have not yet been able to find a criterion for the separatrices.

We have then turned to the generalized Riemann Hypothesis, which is supposed to be true also for the Dirichlet L-functions Λ. In order to compare and contrast ξ and Λ, we have discussed in great detail their analytic continuations. We have emphasized that the appearance of the quadratic polynomial $\wp$ originating from the term $n = 0$ in the functional equation of the Jacobi theta function causes the analytic continuation of ζ and thus of ξ to consist of *three* interfering contributions rather than *two*. The first two terms are the truncated Mellin transforms of the Jacobi theta function, evaluated at s and $1 - s$, whereas the last one is a constant. In the case of the Dirichlet L-functions, it is only the sum of two truncated generalized Mellin transforms of a generalized Jacobi function, again taken at s and $1 - s$.

A closer comparison of the asymptotic behaviors of ξ and Λ at the right edge of the complex plane reveals that they result solely from the exponential product given by a combination of an exponential function and the Gamma function. However, $\wp$ emerging from the absence of the term $n = 0$ in the Dirichlet sum of ζ leads to a slight modification of the corresponding expression for Λ.

We also studied the Titchmarsh counterexample ξ_T which is a superposition of two Dirichlet L-functions and satisfies a functional equation identical to the one of ξ but is known to have zeros off the critical line. Here we have also considered the asymptotic limit at the right edge of the complex plane. Although the expressions for ξ_T and Λ are rather similar, there is a dramatic difference due to the Euler product which exists for ξ as well as Λ but not for ξ_T. As a result, we now have the logarithm of a sum rather than the sum of logarithms in the exponent of ξ_T.

Our considerations may provide a new path towards verifying the Riemann Hypothesis by employing the lines of constant phase. However, more work needs to be done.

Appendix 1: Functional Equation of Jacobi Theta Function

In this Appendix we briefly rederive the functional equation, Eq. (26), of the Jacobi theta function

$$\omega(x) \equiv \sum_{n=1}^{\infty} e^{-\pi n^2 x} \tag{194}$$

defined by Eq. (21), in order to bring out most clearly the origin of the difference to the corresponding relation, Eq. (155), for the generalized Jacobi theta function given by Eq. (154).

Since the summation index n appears in ω as a square, we can immediately include negative values of n together with a factor of $1/2$. However, the term corresponding to $n = 0$ is unity. Thus, we have to subtract this contribution, and arrive at the representation

$$\omega(x) = \frac{1}{2} \left[\sum_{n=-\infty}^{\infty} e^{-\pi n^2 x} - 1 \right] \tag{195}$$

of ω.

Next we employ the Poisson summation formula

$$\sum_{n=-\infty}^{\infty} b_n = \sum_{m=-\infty}^{\infty} \int_{-\infty}^{\infty} dv \, b(v) e^{2\pi i v m} \tag{196}$$

where $b = b(v)$ is a continuous extension of the discrete coefficients b_n. We emphasize that the specific form of the extension is of no consequence as long as $b(n) \equiv b_n$.

Indeed, this fact is a result of the identity

$$\sum_{m=-\infty}^{\infty} e^{2\pi i v m} = \sum_{n=-\infty}^{\infty} \delta(v - n) \tag{197}$$

which is at the very heart of the Poisson summation formula.

For the choice

$$b(v) \equiv e^{-\pi x v^2}, \tag{198}$$

and with the help of the integral relation

$$\int_{-\infty}^{\infty} dv \, e^{-\pi x v^2} e^{2\pi i v m} = \frac{1}{\sqrt{x}} e^{-\pi m^2/x} \tag{199}$$

we arrive at the identity

$$\sum_{n=-\infty}^{\infty} e^{-\pi n^2 x} = \frac{1}{\sqrt{x}} \sum_{n=-\infty}^{\infty} e^{-\pi n^2/x} , \tag{200}$$

which leads us with the representation, Eq. (195), of ω to

$$2\,\omega(x) + 1 = \sum_{n=-\infty}^{\infty} e^{-\pi n^2 x} = \frac{1}{\sqrt{x}} \left(2\omega \left(\frac{1}{x} \right) + 1 \right) , \tag{201}$$

that is, the familiar functional equation

$$\omega(x) = \frac{1}{\sqrt{x}} \omega \left(\frac{1}{x} \right) + \frac{1}{2} \left[\frac{1}{\sqrt{x}} - 1 \right] \tag{202}$$

of ω.

We emphasize that the second term in Eq. (202), which is independent of ω, is a consequence of the absence of the term $n = 0$ in the definition, Eq. (194), of the Jacobi theta function ω leading to the subtraction of unity in Eq. (195).

Appendix 2: Equivalent Condition for Non-trivial Zeros

In Sect. 2.4 we have derived the condition, Eq. (39), for a zero on the critical line given by the identity of an integral

$$J(\tau) \equiv 2 \int_1^{\infty} dx\, \omega(x) x^{-3/4} \cos \left(\frac{\tau}{2} \ln x \right) \tag{203}$$

containing an oscillatory integrand with a Lorentzian. We devote this appendix to the derivation of an alternative but equivalent formulation by casting the integral into a different form.

In particular, we analyze the asymptotic behavior of J. For this purpose, we first derive an exact alternative expression for J, and then consider the limit $\tau \to \infty$.

We start by introducing in J the new integration variable $y \equiv (1/2) \ln x$, or $x = \exp(2y)$, which with the identity $dy = dx/(2x)$ leads us to the form

$$J = 4 \int_0^{\infty} dy\, \omega \left(e^{2y} \right) e^{y/2} \cos (\tau y) \tag{204}$$

of J.

Next, we obtain by integration by parts the equivalent representation

$$J = -\frac{1}{\tau}4 \int_0^\infty dy \frac{d}{dy}\left[\omega\left(e^{2y}\right)e^{y/2}\right]\sin(\tau y) , \tag{205}$$

where we have used the fact that the boundary terms vanish, and find performing the differentiation

$$J = -\frac{1}{\tau}8 \int_0^\infty dy \left[\left.\frac{d\omega}{dx}\right|_{x=e^{2y}} e^{2y} + \frac{1}{4}\omega\left(e^{2y}\right)\right]e^{y/2}\sin(\tau y) . \tag{206}$$

Finally, we introduce the new integration variable $\theta \equiv \tau y$, and arrive at the expression

$$J = \frac{1}{\tau^2}8 \int_0^\infty d\theta \left[-\left.\frac{d\omega}{dx}\right|_{x=e^{2\theta/\tau}} e^{2\theta/\tau} - \frac{1}{4}\omega\left(e^{2\theta/\tau}\right)\right]e^{\theta/(2\tau)}\sin\theta . \tag{207}$$

When we recall from Eq. (21) the definition

$$\omega(x) \equiv \sum_{n=1}^\infty e^{-\pi n^2 x} \tag{208}$$

of ω, the differentiation yields the explicit formula

$$J = \frac{1}{\tau^2}8 \int_0^\infty d\theta \sum_{n=1}^\infty \left(\pi n^2 e^{2\theta/\tau} - \frac{1}{4}\right)\exp\left[-\pi n^2 e^{2\theta/\tau} + \frac{\theta}{2\tau}\right]\sin\theta \tag{209}$$

which decays for increasing τ as $1/\tau^2$ which is identical to the decrease of the Lorentzian on the right-hand side of Eq. (39).

When we now compare this representation of J to the initial one, Eq. (203), we find, apart from the decay with $1/\tau^2$, three distinct features: (i) The oscillatory function in the integral is independent of τ. (ii) We have a double-exponential decay, but (iii) this rapid decay only sets in for integration variables $\theta > \tau/2$.

Hence, the expression, Eq. (209), for J shows that the main contribution to this integral results from the decays characterized by the summation index n that are fast compared to the oscillation period, in complete agreement with the discussion of Sect. 2.4.

Appendix 3: Exponential Product

In this appendix we cast the product

$$\lambda(s; q, \kappa) \equiv \left(\frac{q}{\pi}\right)^{s/2}\Gamma\left(\frac{s+\kappa}{2}\right) , \tag{210}$$

where q is an integer and $\kappa = 0, 1$, into an exponential form, which allows us to derive in a straight-forward way an asymptotic limit of λ. This term is not only a factor in the definition, Eq. (151), of the Dirichlet L-function Λ but also appears in the Riemann function ξ, defined by Eq. (40) for the special case $q = 1$ and $\kappa = 0$.

We first derive a general expression for λ and then perform the asymptotic limit. Finally, we address the special case of the Riemann function.

General Expression

We start from the definition, Eq. (210), of λ in the exponential form

$$\lambda = \exp\left[\frac{s}{2}\ln\left(\frac{q}{\pi}\right)\right]\exp\left[\ln\Gamma\left(\frac{s+\kappa}{2}\right)\right], \tag{211}$$

and recall the representation [17]

$$\ln\Gamma(s) = \left(s - \frac{1}{2}\right)\ln s - s + \frac{1}{2}\ln(2\pi) + \bar{R}(s) \tag{212}$$

of the logarithm of the Gamma function with the remainder

$$\bar{R}(s) \equiv \int_0^\infty dt \left[\frac{1}{2} - \frac{1}{t} + \frac{1}{e^t - 1}\right]\frac{e^{-ts}}{t} \tag{213}$$

which leads us for the argument $(s + \kappa)/2$ to the expression

$$\lambda = \sqrt{\pi}\exp\left\{\frac{s}{2}\left[\ln\left(\frac{qs}{2\pi}\right) - 1\right] + \frac{\kappa - 1}{2}\ln s + \mathcal{R}\right\}. \tag{214}$$

Here we have introduced the abbreviation

$$\mathcal{R}(s;\kappa) \equiv \frac{s + \kappa - 1}{2}\ln\left(1 + \frac{\kappa}{s}\right) - \frac{\kappa}{2} + \left(1 - \frac{\kappa}{2}\right)\ln 2 + \bar{R}\left(\frac{s+\kappa}{2}\right) \tag{215}$$

and $\bar{R}$ is defined by Eq. (213).

Asymptotic Limit

Next we consider the limit $s \to \infty$. In this case we find with the help of the expansion

$$\ln(1 + y) \cong y - \frac{1}{2}y^2 \tag{216}$$

the relation

$$\frac{s+\kappa-1}{2}\ln\left(1+\frac{\kappa}{s}\right)-\frac{\kappa}{2} \cong \frac{\kappa(\kappa-2)}{4s}. \tag{217}$$

Moreover, the generating function [1]

$$\frac{t}{e^t-1} = \sum_{n=0}^{\infty} B_n \frac{t^n}{n!} = B_0 + B_1 t + \frac{1}{2}B_2 t^2 + \frac{1}{6}B_3 t^3 + \frac{1}{24}B_4 t^4 + \dots \tag{218}$$

of the Bernoulli numbers B_n with

$$B_0 \equiv 1, \quad B_1 \equiv -\frac{1}{2}, \quad B_2 \equiv \frac{1}{6}, \quad B_3 \equiv 0, \quad \text{and} \quad B_4 \equiv -\frac{1}{30}, \tag{219}$$

that is,

$$\frac{1}{e^t-1} = \frac{1}{t} - \frac{1}{2} + \frac{1}{12}t - \frac{1}{720}t^3 \dots \tag{220}$$

allows us to expand the remainder $\bar{R}$ given by Eq. (213) into inverse powers of s with the leading contribution

$$\bar{R}(s) = \frac{1}{12}\frac{1}{s} + \mathcal{O}\left(\frac{1}{s^3}\right). \tag{221}$$

Hence, we arrive in the lowest order of $1/s$ at the asymptotic expression

$$\mathcal{R}(s;\kappa) \cong \left(1-\frac{\kappa}{2}\right)\ln 2 + \frac{\kappa(\kappa-2)}{4s} + \frac{1}{6}\frac{1}{s+\kappa} \tag{222}$$

for $\mathcal{R}$, which apart from the constant term decays as $1/s$.

Special Case: Riemann Function

Finally, we discuss the example $q = 1$ and $\kappa = 0$ corresponding to the Riemann function ξ. In this case, Eq. (214) reduces to

$$\lambda(s; q = 1, \kappa = 0) = \sqrt{\pi} \exp\left\{\frac{s}{2}\left[\ln\left(\frac{s}{2\pi}\right) - 1\right] - \frac{1}{2}\ln s + \mathcal{R}\right\}, \tag{223}$$

where $\mathcal{R}$ given by Eq. (215) simplifies to

$$\mathcal{R} = \ln 2 + \bar{R}\left(\frac{s}{2}\right). \tag{224}$$

With the help of the asymptotic limit Eq. (221), we find the expression

$$\bar{R}\left(\frac{s}{2}\right) \cong \frac{1}{6}\frac{1}{s} \tag{225}$$

for the remainder, which together with Eq. (224) is consistent with Eq. (222) for $\kappa = 0$.

Appendix 4: Lines of Constant Height and Constant Phase

In Sect. 3.6.3 we have derived an asymptotic expression for ξ which arises solely from the product $(\wp/2)\pi^{-s/2}\Gamma(s/2)$ since we have made the approximation $\zeta = 1$. This formula is in terms of the complex variable s.

In this appendix, we first cast this expression in terms of amplitude and phase, and then consider two special cases. We conclude by verifying these approximations using the Cauchy-Riemann differential equations in amplitude and phase discussed in the Chapter "Insights into Complex Functions" of this volume.

Decomposition in Amplitude and Phase

We start by decomposing the approximation

$$\alpha_\xi(s) \cong \frac{s}{2}\left[\ln\left(\frac{s}{2\pi}\right) - 1\right] + \frac{3}{2}\ln s \tag{226}$$

of α_ξ into its real and imaginary parts Σ and Φ, that is,

$$\alpha_\xi \equiv \Sigma + i\Phi \tag{227}$$

determining the absolute value

$$|\xi| = \sqrt{\pi}e^\Sigma \tag{228}$$

and the phase Φ of ξ. Here, we first derive general expressions for Σ and Φ, and then consider the two extreme limits $1 < \sigma \ll \tau$ and $0 \le \tau \ll \sigma$.

With the identity

$$s \equiv \sigma + i\tau \equiv \left(\sigma^2 + \tau^2\right)^{1/2}\exp\left[i\arctan\left(\frac{\tau}{\sigma}\right)\right] \tag{229}$$

leading us to

$$\ln s = \ln\left(\sigma^2 + \tau^2\right)^{1/2} + i\arctan\left(\frac{\tau}{\sigma}\right) \tag{230}$$

we find the explicit expressions

$$\Sigma(\sigma, \tau) = \frac{\sigma}{2}\left[\ln\frac{(\sigma^2 + \tau^2)^{1/2}}{2\pi} - 1\right] - \frac{\tau}{2}\arctan\left(\frac{\tau}{\sigma}\right) + \frac{3}{2}\ln\left(\sigma^2 + \tau^2\right)^{1/2}$$

(231)

and

$$\Phi(\sigma, \tau) = \frac{1}{2}\left[\ln\frac{(\sigma^2 + \tau^2)^{1/2}}{2\pi} - 1\right]\tau + \frac{1}{2}(\sigma + 3)\arctan\left(\frac{\tau}{\sigma}\right).$$

(232)

We emphasize that the decomposition, Eq. (227), of the approximation, Eq. (226), of α_ξ into Σ and Φ given by Eqs. (231) and (232) is exact.

Special Limits

In order to gain insight into the dependence of Σ and Φ on σ and τ, we now consider two extreme limits, and derive approximate but analytic expressions for Σ and Φ.

At the Center of the Complex Plane

We start with the case of $1 < \sigma \ll \tau$, and arrive with the help of the approximations

$$\left(\sigma^2 + \tau^2\right)^{1/2} \cong \tau\left[1 + \left(\frac{\sigma}{\tau}\right)^2\right]^{1/2},$$

(233)

that is,

$$\ln\frac{(\sigma^2 + \tau^2)^{1/2}}{2\pi} \simeq \ln\left(\frac{\tau}{2\pi}\right) + \frac{1}{2}\left(\frac{\sigma}{\tau}\right)^2$$

(234)

and

$$\arctan\left(\frac{\tau}{\sigma}\right) \cong \frac{\pi}{2} - \frac{\sigma}{\tau} + \mathcal{O}\left(\left(\frac{\sigma}{\tau}\right)^3\right)$$

(235)

at the expressions

$$\Sigma(\sigma, \tau) = \frac{\sigma}{2}\ln\left(\frac{\tau}{2\pi}\right) - \tau\frac{\pi}{4} + \frac{3}{2}\ln\tau + \mathcal{O}\left(\left(\frac{\sigma}{\tau}\right)^2\right)$$

(236)

and

$$\Phi(\sigma, \tau) = \frac{\tau}{2}\left[\ln\left(\frac{\tau}{2\pi}\right) - 1\right] + (\sigma + 3)\frac{\pi}{4} + \mathcal{O}\left(\frac{\sigma^2}{\tau}\right).$$

(237)

In this limit, the lines $\tau_\Phi = \tau_\Phi(\sigma)$ of constant phase Φ satisfy an interesting symmetry relation. Indeed, for a given value τ the lines of constant phase corresponding to Φ and $\Phi + \pi$ are separated in their values of σ and σ' by 4. This property follows from Eq. (237) which for $\Phi + \pi$ reads

$$\Phi(\sigma, \tau) + \pi = \Phi(\sigma', \tau) = \frac{\tau}{2}\left[\ln\left(\frac{\tau}{2\pi}\right) - 1\right] + (\sigma' + 3)\frac{\pi}{4}, \tag{238}$$

and a comparison with Eq. (237) immediately yields the connection

$$\sigma' = \sigma + 4, \tag{239}$$

and thus the identity

$$\tau_{\Phi+\pi}(\sigma + 4) = \tau_\Phi(\sigma). \tag{240}$$

Next, we derive an explicit expression for τ_Φ. Unfortunately, Eq. (237) is a transcendental equation, that is, the Lambert equation, and cannot be solved directly. However, in the limit of large τ we obtain an approximate expression for τ_Φ by iteration.

For this purpose, we first cast Eq. (237) into the form

$$\tau_\Phi = \frac{2\Phi - (\sigma + 3)\pi/2}{\ln\left(\frac{\tau_\Phi}{2\pi}\right) - 1}, \tag{241}$$

and replace τ_Φ in the slowly varying logarithm by the right-hand side leading us to the formula

$$\tau_\Phi \cong \frac{2\Phi - (\sigma + 3)\pi/2}{\ln\left[\Phi/\pi - (\sigma + 3)/4\right] - 1}. \tag{242}$$

We conclude by noting that this expression obviously also satisfies the periodicity property, Eq. (240).

At the Right Edge of the Complex Plane

Next we consider the case $0 \leq \tau \ll \sigma$ and arrive with

$$\left(\sigma^2 + \tau^2\right)^{1/2} \cong \sigma\left[1 + \left(\frac{\tau}{\sigma}\right)^2\right]^{1/2}, \tag{243}$$

that is,

$$\ln\frac{\left(\sigma^2 + \tau^2\right)^{1/2}}{2\pi} \cong \ln\left(\frac{\sigma}{2\pi}\right) + \frac{1}{2}\left(\frac{\tau}{\sigma}\right)^2 \tag{244}$$

and

$$\arctan\left(\frac{\tau}{\sigma}\right) \cong \frac{\tau}{\sigma} + \mathcal{O}\left(\left(\frac{\tau}{\sigma}\right)^3\right) \tag{245}$$

at the expressions

$$\Sigma(\sigma, \tau) = \frac{\sigma}{2}\left[\ln\left(\frac{\sigma}{2\pi}\right) - 1\right] - \frac{1}{4}\frac{\tau^2}{\sigma} + \frac{3}{2}\ln\sigma + \mathcal{O}\left(\left(\frac{\tau}{\sigma}\right)^2\right) \tag{246}$$

and

$$\Phi(\sigma, \tau) = \frac{1}{2}\left[\ln\left(\frac{\sigma}{2\pi}\right) + \frac{3}{\sigma}\right]\tau + \mathcal{O}\left(\left(\frac{\tau}{\sigma}\right)^2\right). \tag{247}$$

In this limit Σ, and thus $|\xi|$ grows exponentially with $\sigma\ln(\sigma/2\pi)$ for increasing σ, but decays like a Gaussian due to the term $-\tau^2/(4\sigma)$ for τ increasing from zero. Moreover, Φ increases linearly with τ with a rate mainly given by $\ln(\sigma/2\pi)$.

According to Eq. (247), in this asymptotic limit the lines $\tau_\Phi = \tau_\Phi(\sigma)$ of constant phase Φ read

$$\tau_\Phi = \frac{2\Phi}{\ln\left(\sigma/(2\pi)\right) + 3/\sigma} \cong \frac{2\Phi}{\ln(\sigma/(2\pi))}, \tag{248}$$

and decay with a rate, that is, inversely proportional to $\ln(\sigma/(2\pi))$. In particular, $\tau = 0$, that is, the real axis, corresponds to the phase $\Phi = 0$, and all phase lines approach it as $\sigma \to \infty$.

In the last step of Eq. (248), we have neglected the correction term $3/\sigma$ since in obtaining the general expression, Eq. (226), of α_ξ we have already neglected terms of the order $1/s$.

Cauchy-Riemann Differential Equations

Next we probe the consistency of our approximations, Eqs. (236) and (246), as well as Eqs. (237) and (247) for the amplitude Σ as well as the phase Φ of the exponential representation

$$\xi \equiv e^{\alpha_\xi} \equiv e^{\Sigma + i\Phi} \tag{249}$$

of ξ by applying the Cauchy-Riemann differential equations

$$\frac{\partial\Sigma}{\partial\sigma} = \frac{\partial\Phi}{\partial\tau} \tag{250}$$

and

$$\frac{\partial \Sigma}{\partial \tau} = -\frac{\partial \Phi}{\partial \sigma} \tag{251}$$

discussed in the Chapter "Insights into Complex Functions" of this volume.

We start with the asymptotic expressions

$$\Sigma(\sigma, \tau) \equiv \frac{\sigma}{2} \ln\left(\frac{\tau}{2\pi}\right) - \tau\frac{\pi}{4} + \frac{3}{2}\ln \tau \tag{252}$$

and

$$\Phi(\sigma, \tau) \equiv \frac{\tau}{2}\left[\ln\left(\frac{\tau}{2\pi}\right) - 1\right] + (\sigma + 3)\frac{\pi}{4} \tag{253}$$

obtained in the previous section for $1 < \sigma \ll \tau$.

Indeed, by direct differentiation we find

$$\frac{\partial \Sigma}{\partial \sigma} = \frac{1}{2}\ln\left(\frac{\tau}{2\pi}\right) \tag{254}$$

and

$$\frac{\partial \Phi}{\partial \tau} = \frac{1}{2}\ln\left(\frac{\tau}{2\pi}\right) = \frac{\partial \Sigma}{\partial \sigma}, \tag{255}$$

in complete agreement with the Cauchy-Riemann differential equation, Eq. (250).

Moreover, we also obtain the result

$$\frac{\partial \Sigma}{\partial \tau} = \frac{\sigma}{2\tau} - \frac{\pi}{4} + \frac{3}{2\tau}, \tag{256}$$

which is consistent with the expression

$$-\frac{\partial \Phi}{\partial \sigma} = -\frac{\pi}{4}, \tag{257}$$

that is, with Eq. (256) when we neglect the terms proportional to $1/\tau$. Indeed, in the derivation of the expression, Eq. (253) for Φ we have already neglected contributions of this order.

Next, we address the approximations

$$\Sigma(\sigma, \tau) \equiv \frac{\sigma}{2}\left[\ln\left(\frac{\sigma}{2\pi}\right) - 1\right] - \frac{1}{4}\frac{\tau^2}{\sigma} + \frac{3}{2}\ln \sigma \tag{258}$$

and

$$\Phi(\sigma, \tau) \equiv \left[\frac{1}{2} \ln \left(\frac{\sigma}{2\pi} \right) + \frac{3}{2\sigma} \right] \tau \tag{259}$$

valid for $0 \leq \tau \ll \sigma$, and find immediately

$$\frac{\partial \Sigma}{\partial \sigma} = \frac{1}{2} \ln \left(\frac{\sigma}{2\pi} \right) + \frac{1}{4} \left(\frac{\tau}{\sigma} \right)^2 + \frac{3}{2\sigma} = \frac{\partial \Phi}{\partial \tau}, \tag{260}$$

where in the last step we have neglected the term $(\tau/\sigma)^2 \ll 1$. Hence, we have verified the Cauchy-Riemann differential equation, Eq. (250).

Moreover, we obtain from Eq. (258) the relation

$$\frac{\partial \Sigma}{\partial \tau} = -\frac{1}{2} \frac{\tau}{\sigma}, \tag{261}$$

and Eq. (259) yields

$$\frac{\partial \Phi}{\partial \sigma} = \frac{1}{2\sigma} \tau - \frac{3}{2} \frac{\tau}{\sigma^2}. \tag{262}$$

Hence, we satisfy the Cauchy-Riemann differential equation, Eq. (251), in the approximation $\sigma \ll 1$.

Appendix 5: Special Examples of Normalized Gauss Sums

In Sect. 5 we have shown that the normalized Gauss sum

$$G(\chi) \equiv i^{-\kappa} \frac{1}{\sqrt{q}} \sum_{n=1}^{q} \chi(n) e^{2\pi i n/q} \tag{263}$$

is given by a phase factor whose phase β is determined by the Dirichlet character χ. We now illustrate this result by evaluating the normalized Gauss sum of three different characters.

For this purpose, we first consider the normalized Gauss sums associated with the Dirichlet characters forming the Titchmarsh counterexample. In these examples, the phase is non-vanishing. We then calculate the normalized Gauss sum of a character where the phase does vanish.

Non-vanishing Phase

We start by evaluating the Gauss sums G_1 and G_2 corresponding to the characters χ_1 and χ_2 defined mod 5 with the values

$$\chi_1(1) = 1, \quad \chi_1(2) = i, \quad \chi_1(3) = -i, \quad \chi_1(4) = -1, \quad \chi_1(5) = 0 \qquad (264)$$

and

$$\chi_2(1) = 1, \quad \chi_2(2) = -i, \quad \chi_2(3) = i, \quad \chi_2(4) = -1, \quad \chi_2(5) = 0, \qquad (265)$$

and demonstrate that they are phase factors, in complete accordance with Eq. (143). Since χ_1 and χ_2 are the complex conjugate of each other, the corresponding phases must satisfy the symmetry relation, Eq. (144).

We start by noting that in both characters $q = 5$. Thus, we find with $\chi_j(-1) = \chi_j(4-5) = \chi_j(4)$ for $j = 1, 2$, and the definitions Eqs. (264) and (265) of χ_1 and χ_2, the result $\chi_1(-1) = \chi_2(-1) = -1$ which yields with Eq. (127) the parameter

$$\kappa = 1.$$

Hence, the Gauss sum G_1, corresponding to the character χ_1 reads

$$G_1 \equiv G(\chi_1) \equiv i^{-1} \frac{1}{\sqrt{5}} \sum_{n=1}^{5} \chi_1(n) e^{2\pi i n/5}, \qquad (266)$$

and takes with the help of the definition, Eq. (264), of χ_1, the form

$$G_1 = \frac{1}{\sqrt{5}} \frac{1}{i} [e^{2\pi i \frac{1}{5}} + i e^{2\pi i \frac{2}{5}} - i e^{2\pi i \frac{3}{5}} - e^{2\pi i \frac{4}{5}}], \qquad (267)$$

or

$$G_1 = \frac{1}{\sqrt{5}} \frac{1}{i} [e^{2\pi i \frac{1}{5}} - e^{-2\pi i \frac{1}{5}} + i(e^{2\pi i \frac{2}{5}} - e^{-2\pi i \frac{2}{5}})],$$

where in the last step we have taken advantage of the 2π-periodicity of the Fourier factors.

In terms of trigonometric functions, we find

$$G_1 = \frac{2}{\sqrt{5}} [\sin \alpha + i \sin(2\alpha)],$$

where

$$\alpha \equiv \frac{2\pi}{5}.$$

With the identity

$$\sin(2\alpha) = 2 \sin\alpha \cos\alpha$$

we obtain

$$G_1 = \frac{2}{\sqrt{5}} \sin\alpha (1 + 4\cos^2\alpha)^{1/2} \exp[i \arctan(2\cos\alpha)], \tag{268}$$

and the values

$$\sin\left(\frac{2\pi}{5}\right) = \frac{1}{4}\sqrt{10 + 2\sqrt{5}}$$

and

$$\cos\left(\frac{2\pi}{5}\right) = \frac{1}{4}(\sqrt{5} - 1),$$

finally yield the expression

$$G_1 = e^{2i\theta}, \tag{269}$$

for the Gauss sum G_1 as a phase factor where

$$\tan(2\theta) = \frac{\sqrt{5} - 1}{2}.$$

We conclude by briefly addressing the Gauss sum

$$G_2 \equiv G(\chi_2) = i^{-1} \frac{1}{\sqrt{5}} \sum_{n=1}^{5} \chi_2(n) e^{2\pi i n/5} \tag{270}$$

associated with

$$\chi_2 \equiv \chi_1^*. \tag{271}$$

Indeed, this symmetry enforces that in G_2 the second and the third term in Eq. (267) change their signs, but everything else remains the same.

As a result, we arrive at a *negative* rather than a positive phase, that is,

$$G_2 = e^{-2i\theta}. \tag{272}$$

We emphasize that the sign change in the phase due to the transition from G_1 to G_2, Eqs. (269) and (272), can also be viewed as a consequence of the connection, Eq. (271), between χ_1 and χ_2, and the symmetry relation, Eq. (144).

Vanishing Phase

We conclude by discussing an example of a character χ where the phase of the corresponding normalized Gauss sum vanishes. For this purpose, we consider the real-valued character

$$\chi(n) \equiv \begin{cases} 1 & \text{for } n = 4k + 1 \\ -1 & \text{for } n = 4k + 3 \\ 0 & \text{for } n \text{ even} \end{cases} \tag{273}$$

with an integer k.

Since χ is mod 4 we find $q = 4$ and with $\chi(-1) = \chi(3 - 4) = \chi(3) = -1$ following from the definition, Eq. (273), of χ, we obtain with the help of Eq. (127) the value $\kappa = 1$.

Hence, the normalized Gauss sum, Eq. (263), takes the form

$$G(\chi) = \frac{1}{2i} \sum_{n=1}^{4} \chi(n) e^{2\pi i n/4}, \tag{274}$$

which with the definition, Eq. (273), of the character χ reads

$$G = \frac{1}{2i} \left(e^{2\pi i \frac{1}{4}} - e^{2\pi i \frac{3}{4}} \right) = \frac{1}{2i} \left(e^{\frac{i\pi}{2}} - e^{-\frac{i\pi}{2}} \right),$$

or

$$G = \sin\left(\frac{\pi}{2}\right) = 1.$$

For the character χ, given by Eq. (273), we obtain indeed a vanishing phase β for G.

Appendix 6: Functional Equation of Generalized Jacobi Theta Function

Since the functional equation

$$\omega(x, \chi) = G(\chi) x^{-\kappa - 1/2} \omega\left(\frac{1}{x}, \chi^*\right) \tag{275}$$

of the generalized Jacobi theta function plays a crucial role in the analytic continuation of the Dirichlet L-functions, as discussed in Sect. 6, we devote the present Appendix to a brief summary of the derivation. This analysis also brings out most clearly the origins of the Gauss sum G and the character in its complex conjugate form.

Here we proceed in two steps: (i) We first show that due to the appearance of the character χ and the term n^κ in the definition of ω we can express the summation in ω from unity to infinity, into one from minus infinity to plus infinity by merely introducing a factor $1/2$. This additional symmetry is the deeper reason for the simplicity of the functional equation, Eq. (275), of the Jacobi theta function for Λ, compared to the one for ξ. (ii) Next we take advantage of the periodicity of the character together with the Poisson summation formula, and arrive at the functional equation, Eq. (275), of ω corresponding to Λ.

We conclude by briefly comparing the derivation of the functional equation of the Jacobi theta function ω corresponding to Λ to that for ξ. In particular, we show that the origin of the quadratic polynomial $s(s-1)$ and the off-set $1/2$ in the analytic continuation, Eq. (43), of ξ is the fact that the term $n = 0$ in ω corresponding to ξ is nonzero.

Extension of Summation

We start by verifying the representation

$$\omega = \frac{1}{2} \sum_{n=-\infty}^{\infty} \chi(n) n^\kappa e^{-\pi n^2 x/q} \tag{276}$$

of

$$\omega \equiv \sum_{n=1}^{\infty} \chi(n) n^\kappa e^{-\pi n^2 x/q} . \tag{277}$$

For this purpose we consider the sum

$$S \equiv \sum_{n=-\infty}^{\infty} \chi(n) n^\kappa e^{-\pi n^2 x/q} \tag{278}$$

and decompose S into a sum over all positive, and one over all negative integers including zero, that is,

$$S = \sum_{n=1}^{\infty} \chi(n) n^\kappa e^{-\pi n^2 x/q} + \chi(0) 0^\kappa + \sum_{n=1}^{\infty} \chi(-n)(-n)^\kappa e^{-\pi n^2 x/q} ,$$

where in the sum over negative n we have replaced $-n$ by n.

When we recall the connection, Eq. (128), between $\chi(-n)$ and $\chi(n)$ together with $\chi(0) = 0$, Eq. (119), we find the expression

$$S = \left[1 + \chi(-1)(-1)^\kappa\right] \sum_{n=1}^{\infty} \chi(n) n^\kappa e^{-\pi n^2 x/q}. \tag{279}$$

The values, Eq. (127), of κ together with the relation $\chi(-1) = \pm 1$ yield the identity

$$1 + \chi(-1)(-1)^\kappa = 1 + (\pm 1)(-1)^{\frac{1}{2}(1\mp 1)} = 2,$$

and the definition, Eq. (277), of ω the desired representation, Eq. (276).

Emergence of Gauss Sum

Since the character χ satisfies the periodicity condition, Eq. (116), it is useful to apply the summation formula

$$\sum_{n=-\infty}^{\infty} d(n) = \sum_{j=1}^{q} \sum_{r=-\infty}^{\infty} d(j + rq) \tag{280}$$

valid for any appropriately converging sum of coefficients $d(n)$, to S in the form

$$S = \sum_{n=-\infty}^{\infty} \chi(n) h(n) \tag{281}$$

where we have introduced the abbreviation

$$h(n) \equiv n^\kappa e^{-\pi n^2 x/q}. \tag{282}$$

Indeed, with the help of Eq. (116) we immediately find the representation

$$S = \sum_{j=1}^{q} \chi(j) \sum_{r=-\infty}^{\infty} h(j + rq) \tag{283}$$

of S which with the Poisson summation formula

$$\sum_{r=-\infty}^{\infty} d(r) = \sum_{m=-\infty}^{\infty} \int_{-\infty}^{\infty} d\rho \, d(\rho) e^{-2\pi i \rho m} \tag{284}$$

takes the form

$$S = \sum_{j=1}^{q} \chi(j) \sum_{m=-\infty}^{\infty} \int_{-\infty}^{\infty} d\rho \, h(j + \rho q) e^{-2\pi i \rho m} . \qquad (285)$$

Finally, the new integration variable

$$y \equiv j + \rho q \qquad (286)$$

yields the expression

$$S = \sum_{m=-\infty}^{\infty} \tilde{G}(\chi, m) \frac{1}{q} \int_{-\infty}^{\infty} dy \, h(y) e^{-2\pi i y m/q} , \qquad (287)$$

where we have interchanged the two summations, and have recalled the definition, Eq. (130), of the Gauss sum $\tilde{G}(\chi, m)$.

Next we use the reduction formula, Eq. (129), which allows us to factor $\tilde{G}(\chi, 1)$ out of the sum over m and to find

$$S = \frac{1}{\sqrt{q}} \tilde{G}(\chi, 1) \sum_{m=-\infty}^{\infty} \chi^*(m) \frac{1}{\sqrt{q}} \int_{-\infty}^{\infty} dy \, h(y) e^{-2\pi i y m/q} . \qquad (288)$$

With the definition of h, Eq. (282), and the integral formula

$$\int_{-\infty}^{\infty} dy \, y^\kappa \exp\left(-\frac{\pi}{q} x y^2\right) \exp\left(-2\pi i \frac{n}{q} y\right)$$
$$= \sqrt{q} \, (-i)^\kappa n^\kappa x^{-\kappa - 1/2} \exp\left(-\frac{\pi}{q} n^2 \frac{1}{x}\right) \qquad (289)$$

we finally arrive at

$$S = G(\chi) \, x^{-\kappa - 1/2} \sum_{n=-\infty}^{\infty} \chi^*(n) n^\kappa e^{-\frac{\pi}{q} n^2 \frac{1}{x}} , \qquad (290)$$

that is, the functional equation, Eq. (275), of ω when we use the relation, Eq. (276). Moreover, we have recalled the definition, Eq. (142), of the Gauss sum.

Comparison to Jacobi Theta Function of ξ

We conclude by briefly comparing and contrasting this derivation of the functional equation of the Jacobi theta function of Λ to the corresponding one for the Jacobi theta function of ξ, defined by Eq. (21) and discussed in Appendix 1.

In contrast to the generalized Jacobi theta function, Eq. (154), where the term $n = 0$ vanishes due to $\chi(0) = 0$, Eq. (119), the contribution in the one for ξ is unity. This difference is important since the term in Eq. (202) independent of ω is a consequence of the subtraction of the term $n = 0$ in Eq. (195). As a result, in the corresponding functional equation, Eq. (275), for the generalized Jacobi theta function only the contribution proportional to $\omega(1/x)$ enters.

Acknowledgments We thank P. C. Abbott, M. B. Kim, H. L. Montgomery, J. W. Neuberger, M. Zimmermann for many fruitful discussions, E. P. Glasbrenner for technical assistance and J. Pohl for her help with the scans of Riemann's article. Moreover, we are most grateful to the Niedersächsische Staats- und Universitätsbibliothek and, in particular, R. B. Röper for allowing us to present excerpts of Riemann's original manuscript. W. P. S. is also grateful to Texas A & M University for a Faculty Fellowship at the Hagler Institute for Advanced Study at Texas A& M University and to Texas A& M AgriLife for the support of this work. The research of the IQST is financially supported by the Ministry of Science, Research and Arts Baden-Württemberg. I. B. is grateful for the financial support from the Technical University of Ostrava, Grant No. SP2018/44.

References

1. M. Abramowitz, I.A. Stegun, *Handbook of Mathematical Functions with Formulas, Graphs, and Mathematical Tables*, vol. 55 (US Government Printing Office, Washington, 1948)
2. C.M. Bender, D.C. Brody, M.P. Müller, Hamiltonian for the zeros of the Riemann zeta function. Phys. Rev. Lett. **118**(13), 130201 (2017)
3. M.V. Berry, J.P. Keating, A new asymptotic representation for ζ (1/2+ it) and quantum spectral determinants. Proc. R. Soc. Lond. A Math. Phys. Sci. **437**(1899), 151–173 (1992)
4. M.V. Berry, J.P. Keating, The Riemann zeros and eigenvalue asymptotics. SIAM Rev. **41**(2), 236–266 (1999)
5. P. Borwein, S. Choi, B. Rooney, A. Weirathmueller, *The Riemann Hypothesis: A Resource for the Afficionado and Virtuoso Alike*. CMS Books in Mathematics Series (Springer, New York, 2008)
6. D. Cassettari, G. Mussardo, A. Trombettoni, Holographic realization of the prime number quantum potential. PNAS Nexus **2**(1), 1–9 (2022)
7. M. Du Sautoy, *The Music of the Primes* (Harper Collins, New York, 2003)
8. H.M. Edwards, *Riemann's Zeta Function* (Academic, New York, 1974)
9. W.J. Ellison, F. Ellison, *Prime Numbers* (Wiley, New York, 1985)
10. C. Feiler, W.P. Schleich, Entanglement and analytical continuation: an intimate relation told by the Riemann zeta function. New J. Phys. **15**(6), 063009 (2013)
11. C. Feiler, W.P. Schleich, Dirichlet series as interfering probability amplitudes for quantum measurements. New J. Phys. **17**(6), 063040 (2015)
12. G. Freiling, V.A. Yurko, *Inverse Sturm-Liouville Problems and Their Applications* (NOVA Science Publishers, Huntington, 2001)
13. F. Gleisberg, W.P. Schleich, Factorization with a logarithmic energy spectrum of a central potential. Acta Phys. Pol. A **143**, S112 (2023)

14. F. Gleisberg, R. Mack, K. Vogel, W.P. Schleich, Factorization with a logarithmic energy spectrum. New J. Phys. **15**(2), 023037 (2013)
15. F. Gleisberg, M. Volpp, W.P. Schleich, Factorization with a logarithmic energy spectrum of a two-dimensional potential. Phys. Lett. A **379**(40–41), 2556–2560 (2015)
16. F. Gleisberg, F. Di Pumpo, G. Wolff, W.P. Schleich, Prime factorization of arbitrary integers with a logarithmic energy spectrum. J. Phys. B **51**(3), 035009 (2018)
17. I.S. Gradstein, I.M. Ryzhik, *Tables of Integrals, Sums, Series and Products* (Academic Press, New York, 1994)
18. R. Grimm, M. Weidemüller, Y.B. Ovchinnikov, Optical dipole traps for neutral atoms. Adv. Atom. Mol. Opt. Phys. **42**, 95–170 (2000)
19. D. Hilbert, Mathematische Probleme. Arch. Math. Phys. **1**, 44–63 and 213–237 (1901). English translation by Mary Newson. Bull. Am. Math. Soc. **8**, 437–479 (1901)
20. H. Iwaniec, E. Kowalski, *Analytic Number Theory* (American Mathematical Society, Providence, 2003)
21. B.M. Levitan, *Inverse Sturm-Liouville Problems* (VNK Science Press, Utrecht, 1987)
22. R. Mack, J.P. Dahl, H. Moya-Cessa, W.T. Strunz, R. Walser, W.P. Schleich, Riemann ζ function from wave packet dynamics. Phys. Rev. A **82**(3), 032119 (2010)
23. G.E. Mitchell, A. Richter, H.A. Weidenmüller, Random matrices and chaos in nuclear physics: nuclear reactions. Rev. Mod. Phys. **82**(4), 2845 (2010)
24. J.W. Neuberger, C. Feiler, H. Maier, W.P. Schleich, Newton flow of the Riemann zeta function: separatrices control the appearance of zeros. New J. Phys. **16**, 103023 (2014)
25. J.W. Neuberger, C. Feiler, H. Maier, W.P. Schleich, The Riemann hypothesis illuminated by the Newton flow of ζ. Phys. Scr. **90**, 108015 (2015)
26. B Riemann, *Monatsberichte der Berliner Akademie* (1859). Transcribed German version and English translation by D. R. Wilkins see http://www.claymath.org/publications/riemanns-1859-manuscript
27. D. Rockmore, *Stalking the Riemann Hypothesis: The Quest to Find the Hidden Law of Prime Numbers* (Pantheon Books, New York, 2005)
28. K. Sabbagh, *The Riemann Hypothesis: The Greatest Unsolved Problem in Mathematics* (Farar, Straus and Giroux, New York, 2003)
29. W.P. Schleich, *Quantum Optics in Phase Space* (VCH-Wiley, Weinheim, 2001)
30. W.P. Schleich, I. Bezděková, M.B. Kim, P.C. Abbott, H. Maier, H.L. Montgomery, J.W. Neuberger, Equivalent formulations of the Riemann Hypothesis based on lines of constant phase. Phys. Scr. **93**, 065201 (2018)
31. D. Schumayer, D.A.W. Hutchinson, Physics of the Riemann hypothesis. Rev. Mod. Phys. **83**(2), 307 (2011)
32. C.L. Siegel, Über Riemanns Nachlaß zur analytischen Zahlentheorie. *Quellen Stud. Geschichte Math. Astron. Phys. Abt. B: Stad 2* (1932), p. 45
33. R. Spira, Some zeros of the Titchmarsh counterexample. Math. Comput. **63**, 747–748 (1994)
34. E.C. Titchmarsh, *The Theory of the Riemann Zeta-Function* (Clarendon Press, Oxford, 1967)
35. J. Twamley, G.J. Milburn, The quantum Mellin transform. New J. Phys. **8**(12), 328 (2006)
36. F. Ullinger, M. Zimmermann, W.P. Schleich, The logarithmic phase singularity in the inverted harmonic oscillator. AVS Quantum Sci. **4**(2), 024402 (2022)
37. J.A. Wheeler, in *Studies in Mathematical Physics: Essays in Honor of Valentine Bargmann.* ed. by E.H. Lieb, B. Simon, A.S. Wightman (Princeton University Press, Princeton, 1976)
38. E.T. Whittaker, G.N. Watson, *A Course of Modern Analysis* (Cambridge University Press, Cambridge, 1996)
39. M. Wolf, Will a physicist prove the Riemann Hypothesis? Rep. Prog. Phys. **83**(3), 036001 (2020)

Printed in the United States
by Baker & Taylor Publisher Services